Interdisciplinarity in the Making

Interdisciplinarity in the Making

Models and Methods in Frontier Science

Nancy J. Nersessian

The MIT Press
Cambridge, Massachusetts
London, England

The MIT Press would like to thank the anonymous peer reviewers who provided comments on drafts of this book. The generous work of academic experts is essential for establishing the authority and quality of our publications. We acknowledge with gratitude the contributions of these otherwise uncredited readers.

This book was set in Stone Serif and Stone Sans by Westchester Publishing Services. Printed and bound in the United States of America.

Library of Congress Cataloging-in-Publication Data

Names: Nersessian, Nancy J., author.
Title: Interdisciplinarity in the making : models and methods in frontier science / Nancy J. Nersessian.
Description: Cambridge, Massachusetts : The MIT Press, [2022] | Includes bibliographical references and index.
Identifiers: LCCN 2021061880 (print) | LCCN 2021061881 (ebook) | ISBN 9780262544665 | ISBN 9780262372268 (epub) | ISBN 9780262372275 (pdf)
Subjects: LCSH: Biotechnology—Methodology. | Bioengineering—Methodology. | Biotechnology—Research—Case studies. | Bioengineering—Research—Case studies. | Biotechnology laboratories. | Scientific surveys. | Interdisciplinary research.
Classification: LCC TP248.24 .N47 2022 (print) | LCC TP248.24 (ebook) | DDC 660.6—dc23/eng/20220720
LC record available at https://lccn.loc.gov/2021061880
LC ebook record available at https://lccn.loc.gov/2021061881

10 9 8 7 6 5 4 3 2 1

Dedicated to the Memory of
Robert "Bob" Nerem
without whom nothing would have been possible

Contents

Acknowledgments

The research in this book began in a conversation with three forward-thinking biomedical engineers: Robert Nerem, Donald Giddens, and Ajit Yoganathan. I am grateful for their willingness to discuss what cognitive science might contribute to their cutting-edge educational vision, and for their whole-hearted embrace and sustained support of our novel proposal to investigate BME laboratory research practices as a means to assist in the design of their educational program. I thank all our BME collaborators, especially Steve Potter, Eberhard Voit, and Melissa Kemp. I appreciate also the support of my colleagues in the College of Computing and the School of Public Policy as I ventured into a novel line of research for a philosopher and cognitive scientist.

Earlier conversations with James G. Greeno primed me to see the opportunities to pursue research on cognitive-cultural integration in scientific practices that the request for assistance by the biomedical engineers presented. Jim guided my philosophical investigations into situated, embodied, and distributed cognitive perspectives and what these might offer to understand scientific practice. I am deeply grateful for these conversations, which carried on throughout our projects, as Jim continued to provide advice to us on analysis and evaluation. John Jungck offered helpful insights when serving as adviser on our research in the systems biology labs. I am also grateful for conversations with Ronald Giere, my fellow traveler in advocating for the fruitfulness of a "cognitive approach" within the philosophy of science, and with Ryan Tweeny, a co-pioneer in psychology of science. Each made valuable contributions to how I framed this research.

The research presented here has been a collaboration from the outset with members of our Cognition and Learning in Interdisciplinary Cultures (CLIC) research group. It is impossible to disentangle their individual contributions, but I do credit specific analyses, based on coauthored publications, within the chapters of the book. The CLIC research group constituted our own complex distributed cognitive-cultural system, which varied in membership and configuration over the course of fifteen years to comprise, ultimately, forty researchers with a wide range and variety of backgrounds. Our research discussions were not only spirited and intellectually stimulating, but also fun. As colleagues often remarked, much laughter floated down the corridor during these meetings. We did, indeed, "click," through many changes of membership. I single out the primary, long-term members from whom I learned so much for special acknowledgement. I thank my co-PI, Wendy Newstetter, for helping to make me a better ethnographer, and our initial postdoc, Elke-Kurz Milcke, for helping us to establish our interview, observational, and coding procedures. As research scientist across all the projects, Lisa Osbeck helped us all learn to use qualitative methods, especially coding and thematic analysis, and to develop evaluations. She also initiated a line of research contributions to social and theoretical psychology. As postdoctoral researchers, Sanjay Chandrasekharan helped to articulate distributed cognition as it might be applied to science, and Miles MacLeod, to develop our contributions to philosophy of biological science and engineering. All these senior researchers are highly creative, talented, insightful, and dedicated researchers, as are our student researchers. Jim Davies was the primary graduate student on the BME project, and Vrishali Subramanian, on the ISB project. Our main undergraduate researchers on the BME project were Ellie Harmon and Christopher Patton, who stayed with us through their MS degrees, and Joshua Jameson, on the ISB project. Each of the student researchers conducted numerous interviews, collected field observations, and taught us the details of the lab technologies they were learning about, all while—it should be noted—they were conducting their own dissertation or thesis research on different topics in their background fields of computer science, public policy, or industrial design. There are no philosophy or cognitive science degrees at Georgia Tech, and so we gathered student researchers from the courses I taught. Wendy and I were responsible for how the project started out, but we could not have envisioned what it would become through the contributions of this creative and energetic group of researchers. Only as I have been able to stand back and reflect, have I come to realize, fully, the range and complexity of the problems they enabled us to attack and how necessary they all were to making progress.

I appreciate feedback from philosophical and cognitive science colleagues on presentations too numerous to list, most recently at various colloquia and workshops associated with the Pittsburgh Center for Philosophy of Science and with the Boston University Center for History and Philosophy of Science, where I am appointed as a visiting researcher, as well as from presentations to bioengineering scientists. I especially appreciate the opportunities to discuss this research, both content and methods, with students and faculty in science education at Vanderbilt University over a period of six years with repeated invitations from Richard Lehrer to participate in his epistemology seminar.

With respect to the manuscript, I am grateful to former research group members for the comments they provided on various chapters, to Catherine Elgin for reading rough drafts of several chapters, and especially to Michael Stuart, who read and commented on the first draft of every chapter. I appreciate the comments of five anonymous reviewers of the manuscript, each of whom approached the book from a different perspective. Their summaries of what they saw as the main contributions give me confidence in the book's intelligibility to an interdisciplinary audience, and their critiques have led me to improve the analysis, hopefully to their satisfaction. My greatest debt is to Floris Cohen, who, as a historian of the scientific revolution, provided the litmus test for the intelligibility of this book for the nonspecialist. Not only did he read and comment on every chapter more than once, he also offered stylistic changes to "enliven" the prose where needed. In addition, he made significant use of his status as an "MS Wordmeister" to help me wrestle the book into proper MIT Press format. I am also deeply grateful for our frequent transatlantic Skype sessions, which kept my spirits up and me at the keyboard during the darkest period of pandemic lockdown in which most of this book was written.

Philip Laughlin and Haley Biermann, MIT Press, and Christine Marra, Westchester Publishing Services, have been a pleasure to work with, and Susan Campbell's careful editing has improved the book.

Some sections of the book draw from previously published material: "Creating Cognitive-Cultural Scaffolding in Interdisciplinary Research Laboratories," in *Beyond the Meme: Development and Structure in Cultural Evolution*, edited by A. C. Love and W. C. Wimsatt, 64–94, Minnesota Studies in the Philosophy of Science 22 (Minneapolis: University of Minnesota Press, 2019); "Interdisciplinarities in Action: Cognitive Ethnography of Bioengineering

Sciences Research Laboratories," *Perspectives on Science* 27 (2019): 553–581; "Hybrid Devices: Embodiments of Culture in Biomedical Engineering," in *Cultures without Culturalism*, edited by K. Chemla and E. F. Keller, 117–144 (Durham, NC: Duke University Press, 2017); "Engineering Concepts: The Interplay between Concept Formation and Modeling Practices in Bioengineering Sciences," *Mind, Culture, & Activity* 19 (2012): 222–239; "Modeling Practices in Conceptual Innovation: An Ethnographic Study of a Neural Engineering Research Laboratory," in *Scientific Concepts and Investigative Practice*, edited by U. Feest and F. Steinle, 245–269 (Berlin: DeGruyter, 2012); M. MacLeod and N. J. Nersessian: "Mesoscopic Modeling as a Cognitive Strategy for Handling Complex Biological Systems," *Studies in the History and Philosophy of the Biological and Biomedical Science* 19 (78): 101201 (2019, with Miles MacLeod); "Modeling Complexity: Cognitive Constraints and Computational Model-Building in Integrative Systems Biology," *History and Philosophy of the Life Sciences* 40 (2018): 70 (with Miles MacLeod); "Building Cognition: The Construction of Computational Representations for Scientific Discovery," *Cognitive Science* 39 (2015): 1727–1763 (with Sanjay Chandrasekharan); "Building Simulations from the Ground Up: Modeling and Theory in Systems Biology," *Philosophy of Science* 80 (2013): 533–556 (with Miles MacLeod); and "Coupling Simulation and Experiment: The Bimodal Strategy in Integrative Systems Biology," *Studies in the History and Philosophy of the Biological and Biomedical Sciences* 44 (online 9/13/13): 572–584 (with Miles MacLeod).

A research project of this scope and extent would not have been possible without the substantial funding provided by the United States National Science Foundation. I appreciate not only the generous funding, but I am equally grateful for the dedicated program officers whose guidance, insight, and support were central to our success over the course of several grants. Edward Hackett and Bruce Seely supported the Scholar's Award from the Science, Technology, and Society Program in the Directorate for Social, Behavioral, and Economic Sciences (SBE9810913) that I received to begin an investigation into how research on situated, distributed, and embodied cognition might be brought to bear on understanding scientific practice. This research, then, provided the cognitive and philosophical framing for the grants received for the bioengineering sciences projects from the Division of Research on Learning in the Directorate on Education and Human Resources from 2001 to 2014 (DRL0106773; DRL0411825; DRL0909971). These grants all proposed both to conduct basic research on frontier interdisciplinary scientific practices, and the challenges these present for learning, and to use design-based research to work with faculty to translate our findings into classrooms and instructional laboratories rooted in authentic practices in these emerging fields. Nora Sabelli and Elizabeth VanderPutten saw the potential in an unusual proposal to conduct basic research on graduate student researchers in laboratories on the frontiers of biomedical engineering. They took a chance on us and awarded a grant deemed "high risk, potentially transformative." Without that grant the project could never have gotten off the ground. Gregg Solomon joined NSF near the end of the first grant and guided us through the next project, in which we proposed to continue to examine frontier experimental practices, this time with the aim of transforming the customary recipe-like instructional lab into one more closely aligned with research practices. Throughout, he provided opportunities for me to promote and showcase our research across NSF, for which I am grateful. Janice Earle provided support for our final grant, which focused on research and learning in a different kind of interdisciplinary frontier, integrative systems biology. The opinions expressed in this book are my own and do not reflect those of the NSF.

In addition, over the course of this research, I appreciate funding in the form of fellowships from the National Endowment for the Humanities, the Harvard Radcliffe Institute (Benjamin White Whitney Fellow), the Institute for Advanced Study on Media Cultures of Computer Simulation (MECS), Leuphana University of Lüneburg, Germany, and the University of Pittsburgh Center for Philosophy of Science (Senior Visiting Fellow). I am grateful to Susan Carey and the Department of Psychology, Harvard University, for my appointment as research associate. The Carey Lab and the Laboratory for Developmental Studies provided not only a home while I was writing this book, but also a look at frontier research in cognitive development.

Finally, ethnographic research is wholly dependent on the cooperation of the members of the communities it investigates. I thank the directors for opening their labs to us and for being willing to discuss their research and ours during the course of these investigations, as well as for inviting us to their celebrations and social gatherings. I thank the lab members for welcoming us into their environment and their openness and generosity while participating in numerous interviews and clarifying discussions, and for their willingness to be observed while they were conducting their research. I remain in awe of and full of admiration for what these student researchers were able to achieve, for the way they went about their daily research with general good cheer in the face of numerous obstacles, and for their communal spirit.

1 Investigating Practice: The Cognitive-Cultural Systems of the Research Lab

Science is one of the most significant creative pursuits of humankind. How can we understand, and account for, the epistemic accomplishments of science given that scientists are limited beings and the natural world is vastly complex? I have been occupied with this problem in various formulations starting from when I was an aspiring theoretical physicist and then as a philosopher of science but also, in addition, as a cognitive scientist. Science is an activity with many dimensions, and for the last twenty years I have been following out my conviction that the answer lies in fathoming how these are integrated in the problem-solving practices scientists create to get a grip on the world.

By "many dimensions" I mean that there is, obviously, a *cognitive* side to science (there's no way to do science without using your mind/brain), but, equally obviously, also a *social* side (lab organizations, academic institutions, how people work together, and so forth), a *material* side (for example, computers, pipettes, instruments, cells, and chemicals), and a *cultural* side (for instance, locally maintained traditions). In each of these dimensions, scientists create resources through which to think. I mark these jointly as "cognitive-cultural" to indicate that in scientific practice these dimensions are integrated.

Scientists' problem-solving practices comprise a range of activities that are of great interest to philosophers as well as to cognitive scientists. Such activities include, among others, the manner in which scientists reason, how they make representations of phenomena, how they construe "understanding," how they use their imagination, and how they work together on a day-to-day basis. The problem-solving practices of scientists on the frontier are exploratory; they are incremental in the sense of moving step-by-step;

and they are nonlinear in that they do not follow any obvious, let alone preexistent, pathway from original problem to final resolution. It is in the processes of developing and using these practices that scientists create complex dynamical investigative systems of research in which cognition and culture are mutually implicated in epistemic practices. Thus, scientific discovery and creativity need to be understood as system phenomena.

The route I have taken toward understanding what is going on in this vast multidimensional realm is to develop fine-grained examinations of those problem-solving practices, their origins and advances over time, and the epistemic principles that guide them. As a professor in philosophy of science *and* cognitive science at Georgia Tech, I was presented with an opportunity to learn what was going on in research undertaken there in laboratories working at the exciting intersection of biology and engineering. It seemed to me from even a cursory initial look that the bioengineering sciences, as frontier research areas, provide an excellent locus for examining cognitive-cultural integration. They are paradigmatic of the interdisciplinary aspirations of frontier twenty-first-century science. Researchers in these fields aim to bring engineering, technological, and computational resources to bear on understanding and controlling complex biological systems. To achieve their ambitions, they would need to "integrate" cognitive-cultural resources in the form of concepts, methods, and materials from the domains of biology and engineering, which requires understanding their affordances and limitations, and to train new kinds of researchers.

I had no prior understanding of the domain, and the selection of labs to investigate was largely a matter of chance. Each specific field was undertaking research on complex biological systems about which there was little scientific understanding. In each lab, the members self-consciously referred to themselves as pioneering researchers on the frontiers of engineering and science. They expressed excitement at being on the forefront of research. As one told us, "*What we are doing right now is the most exciting thing I can think of, and I think most people know that too when they hear about it.*" We began with a lab where engineering techniques are applied to cells and tissues that constitute the blood vessel wall, simply because the director approached me in my role as director of the Program in Cognitive Science. He, along with other senior colleagues, wanted to explore what cognitive scientists might contribute to the development of a new educational vision they had: to train the emerging field's researchers as truly integrative, hybrid "bio-medical-engineers." I immediately saw in this offer the opportunities I had been seeking to both extend my research from historical studies in physics to ethnographic investigations, now, on frontier research in the bioengineering sciences, and to extend my advocacy that scientists should "preach what they practice" (Nersessian 1995) to an innovative educational program as it was being developed. However eager I was to join their endeavor, I responded that I needed first of all to develop some understanding of their research problems and their practices in attempting to solve them. I put together a research group that was (in gradually changing compositions) to investigate their research and learning practices for more than fifteen years.[1] We began to investigate the tissue engineering lab, and three subsequent labs, by conducting a pilot study. We interviewed the lab director about his background and the lab's history and current research, toured the lab while the director and the lab manager described artifacts and activities, and met with the group of researchers who would participate in the investigation. This preliminary research enabled us to focus our research project better.

That is how it began. This book is where it ends. I have two main goals with it. One goal is to establish how the modeling practices of bioengineering research labs exemplify, and work within, the cognitive-cultural framework.[2] I cast research labs and the problem-solving activities within them as what I call *distributed cognitive-cultural systems with epistemic aims*—that is, roughly, a complex system comprising researchers, artifacts, social structures, and practices through which specific epistemic aims are advanced. Within that special frame, which I discuss at some length in section 1.2. below, I examine the building of models (that is, of material and computational artifacts) as what drives the creation and evolution of these systems. On my analysis, models are loci of cognitive-cultural integration. The processes of model-building make a researcher a part of "the lab" socially, culturally, and cognitively.

The other goal is to build on my previous work to further articulate, and to develop the rationale for, the integrative cognitive-cultural framework itself. This framework is applicable to scientific practice, in a broader, more general sense, as well. That is, although my exploration is of the bioengineering sciences that provide my case material, the analytic framework, along with my methodological approach, applies to scientific research as such.

In this introduction, I first lay out the reasoning underlying the integrative framework, and also the methodological approach my research group and I have followed—what I call "cognitive ethnography of research labs." Next, I describe the four labs we have investigated over many years, and how we adapted and used ethnographic methods for our specific purposes. Finally, I sum up the themes of the successive chapters of the book.

1.1 A Cognitive-Cultural Framework for Investigating Practice

Thomas Kuhn's *The Structure of Scientific Revolutions* (Kuhn 1962) was a major impetus in the flourishing of case studies of the investigative practices of scientists since the 1970s. From the start of the "practice turn," accounts of scientific practices have tended to focus on *either* "cognitive/rational" *or* "cultural/social" factors, where these are taken to be separate—often mutually exclusive—interpretive categories. There have been, on the one hand, philosophical and cognitive studies of science, both of which focus on individuals and on the cognitive representations and processes they use in problem-solving and, for philosophy, on the epistemic standards they employ in carrying out experimental or theoretical practices.[3] But there also have been, on the other hand, social and anthropological studies of practice (STS), which focus instead on interests, motivations, and a range of social, cultural, and material factors in play within communities of scientists. Both kinds of analyses have certainly produced valuable insights about scientific practice. Even so, their oppositional stances have prevented fruitful interaction. STS accounts have programmatically constrained relevant explanatory factors to social and cultural, thus downplaying (or denying entirely) the relevance of rational and cognitive factors. On the cognitive side there have been two main objections. Philosophers reject STS accounts for their epistemic relativism. In addition, both philosophers and cognitive scientists are opposed to the social reductionism, likewise characteristic of STS studies, that completely omits cognitive and rational factors from explanations of how science produces knowledge. Such reductionism was most famously announced in the "ten-year moratorium" on cognitive explanations issued first in 1986 by Bruno Latour and Steven Woolgar (Latour and Woolgar 1979, 280; Latour 1987, 247).

The moratorium is long over, but one still needs to ask: What, besides a penchant for rhetorical flourish, could explain such a pronouncement as the ten-year moratorium? One can agree that scientists have interests, motivations, and sociocultural loci in conducting research, but they also have human cognitive capabilities and epistemic aims. Scientists develop and use artifacts ("material culture"), but they also articulate and use evaluative procedures and standards. These "both/ands" look like such truisms that only the investigation of a mix of complex issues can lay bare the roots of the moratorium in twentieth-century intellectual history. I have attempted such an examination elsewhere (Nersessian 2005), but there is no need to detail it here. For our purposes, it suffices to agree that any such divide, though at times analytically useful, is artificial. It imposes boundaries on dimensions that are inherently integrated *in practice*.[4] What twenty-first-century studies of science require is *an analytical framework that facilitates theorizing the cognitive and cultural aspects of practice in relation to one another.*

Here is how I articulate such a cognitive-cultural framework for investigating practice. As with my previous methodological approach of cognitive-historical analysis (Nersessian 1987, 1992a, 2008), I draw, primarily, on resources in *both* the philosophy of science *and* the cognitive sciences. From a philosophical perspective, the project of integration fits into a philosophical tradition of epistemological naturalism (Quine 1969; Nersessian 2008). My construal of that naturalism has the following requirements. First of all, the project needs to be informed by the best available scientific understanding of humans that the biological, cognitive, and social sciences have to offer. Next, it needs to be informed by a grasp of the actual investigative practices as they are created and used by scientists, including their epistemic warrant. Finally, we need to avail ourselves of appropriate empirical methods for determining these practices.

Already in my earlier work on the cognitive basis of model-based reasoning I have established the fruitfulness of thinking about scientific practices through the lens of specific research in the cognitive sciences (Nersessian 1984, 1987, 1992a, 2002a,b, 2008). And, indeed, for the case at hand, there is a line of research that can provide important insights of an empirical, but also theoretical and even methodological, nature—insights that are most helpful if we want to attain an integrative study of scientific problem-solving practices as located with complex cognitive-cultural systems.

Within contemporary cognitive science, there is a movement across the disciplines it comprises that has advanced and persuasively argued for an integrated understanding of cognition and culture. This understanding did not begin with science as a specific object of investigation—that is where much of my own work over past decades, and also the present book, comes in. Even so, this specific direction in cognitive science has much to offer by way of a starting point from which to tackle the problem of integration in science. Indeed, participants in the movement have, themselves, advocated that cognitive science study real-world practices because, "it is in real practice that culture is produced and reproduced. In practice we see the connection between history and the future and between cultural structure and social structure" (Hutchins 1995a, xiv).

The argument in cognitive science begins from a critique of traditional research that locates and studies cognition in terms of what goes on "in the head," without taking into account the role of environmental resources (social, cultural, and material) in shaping and participating in cognitive processes. What I have called "environmental perspectives" (Nersessian 2005) are grounded in empirical evidence from a range of research in the cognitive, biological, and social sciences. Environmental perspectives comprise accounts of cognition as an embodied, artifact-using, and situated process that the social, cultural, and material environment does not just scaffold, but also provides resources that are integral to cognitive process.[5] Advocates for environmental perspectives contend that analyses of cognitive processes need to determine and incorporate salient dimensions of the contexts and activities in which cognition occurs.

Traditionally, the environment is construed as providing mental content on which "internal" cognitive processes operate. In contrast, pioneers of the new approach have argued that "cognition" and "culture" both need to be seen as processes that are integrated in human activity (Hutchins 1995a; Shore 1997). This construal points away from the traditional perspective that regards cognitive and cultural *factors* as independent variables in an explanation of intelligent behavior. Instead, human cognition is understood to be inherently cultural, as being shaped in ongoing processes of the evolutionary history of the human species, of the development of the human child, and of the various sociocultural environments in which learning and work take place—*processes* that are integral to intelligent behavior.

The route to attaining analytical integration for scientific practice follows a path similar to environmental perspectives in that it, too, moves the boundaries of scientific problem-solving beyond the individual to comprise complex cognitive-cultural systems. The grain size of the system—individual, group, groups of groups—depends on the focus of analysis. Here the analytical framework of "distributed cognition" proves particularly useful, in that it incorporates all of the dimensions of an integrated analysis advanced by the range of environmental perspectives. From that perspective, scientific researchers are understood to be embodied agents situated in problem-solving contexts. That is why I have adopted this analytical framework as the starting point of a cognitive-cultural analysis of scientific practice. In this project, we analyze scientific problem-solving as situated within contexts and distributed across researchers and specific artifacts. Extremely fruitful as it is for starters, the framework of distributed cognition is itself, however, in need of further development. What, for present purposes, it needs in particular is to think about what adjustments and enrichments are required to accommodate scientific practice.

1.2 Broadening the Framework of Distributed Cognition

The framework of distributed cognition (from here on, D-cog) has its roots in concerns that arose, starting in the mid-1980s, from the sense that traditional cognitive science had put too much of what goes on in cognition "in the head." Clearly humans are able to do some simple reasoning and problem-solving in the absence of external resources, but we are able to accomplish much more by drawing on those resources—either using or creating them as needed. As Daniel Dennett has remarked, "Just as you cannot do very much carpentry with your bare hands, there's not much thinking you can do with your bare mind" (Dennett 2000, 17). Thinking requires doing with resources in the mind and in the world. One question is how the mind incorporates external resources in thinking. Several streams of research, much of it using ethnographic methods, converge on the notion of cognitive processes as distributed across people and artifacts, including distributed cognition, distributed intelligence, activity theory, situated action, and extended mind theory.[6] Of these various approaches, Edwin Hutchins's research has been most influential on my project of integration because he has focused on the roles of representational artifacts in problem-solving processes ("tasks") situated in technologically rich environments ("socio-technical systems"). Although not themselves science, these systems have features that do align well with science. Further, I agree with Hutchins in that I understand D-cog to be a framework for analyzing cognitive-cultural processes, not an ontological thesis.[7]

The primary unit of analysis is what Hutchins calls a distributed socio-technical system. It consists of people working either together or individually to accomplish a task, along with the specific artifacts they use in the process. *Cognitive artifacts* are the artifacts that have been designed to perform cognitive functions and are used in problem-solving tasks by groups of people (e.g., piloting a ship, Hutchins 1995a) or individuals (e.g., piloting a plane, Hutchins 1995b). The classic exemplar is that of the speed bug on the airplane cockpit instrumentation panel. It eliminates the need for the pilots to remember the range of minimum maneuvering speeds while taking off and landing a plane and, when there are copilots, enables them to have a coordinated perspective (Hutchins 1995b). What makes the speed bug artifact "cognitive" is that it provides a representation that "remembers" the speed constraints for the pilots, thus reducing their cognitive load at a critical juncture in flying a plane. On Hutchins's account, cognitive artifacts, generally, are material media that possess the cognitive properties of generating or manipulating or propagating representations. D-cog analyses, then, focus on the functions of these representational artifacts in human activities.

Science, however, differs in significant ways from the socio-technical systems Hutchins has investigated. Therefore, to accommodate it to our purposes, the D-cog framework needs to be broadened.[8] The framework of distributed cognition has been developed largely from investigations of contexts quite different from research laboratories and of practices quite different from those of scientific problem-solving. In the sections that follow I elaborate on these differences.

First, much of the research that has contributed to the initial framework of D-cog has focused on highly defined task environments with well-defined problems and goals and ready-to-hand cognitive artifacts. Scientific environments, in contrast, are continually evolving along all dimensions.

Second, initial D-cog research has focused on the way artifacts/technologies, as "external" representations, contribute to accomplishing cognitive tasks, especially how they reduce internal cognitive load by "off-loading" cognitive processes, such as memory, to the environment. Scant attention has been paid to the nature of the resources the mind/brain, especially "internal" (human memory) representations and processes, contribute to problem-solving tasks.[9] If we wish to understand scientific problem-solving, we need to pay attention to how these two kinds of representations—artifact and memory—interact.

Third, unlike other practices that have been investigated in a D-cog framework, science is an epistemic practice that needs to justify its outcomes. A scientist cannot just say, "I have discovered this or that." She is bound to offer arguments grounded in the observations made, in the theories employed, and in the methods followed, for why she regards her discovery claims as justified. In D-cog analyses so far, attention to this dimension has, by and large, been lacking.[10]

In these three distinct (though naturally overlapping) regards, then, the D-cog framework needs to be extended: science as a process of ongoing growth and change; science as marked by ongoing interaction between memory and artifactual representations; science as standing in ongoing need of an epistemic warrant. I now successively discuss in more detail these three dimensions to be incorporated into the D-cog framework, always with a view to making it the best possible analytical tool for our investigation of present-day science, in general, and, in particular, bioengineering epistemic practices.

1.2.1 Scientific Task Environments

The distributed cognition framework was developed to analyze cognitive processes in complex problem-solving environments in which there are well-defined problems and tasks for which the requisite salient artifacts and technologies for problem-solving are already at hand. In these studies of environments, for instance the cockpit (Hutchins 1995b) or a naval ship (Hutchins 1995a), the problem-solving situations change in time, but, for instance, how to land a plane is a well-defined problem. Situational features of the problem faced by the pilot or crew can change in the process of landing the plane or bringing a ship into the harbor, and creative solutions can arise in novel situations. However, the technology (cognitive artifacts) required to accomplish the problem-solving, such as the alidade or the speed bug, the practices surrounding them, and the knowledge the pilot and crew bring to bear in those processes, are relatively stable. Even though the artifacts have a history in the field, as Hutchins documents for the instruments of ship pilotage, they do not change in the day-to-day problem-solving processes on board. Thus, this kind of cognitive system is dynamic but largely *synchronic*. By contrast, the cognitive systems of scientific problem-solving are not only dynamic but also *diachronic*.

Scientific problem-solving environments have open, ill-formed problems and goals, and tasks for which methods and artifacts for problem-solving often need to be created, depending on the science and its state of development. Although there are loci of stability, the salient artifacts, methods, knowledge, and the problem formulations themselves can all undergo development and change during problem-solving processes. This is especially so in frontier science environments. A bioengineering research lab provides a good example. Not only does this environment have open, ill-defined problems and goals, but novel methods are under development, and the most salient cognitive artifacts—in this case, physical and computational simulation models—need to be created. Indeed, one could say the lab itself needs to be built in the process of articulating its problems and goals (Nersessian 2012).

Further, university labs are sites of situated learning within a community of practice (Lave and Wenger 1991). In the labs we investigated, the researchers are primarily graduate students, who are "researcher-learners" with developmental trajectories that intersect with the trajectories of development of the technological resources and other dimensions of the problem space. Of course, problem-solving on the frontier requires everyone—including lab directors and managers—to be learning continuously, but student researchers have the additional challenge of acquiring the knowledge and skills to become scientists. A further challenge for everyone in interdisciplinary communities is that interactions among concepts, knowledge, practices, norms, and so forth from more than one field need to be formed into new cognitive-cultural practices. The dynamic and diachronic nature of research labs has led us to characterize them as *evolving distributed cognitive-cultural systems* (from here on to be abbreviated as cognitive-cultural systems or, simply, D-cog systems). Research labs, as problem-solving environments, provide the framework of D-cog with the opportunity to attain what Hutchins has called its "most ambitious goal": to investigate "a system in which adaptive processes that are continually operating are responsible for the production of both stability and change" (1996, 67). Lab researchers build their cognitive-cultural systems as the research moves along, especially through building the artifacts necessary for the activities he singles out: generating, manipulating, and propagating representations within the system.

Much of the work within D-cog has focused on constructing detailed descriptions of the way artifactual representations are used and how they change the nature of the cognitive tasks, especially by reducing cognitive load through "off-loading" memory to the environment. Less understood are the processes of generating/building artifacts to alter task environments in the course of problem-solving.[11] Indeed, as Daniel Schwartz and Thomas Martin have observed, "most cognitive research has been silent about the signature capacity of humans for altering the structure of their social and physical environment" (Schwartz and Martin 2006, 314). Rogers Hall et al. have also noted the lack of attention paid to the practices through which people actively *distribute* cognitive processes to the environment (Hall et al. 2002; Hall et al. 2010). Yet, a central premise of D-cog is precisely that. As Hutchins has succinctly stated, "Humans create their cognitive powers by creating the environments in which they exercise those powers" (1995a, 169). Scientific practices provide an especially good locus for examining the human capability to create or extend cognitive powers. After all, distributing cognition through creating problem-solving environments is a major component of scientific research. The problem-solving environments scientists create include material and conceptual artifacts, methodological practices, and communities of researchers (whether working together or alone), such as labs. By examining the processes through which these environments are built, we can begin to understand how the artifactual resources, in particular, are *incorporated into* problem-solving systems, and are not just external representations "used by" the mind/brain. In our analyses of the interactions between the mental and the artifactual components of the D-cog system in model-based reasoning processes, we cast the relationship as one of "coupling," rather than "off-loading" and then "using." This distinction will become clearer in the chapters of this book, but for present purposes, in general, we use the notion of coupling to indicate that in problem-solving processes, each component of the D-cog system—mental model and material model—interacts in a manner that can create change in the other. But to extend the D-cog framework in this way requires we attend to the nature of the resources the mind/brain contributes to problem-solving. That is our second missing dimension, to be taken up now.

1.2.2 Mental Modeling

To date in D-cog research there has been scant discussion on the nature of the mental resources the human component of the system contributes beyond attending to how the people involved perform coordinating functions. There are, however, many cognitive capacities that come into play in problem-solving. Important examples are memory, representation, reasoning, imagination, and executive functions. Hutchins is explicit that his is a representational view of cognition, in which intelligent behavior results from interactions among "internal" (human) and "external" (artifact) representations in cognitive processes. His analyses track what he calls the generation, manipulation, and propagation of representations in the D-cog system. However, Hutchins focuses primarily on the nature and function of the *artifact* representations, while remaining silent about the *human* representations.

Jaijie Zhang and Donald Norman (Zhang and Norman 1995) provide a rare analysis from a D-cog perspective of the interactions among external and internal representations, in this case, solving insight problems (e.g., tower of Hanoi) and reasoning in games (e.g., tic-tac-toe). They assume, on the basis of a substantial experimental literature, that the internal representations are mental models; however, they do not elaborate on their nature. In earlier work I have developed an account of the capacity for mental modeling, as it provides a cognitive basis for model-based reasoning practices in science (Nersessian 1992a,b, 2002, 2008). My pathway to constructing that account was to synthesize a wide range of cognitive science research from what are usually viewed as separate areas, but which I came to see as interrelated from the perspective of scientific thinking, and to add data from investigations of scientific model-based reasoning practices. I have argued that the cognitive basis for these practices lies in the mental ability to imagine real-world, counterfactual, and impossible situations, and to make inferences about future states of these situations through manipulation of a model of the situation. The remainder of this section provides the basic features of that account, by way of a brief summary of the full analysis to be found in chapters 4 and 5 of Nersessian (2008).

Several important lines of theoretical and experimental research concerning mental models started with the 1967 reissue of the book on explanation by Kenneth Craik (Craik 1943). One specific line focuses on working-memory processes of dynamic and mechanistic mental modeling. My analysis draws on this strand of research in the literature that, like Craik's, examines the processes of constructing and manipulating a mental model during reasoning and problem-solving. Inferences, on this account, are made by means of manipulating both static and dynamic features of the model. That literature, including my own contribution, addresses working memory representations and does not make any claims about the nature of long-term memory representations, which some have claimed to be mental models as well. I have argued that to accommodate the complex nature of scientific reasoning requires an account of model-based inference to move beyond the mental modeling literature per se, so as to construct a synthesis of an extensive range of experimental literature that consists of research on a whole range of issues that implicates mental modeling. These issues comprise discourse and situation modeling, mental animation, mental spatial simulation, and, finally, perceptual simulation in embodied mental representation.[12] When taken together, this research supports a "minimalist hypothesis" that "in certain problem-solving situations humans reason by constructing a mental model . . . in working memory that in dynamic cases can be manipulated by simulation" (Nersessian 2002, 143). That is, people have the capacity to perform what I call simulative model-based reasoning. Such a mental model "is an organized unit of knowledge that embodies representations of spatio-temporal relations of situations, entities, and processes, as well as representations of other pertinent information, such as causal structure" (Nersessian 2008, 128).

Mental modeling is subject to the representational and processing constraints/capacities of the brain, chief among which is memory. Thus, mental model representations are limited in how much detail they can contain, and for how long. The experimental research associated with mental model simulations of physical and mechanistic representations (for instance, pulley systems) identifies specific features of these as being qualitative (DeKleer and Brown 1983; Roschelle and Greeno 1987) and with animation/simulation being piecemeal (Roschelle and Greeno 1987; Hegarty 1992, 2004; Schwartz and Black 1996). Long-term memory provides background knowledge to support mental model-building and simulation processes (Roschelle and Greeno 1987). Information in various formats can be used to construct, manipulate, and revise a model. Such formats may cover language but also formulae, pictures, sounds, and kinesthetic phenomena.

In line with Dennett's comment, simple mental simulations might be possible "in the head" alone, yet mental modeling of any complexity needs to be carried out in the presence of real-world resources. For example, consider a mundane case. It is much easier to mentally simulate how to get an awkward piece of furniture through the door when it is in front of the reasoner in the doorway, rather than recalled from a visit to the furniture store. This is even more true in the case of science. A wide range of data, which comes from historical records, from protocol studies, and from ethnographic investigations, establish that many kinds of external representations are integral to reasoning in scientific problem-solving processes. Such representations can be linguistic (descriptions, narratives, written and oral communications), or mathematical (equations), or visual (diagrams, graphs, sketches, computer). They also include gestures, physical models, and computational models.

How might the capacity for mental modeling interface with the relevant resources in the external world? Much of the experimental research on this interface has been directed toward the use of diagrams and other visual representations. The research noted above by Zhang and Norman, for instance, analyzes diagrams as external representations that are coupled as information sources with mental models in problem-solving. Mary Hegarty (Hegarty 2004) has argued that the corpus of research on mental animation in the context of external visual representations leads to the conclusion that these and internal representations form a coupled system in inferential processing (see also, Gorman 1997; Greeno 1989b). Although limited, the experimental research on scientific reasoning, in particular with computational representations, promotes the coupled system view as well (Christensen and Schunn, 2008; Trafton et al. 2005; Trickett and Trafton 2007). Our research extends the notion of mental model and external representational coupling to include physical and computational models in processes of simulative model-based reasoning (Osbeck and Nersessian 2006; Nersessian 2008, 2009; Chandrasekharan and Nersessian 2015; Nersessian et al. 2003). Diagrams and other visual representations (with the exception of computational) considered in the experimental literature are static, but our analyses concern also the interface between dynamic representations: physical model simulations, computational model simulations, and mental model simulations. Unlike most cases examined in the experimental literature, our research looks at cases in which each part of the system is subject to change in inferential processes—mental models can improve from interaction with material models and vice versa. We understand "coupling" on analogy with how it is understood, generally, in mechanics as heterogeneous components interacting dynamically in a feedback loop to improve the function of a system, in our case, model-based reasoning in a D-cog system.

One way to accommodate this notion of coupling would be to expand what is understood as memory to encompass external representations and cues. If memory is so distributed, problem-solving affordances and constraints in the environment are ab initio part of cognitive processes, which now incorporate both kinds of representations (see, e.g., Donald 1991). In this case, during model-building processes, mental and real-world models iteratively develop correspondences among - their features, and in reasoning, information is co-processed in human memory and in the environment. Although cognitive science stands in need of an account of the nature of the mechanisms of internal/external representational coupling (see, e.g., Chandrasekharan and Stewart 2007; Rahaman et al. 2018), we contend that coupling provides a better metaphor than off-loading (itself in need of such an account of mechanisms), since coupling intimates that cognitive artifacts become incorporated into a D-cog system. We would not want to say that reasoning processes are off-loaded to an artifact model as memory processes are to the speed bug; rather the model and modeler form a dynamic, coupled system that performs reasoning, and, over time, and in interaction with other elements in the system, changes in one lead to changes in the other. Indeed, we have been arguing and I will show in this book, building coupling between mental and artifact models in a D-cog system is a major means through which scientists "create cognitive powers," which extend beyond enhanced memory. To paraphrase Hutchins, *one way in which scientists create their cognitive powers is by creating modeling environments*.

1.2.3 Scientific Practice Is Epistemic

Finally, unlike most other practices that have been investigated within the D-cog framework, science is an epistemic practice. Research labs, and other configurations of scientific practice, can be cast as what Karin Knorr Cetina has called epistemic cultures (1999). These "are cultures that create and warrant knowledge" (Cetina 1999, 1). She chose that designation to contrast with the customary terms of "discipline" and "specialty," which typically refer, in the social sciences, to institutional organizations of knowledge. "Epistemic culture" is used to shift the focus of attention to "knowledge-in-action" (Cetina 1999, 3), or practice. As an approach to the study of science, according to Cetina, to analyze an epistemic culture requires one attend to the differences of "knowledge-making machineries" in different scientific cultures and subcultures (Cetina 1999, 3). Her case studies of practices in particle physics and molecular biology make clear that "machineries" comprise sociocultural practices as well as the technologies of research. Her analysis of these epistemic practices, however, is too limited. It largely omits the cognitive and rational dimensions of the practices.[13] In particular, she does not attend to how a culture provides, or develops, warrant for its practices, including its "machineries." What makes science epistemic, however, is not only that it makes claims to create knowledge, but also that it provides warrant for those claims and the investigative methods leading to them. Nor does she attend to differences in epistemological assumptions, norms, and values among cultures, which Evelyn Fox Keller has pointed out are equally significant for individuating scientific cultures and understanding their practices (Keller 2002). My understanding of an "epistemic culture," and use of that designation, include all of these dimensions.

In the context of extending the framework of D-cog to accommodate science, it is important underscore the normative dimension of a practice. Joseph Rouse's general analysis of practice is useful to understand what this means. Rouse characterizes practices as "situated patterns of activity" (Rouse 1996, 150). Such a pattern "constitutes a practice rather than some other kind of regularity to the extent that it is a pattern of correct or appropriate performance" (Rouse 1996, 137). So, the idiosyncratic activities of a pilot or a scientist do not constitute a practice. Rather, a practice has correct and incorrect performance standards, and is evaluated in light of these. Unlike other practices investigated by D-cog, however, science makes epistemic claims on the basis of its practices. Just as a claim to scientific discovery is not a "discovery" until it has been accepted as justified by something approaching consensus in the relevant field (see, e.g., Arabatzis 2006), a scientific practice is not a "practice" until it is acknowledged in the field as a warranted means of investigation. As Helen Longino has pointed out in her analyses aimed at bridging what she calls the "rational-social" divide, the normative dimension of practice is inherently social, in that it relies on the critical scrutiny of the relevant scientific community (Longino 1990, 2001).

The decision to warrant a scientific practice or system of practices is based on an assessment of the reasons advanced in favor of the practice or the system, and on a history of reliable success at achieving its epistemic aims. In developing novel practices, researchers are required to advance reasons for why their specific application is warranted, and also for their more general applicability in the field. For instance, the computational systems biologists we studied provided reasons in support of the abstractions they used to represent biological interactions while they were building specific models. They came up, likewise, with arguments to account for why their practice of building certain midlevel ("mesoscopic") models with less fidelity to the biological details than the field customarily aims at is warranted. They argued, and demonstrated, that such models can produce significant understanding and possibilities for intervention (e.g., to create biofuels or kill cancer cells), while also providing insight into how one might build in fidelity as capabilities for modeling develop in the field.

In our investigations of problem-solving in research labs, we consider "normativity," in its most general sense, to mean that there are specific constraints on practice. These constraints are of three kinds. Some are material, in that they are tied to the composition and behavior of entities and objects. Others, of a cognitive nature, are imposed by the processes and structures by which humans, for instance, categorize and make inferences. Again, others are imposed socially, in accordance with standards of research and of professional conduct. All these constraints can be shared by other kinds of practices, but science alone is subject, in addition, to constraints that are imposed by the epistemic aims and justification of the practice. That is why our examinations of innovative methodological practices for modeling attend to how warrant is built for novel practices in the course of their development, as well as for specific models, which are intended to function epistemically.

We, of course, realize that the wider community is always implicated in the development a practice. Even so, for practical reasons we limited the scope of our investigations mainly to the participants in our labs, including external collaborators, where present and feasible. We did nonetheless attend to how the researchers expressed the norms and values of their communities, and also to how their work was received and critiqued in community responses, notably in the context of conference presentations and of reviewer responses to publication submissions and grant proposals (see, e.g., Osbeck et al. 2011, chapters 5 and 6).

1.3 Cognitive Ethnography of Research Labs

To determine how warrant is built for practices, and indeed to move beyond the perceived cognitive-cultural divide more broadly, requires more than theoretical argumentation of the kind broached above. It requires likewise fine-grained empirical investigations of the epistemic practices of scientists to determine how integration takes place in these practices. Until quite recently (and definitely when we began our investigations), philosophers out to determine practices have been relying primarily on historical data of two kinds—archival records and publications. Historical records, however, leave us at the mercy of "what's left behind"—a problem even more significant with late twentieth and early twenty-first-century science—and rarely are they sufficient to provide details on the day-to-day research problems and paths to solution implicated in what become discoveries. Established canons of modern science also require that such details be omitted from published accounts of research. Impasses and obstacles encountered along the way, in particular, rarely make an appearance in published accounts, which frame the problem-solving processes and reasoning as linear and relatively straightforward (Bazerman 1988). Historical data do afford a means to contextualize the practices of scientists with respect to their historical situatedness from the viewpoint of the problem situations of the traditions in which scientists carried out their work, and so can advance the project of integration (Nersessian 2008). But full analysis of integration requires more. It necessitates that we move beyond exclusive use of historical records, which by their nature place limitations on thinking about the interplay of cognition and culture, to carrying out empirical investigations of practices *as they are enacted in situ* in order to determine the range and kind of resources that contribute to the epistemic accomplishments of science. In other words, we need to move beyond historically informed philosophy to ethnographically informed philosophy.

What understanding philosophers of science have of ethnography derives, primarily, from how it has been portrayed and used in STS. STS researchers have established that ethnography, which comprises field observations and interviews, provides a fruitful means of collecting data on day-to-day scientific practices in research labs.[14] The main objection of philosophers to STS accounts has been, and still is, their tendency toward epistemic relativism, which stems from their alignment with the so-called strong program of the "sociology of scientific knowledge (SSK)" (see, e.g., Bloor 1991). Since their inception and in line with the "ten-year moratorium" mentioned above, these accounts have programmatically constrained relevant explanatory factors in their analyses to social and cultural factors, including personal motivations and interests, while downplaying or denying entirely the relevance of rational and cognitive factors.[15] There is nothing in principle, however, about ethnographic methods as such that would prohibit addressing those aspects of practice. Indeed, a major contribution of STS ethnographies of research labs is that they have demonstrated repeatedly the value of ethnographic methods for investigating scientific practices across a range of sciences. I began the project presented here with the conviction that philosophers need not cede this important methodological tool to sociocultural science studies fields, but that it could be adapted and used to address philosophical issues (Nersessian and Macleod 2022). This conviction was supported by its use by cognitive scientists to investigate other kinds of problem-solving practices in D-cog. Ethnography conducted for philosophical objectives can be placed within the perspective of what is called "cognitive ethnography" within the D-cog framework.

1.3.1 Cognitive Ethnography of Scientific Practices as a Method

D-cog has been using ethnographic methods to move the study of human cognitive processes out of the experimental psychology lab and into real-world settings, ranging from ordinary activities to sophisticated work practices. As Edwin Hutchins argued, "We need to look in the wild, not because that is where cognition is, but because it is a place where it is easier to see the cultural nature of cognition" (Hutchins 1996, 67).[16] Originally, ethnography was developed by anthropologists as a method by which to study and interpret cultural and social practices of indigenous communities. In the late 1970s, it began to be used to study the practices of other kinds of communities as situated in their natural settings. As methods of "qualitative analysis" began to develop, Egon Guba, a pioneer in promoting ethnographic educational research, characterized ethnography, generally, as a form of "naturalistic inquiry" (1978; Lincoln and Guba 1985). This characterization was introduced to contrast it, and qualitative methods, with empirical inquiry by means of controlled experimental design. A naturalistic inquiry aims to collect in situ data and extract information on practices and their relations to context through an intensive and detailed description and systematic analysis of those practices and their contextual relations. A naturalistic inquiry is ecologically valid in that there is at most minimal manipulation of existing settings, and no strict constraints, such as predetermined categories of interpretation, are placed on outcomes. Such studies are principally inductive rather than hypothetico-deductive.

An ethnographic investigation is geared toward the open exploration of practices, rather than the testing of hypotheses. The scope and focus of an inquiry are, of course, framed by its research questions, which serve to focus the ethnography.

Ethnography is interpretive research. The anthropologist Clifford Geertz characterized a main objective of ethnographic analysis as "thick description"—a term he claims to have borrowed from the philosopher Gilbert Ryle (Geertz 1973). Thick descriptions interweave description and explanation of an observed phenomenon or practice by unpacking it layer by layer with respect to its context. In general, ethnographic investigations are built around a family of tools for gathering data, mainly field observations and interviews, and around interpretive data analysis methods, such a grounded coding (Corbin and Strauss 2008) and thematic analysis (Braun and Clark 2006). Ethnography provides systematic methods of data collection and analysis, some of which are discussed in section 1.3.2, in order to establish that the interpretations are robust and consistent across a range of evidential sources ("triangulation"), thereby establishing warrant for the claims advanced from the investigation (see, e.g., Guba 1981).

"Cognitive ethnography" (dubbed thus by Hutchins 1995a) is used to gather data in real-world settings on how conceptual, social, and material resources are integrated in cognitive processes. What makes this "cognitive" is, among other things, that the focus of the ethnography is on how individuals and communities solve problems by reasoning about them, by seeking to understand them, by altering the concepts they use, by working together, by using their imaginations, and by learning. These practices are investigated as situated in contexts, with their attendant resources, which include, importantly, material artifacts. For example, one pioneer in the approach, Jean Lave (1988), rooted her critique of traditional experimental cognitive science in her own ethnographic investigations focused on mathematical problem-solving by "just plain folks" in their natural environments, such as home and the grocery store. Her studies showed in what ways people integrate environmental resources into their mathematical reasoning and problem-solving, which has the effect of making them generally much more competent at these tasks in the real world than they demonstrate as subjects in experimental psychology labs or in traditional school settings, where they are usually deprived of such resources. Meanwhile the most widely influential cognitive ethnographic research, and the one best known among philosophers, is Hutchins's research on technologically rich and well-defined problem-solving environments, where he extends "natural" to comprise specific work contexts. In keeping with the traditional cognitive science framing, Hutchins conceives of problem-solving as a form of information processing that uses representations and reasoning in pursuit of goals. However, his conception diverges from the traditional framing in that, from the D-cog perspective, the relevant representations and reasoning processes are located not only "in the head" of an individual, but also situated in the problem-solving environment and distributed across one or more individuals and select artifacts.

"Naturally occurring culturally constituted human activity" (Hutchins 1995a, xiii) of any kind can, in principle, be investigated with ethnographic methods. Research labs like the ones we have investigated certainly constitute such natural environments of scientific practice. In a cognitive ethnographic investigation, philosophers of science, too, are likely to focus on problem-solving contexts. These contexts can provide detailed information on many issues of interest to philosophers, including the nature and structure of scientific problems; how these are modified in the course of research; how scientists develop and use methods and concepts; how they create and evaluate claims and explanations; and how they communicate results.

Another way in which D-cog ethnography is "cognitive" is that a central aim of analysis besides providing richly nuanced thick descriptions of the particularities of a given case is to advance a more general, theoretical account of cognitive processes. As Hutchins has framed this objective, "There are powerful regularities to be described at the level of analysis that transcends the details of the specific domain. It is not possible to discover these regularities without understanding the details of the domain, but the regularities are not about the domain specific details, they are about the nature of cognition in human activity" (quoted in Woods 1997, 177).[17] Cognitive ethnography, although rooted in the concrete, can make use of several kinds of qualitative data analysis methods to abstract to the extent warranted beyond the details, such as "grounded theory" and "thematic analysis" (see, e.g., Corbin and Strauss 2008; Patton 2002; Braun and Clarke 2006). Such analyses aim to move from the specificity of the case to build a broader interpretive account by using systematic procedures to abstract and coalesce interpretive categories and, where appropriate, formulate candidate hypotheses to transfer and assess across cases, using multiple cases to work back and forth between data and theory to attain a warranted degree of generality. This is a different kind of process from inductive generalization. Insights from philosophy of science on how to use case-study material to build theory help to illuminate the difference.

As with cognitive science, philosophy of science, too, is interested in using empirical insights from data on scientific practices to develop or examine theoretical notions, while avoiding unwarranted generality. Early critiques, including my own, of a simple inductivist perspective on case data advocated that the way to understand the relation between specific cases and theory is as a bootstrapping method customarily used in the sciences (Nersessian 1991a), which we will see ample examples of in the case studies developed in this book. Roughly, in such bootstrapping processes, "hypotheses are made within a background of beliefs and problems. . . . They are refined, made more specific, modified, or rejected in light of more constraining data (a detailed case study). Surviving hypotheses are then tested against other data and other hypotheses to determine the extent of their validity" (Nersessian 1991a, 683). Bootstrapping entails working back and forth between data and theory, until a satisfactory accommodation is achieved. It is an iterative and incremental, open-ended process.

Recently, in thinking about the use of qualitative data on scientific practices, such as from ethnographies or interview studies, Erika Mansnerus and Susann Wagenknecht (Mansnerus and Wagenknecht 2015) follow up on a recommendation of Hasok Chang (2012, 111) to construe the relation between historical case studies and philosophical theorizing in terms of the "concrete" and the "abstract" instead of the customary inductive categories of the "particular" and the "general." They use this suggestion to further articulate the bootstrapping account and argue that the way philosophers can arrive at "limited generalizations," while they "avoid unwarranted generality," is to "create a *dialogue between the abstract and the concrete*" (emphasis original). That is, to work back and forth between data and theory, which they, too, note is a bootstrapping procedure. Further, they contend, such "productive interplay" (Mansnerus and Wagenknecht 2015, 40) makes it possible to examine and further develop philosophical concepts and theories with qualitative case study data, while avoiding the pitfalls philosophy has often succumbed to of fitting the data to the theory. With ethnographic studies of scientific practice, the context for "productive interplay" is established in the way philosophers of science frame the investigation, how and what data are collected, and how analysis proceeds.

Such dialogue is in line with traditional practices in ethnographic analysis. As Geertz has emphasized, "one does not start intellectually empty-handed. Theoretical ideas are not created wholly anew in each study. . . . They are adopted and refined from other, related studies, and, refined in the process, applied to new interpretive problems. If they cease being useful . . . they stop being used. . . . If they continue to be useful, throwing up new understandings, they are further elaborated and go on being used" (1983, 57).

To consider here, in general, how ethnography can be adapted from its social science roots to serve as a method of philosophical investigation, in line with philosophers' interests, goals, and values, would take us too far afield from the subject of this book. I have undertaken that analysis elsewhere in order to promote an ethnographic approach in philosophy of science (Nersessian and Macleod 2022). Instead, I next detail, specifically, how we framed and developed the multiyear ethnographic investigation of epistemic practices in bioengineering sciences research labs that is discussed in this book. The chapters that follow provide the fruits of our investigation.

1.3.2 Our Own Cognitive Ethnography of Bioengineering Labs

The "wild" of our ethnographic investigations is the university bioengineering sciences research lab and the researchers within it. We have been making use of cognitive ethnography and the broadened framework of D-cog to investigate scientific labs and the problem-solving practices within them as distributed cognitive-cultural systems. As such, "the lab" is not simply a physical space existing in the present, but a dynamic problem space that reconfigures itself as the research program moves along in time and takes new directions in response to what occurs both within the lab and in the wider community of which the research is part.[18] My choice to study specific research labs was largely serendipitous, but once the opportunity presented itself it was apparent they were ideal loci in which to investigate how cognitive-cultural integration proceeds in the reality of everyday research. The bioengineering sciences are a hotbed of creativity and innovation, including of ways to go about doing research on complex biological systems. The pioneering, interdisciplinary nature of the research offers a perfect opportunity to study investigative practices as they are created, as well as how they are used.

The central epistemic practice across all labs is building some kind of model of a complex biological system to use as a basis for inference and understanding about the system. Our approach to cognitive-cultural integration was to frame these labs and the problem-solving configurations within them as D-cog systems and to examine the various components these comprise and their interrelationships. The project of cognitive-cultural integration is rich, multidimensional, and difficult. I do not claim that the analyses presented here fully cover this richness or dimensionality in the problem-solving practices of the labs we investigated. Among the specific components of a distributed cognitive-cultural system we attended to are conceptual and methodological resources, artifacts central to the research, and epistemic norms and values. We carried out several kinds of analysis with respect to these, as will be discussed throughout the book.[19] In the analyses I present in the chapters of this book, I first attend to the specific nature and configuration of these aspects, as well as the interactions among them, in each lab. I then examine specific *epistemic affordances* of each D-cog system. Epistemic affordances, as I characterize them here, are those features of the D-cog system that enable or facilitate epistemic access to the phenomena under investigation.

The bioengineering sciences are inherently complex in at least three ways: how researchers think, what technologies they work with, and how they work together. Researchers in these fields make use of a range of conceptual, methodological, theoretical, and material resources drawn not only from various engineering fields but also from the biosciences and from computational sciences—resources they use to conduct groundbreaking, basic biological research in the context of potential application. It is a field initiated largely by engineers, who are recasting specific biological problems as bioengineering problems. For example, they recast the problem of changes in functional properties of endothelial cells in the cardiovascular system in terms of the effects of mechanical forces on them. They aim, ultimately, to *get a grip* on complex biological systems in both senses of the word "grip": to understand and to control. The understanding they aim at is to develop a model of a complex biological phenomenon, for instance, a model of its underlying mechanism or a mathematical model of the interactions among the components of a system. Researchers in bioengineering sciences hope that such a model will offer the possibility to control specific processes, such as disease processes (usually by others, for instance, medical researchers). However, given the frontier nature of the research, the initial aim of the researchers is to develop, understand, and control a physical or computational model that has the potential to be informative about the behavior of the biological system, as was the case in the labs we investigated.

The movement of engineering into biology has given rise to a multifaceted interplay of quite disparate conceptual frameworks, methodological approaches, and epistemic values (see also Boon 2011). Figuring out what resources to draw from and how to adapt them is in no way straightforward. For instance, concepts developed in the context of engineered systems, such as robustness, modularity, and noise, are being transferred to the study of biological systems, but their attempted transfer has required not only modification, but often conceptual innovation (Knuuttila and Loettgers 2011; Loettgers 2007; Nersessian 2012a,b). Further, mathematical engineering theories and frameworks, as well as engineering methodologies, are being imported and adapted to perform as tools of biological representation and analysis (Wimsatt 2007). Along with these tools, bioengineers transfer certain values embedded in the practices of the engineer, such as precision, control, isolation, and abstraction—values that often conflict with assumptions and epistemic values of their collaborators in the biological sciences.

At the outset, bioengineering scientists face a major challenge in that they usually cannot experiment on biological systems directly. The complexity of the real-world (in vivo) phenomena makes experimentation too difficult, or even impossible, to control. Also, to intervene on in vivo systems often presents significant ethical issues. Thus, researchers in these fields need to devise means to model the phenomena in sufficient measure to enable experimentation on the model to yield plausible understanding of the in vivo system. Physical simulation models (in vitro) and computational simulation models (in silico) are artifacts designed to function as epistemic tools (Boon and Knuuttila 2009; Knuuttila 2011). They are part of the epistemic infrastructure through which bioengineering scientists manage and probe the nature of complex biological systems. But to devise an appropriate modeling practice as well as specific models faces all of the challenges of biological-engineering integration noted above. Thus, by investigating specific practices we are also laying out the basic epistemic structure of biological engineering; namely, to bring the conceptual, methodological, technological, and material resources of engineering to bear on the problem of managing the complexity of biological systems so as to be able to study them.

Generally speaking, bioengineering scientists investigate biological systems in either of two ways. They work through iterative and incremental processes of designing, constructing, redesigning, evaluating, and experimenting either with surrogate in vitro physical simulation models, which comprise biological and engineering materials, or with in silico computational simulation models. We refer to these basic processes as *building to discover*. The processes themselves, and the question of how they advance their epistemic goals, have been the focus of our research on cognitive-cultural integration. It would not be far-fetched to conclude that understanding in bioengineering sciences, and engineering sciences in general, derives largely from *building*. It is no accident that the famous saying attributed to Richard Feynman, "what I cannot create, I do not understand," is misquoted by bioengineering scientists as "what I cannot build, I do not understand" (famously encoded in the first synthetic cell by Craig Venter; see also Voit et al. 2012). However, this process of building to discover has not received much attention in either philosophy of science or cognitive science. Our research shows that—and how—building simulation models is a major means through which bioengineering scientists build understanding of complex biological phenomena. Model-building is the means by which they actively distribute cognition in their environment, thereby creating complex distributed cognitive-cultural systems of people, practices, and artifacts.

The bioengineered models thus function as cognitive artifacts, which participate in the reasoning processes and the representational processes of a distributed cognitive system. They function equally as what researchers in STS studies of science refer to as the "material culture" of communities, which participate in social and cultural processes. Our research demonstrates on a day-to-day basis that it is simply not possible to fathom the epistemic work such models enable by focusing exclusively on one or the other aspect. Bioengineered models are representations, and as such they play a role in reasoning processes. They are central to social practices related to community membership, such as mentoring and identity formation. They are sites of learning. They embed epistemological norms and values. They are repositories of lab history. They perform as cognitive-cultural "ratchets" (Tomasello 1999) in an epistemic community, which enable one generation to build on the results of the previous, and so serve as loci of stability in the context of innovation, while moving the problem-solving forward. In sum, *models are central components of the cognitive-cultural fabric of creative problem-solving in bioengineering sciences*. Examining how that fabric is built in specific cases provides insight into how cognition and culture are intertwined in scientific practice. It is for all these reasons that we have focused our investigations on the modeling practices we have encountered in the research labs.

1.4 Four Cognitive Ethnographies: An Overview of Our Bioengineering Sciences Project

Our investigational settings comprise four pioneering university research laboratories. Two labs work in biomedical engineering (BME)—tissue engineering and neural engineering, respectively. The other two specialize in integrative systems biology (ISB)—one solely computational, the other a combined computational and wet experimentation lab. We chose university labs because they are largely populated by graduate students, who are pioneers in research and at the same time learning to become scientists. Many of the graduate students were in an educational program aimed at moving beyond collaboration between engineers and biologists through producing *hybrid* biomedical engineering researchers. The educational program was itself under development at the time, and we were assisting the faculty in this undertaking. As a central part of our NSF-funded project on the labs, we had proposed not only to investigate the nature of emerging practices in these fields, but also to determine important requirements for learning them, and to work with bioengineering faculty to "translate" findings from our research into educational experiences. Our investigations taught us, and the faculty as well, that the forms of interdisciplinarity practiced in BME and in ISB are quite different, leading to different environments, requirements, and challenges for problem-solving and learning.

We chose to investigate interdisciplinary fields at the intersection of biology and engineering, but the selection of labs was largely a matter of chance, and we had no prior understanding of the fields. Each field was conducting research on complex biological systems about which there was little scientific understanding. In our first pilot research, the tissue engineering lab, we were surprised that its physical space had the look of a biology wet lab, with pipettes, flasks, sterile hoods, and cell cultures, but also included strange-looking engineering artifacts, which members were referring to as "devices." We quickly realized that these artifacts were the focal point of the research life of the lab, as they provided the means of what the director called "*taking the research in vitro.*" These lab-built models are intended to replicate selective processes of complex biological systems. We also quickly realized that conducting physical simulation experiments with these hybrid models—part living tissue and cells and part engineered materials—constitutes a novel modeling practice not previously investigated in the philosophical literature. We made these models and the practices surrounding them our focus in BME. Similar pilots with each lab directed our attention to the salient modeling practices that became the focus of our data collection and analysis.

What labs? Now that we had discovered the novel practice of building in vitro physical simulation models, we wanted a second BME lab with that practice in a different domain. We also thought that a newly established lab would be good for contrast, which is why we chose the neuroengineering lab. We began our ISB study because we wanted to move into a pioneering area in computational modeling and simulation of biological systems, and the director of the purely computational lab was interested in how we might assist research and learning in that area. By then we had established an excellent reputation in the department, and lab directors often asked if we could "do" their lab. In addition to our contributions to the development of their educational programs, the faculty began to notice that the researchers in the labs we studied had become more reflective about their research. Indeed, the researchers, themselves, told us that our interviews always provided an opportunity to reflect on their research problems and, on occasion, served to rekindle motivation when they were tired and lagging. As one researcher told us, "*talking about it is good 'cause it also reinforces what you're doing. So, I can go back and feel motivated about it now.*" For a second ISB lab we sought one that conducted experimental research as well as computational modeling, and also was newly established. When an assistant professor offered her lab, we concluded from her description that it would contain experimentalists as well as modelers. In our pilot we were surprised to discover that the modelers themselves were being trained to conduct the experimental work. Since the challenges of the dual nature of the practice were novel and interesting, we decided to stay with that lab. We coded the practice as that of the "bimodal strategy" (MacLeod and Nersessian 2013).

In both BME labs, the researchers tackled problems of bringing cells and tissues together with engineered materials in processes of building living hybrid models to simulate in vivo processes of complex biological systems.

Their aim was to understand basic processes within the biological systems, with the hope of providing a basis for future application, chiefly in medicine. Their research problems required little collaboration with researchers outside the lab.

Both ISB labs aim to understand biological systems that comprise integrated and interacting complex networks of genes, proteins, and biochemical reactions. When this is achieved in sufficient measure, they expect it will allow collaborators to attempt interventions such as to produce a better biofuel or to make cancer cells receptive to treatment. Solutions to the problems they tackle require that they construct computational simulation models in need of rich experimental data, which creates an essential epistemic interdependence with their collaborators in the biological sciences.

1.4.1 Research Questions and Data Collection

Our objective in data collection can be summarized as follows. Starting from an open and broad stance about what might prove relevant to our research questions, we aimed to conduct a systematic longitudinal investigation involving numerous bioengineering scientists across a broad range of perspectives, problems, and lab organizations. In each case we aimed to collect a range of data from different sources with which to triangulate the analyses.

Our investigations in each lab began with a basic set of questions motivated by our combined philosophical and cognitive science interests:[20]

- What are the representational and reasoning practices used in problem-solving in this community?
- How is epistemic warrant developed for novel practices?
- What are the epistemic assumptions, values, and norms at play in each of these interdisciplinary communities?
- What concepts, methods, and theories are being used from engineering, and how? Ditto from biology? What is the nature, or are the results, of their interaction?
- In what ways might cognitive, social, cultural, and material "factors" be mutually implicated in these epistemic practices?

These questions enabled us to focus our research while remaining open in sufficient measure to guide, but not totally constrain, our data collection and our analysis. Within the course of the investigation, we intended that specific issues to be addressed with respect to these questions would emerge from our findings, as indeed they did.

We investigated each lab for approximately five years. We collected data between 2000 and 2014, and data analysis continues in the present. Data collection in all labs comprised the following main items:

- audio-taped open and semi-structured interviews
- participant field observation with note-taking
- lab tours (given for us and for visitors)
- arranged demonstrations of experimental procedures and technologies involved in the lab researchers' data collection and analysis
- video and audio recorded lab meetings
- "journal club" meetings in which pertinent articles were discussed
- photographs of white boards
- diagrams of the spatial layout of each lab and photographs of how lab space changed over time
- artifact collection: grant proposals, paper drafts, presentations, dissertation proposals, emails, diagrams/sketches, and so forth

For each lab we compiled an extensive "technology document" that surveyed all the technologies in the lab, which researchers used them, and what their functions were within the research.

The extent of our interview and observational data is summarized in table 1. All interviews and some (or parts of) meetings we thought especially significant have been transcribed. In our interviews and discussions with the researchers, we probed the nature of their research, how they went about doing their research, what problems they were encountering, how they responded to those, how and what kind of learning was needed as they went along with their research, and how they positioned their research with respect to the broader field. We began with unstructured interviews, with the initial ones focused on their background, motivation for choosing that bioengineering science, and an overview of their research. As we gathered more information on their projects and as we developed a better understanding of the scientific/engineering content and methods, we conducted more targeted interviews both to probe their reasoning as they were working on specific problems and to probe specific issues that arose as we began to analyze the transcripts and other data. We learned a great deal about their practices and about the content and context of the research. Often, when making field observations and when it would not interfere with the research, we were able to ask questions about what they were doing at that time and why. They also were willing to set up meetings with us to demonstrate procedures they used in model-building and data analysis. In return, as I noted, our probing provided the opportunity for them to articulate and reflect on their research. One student likened our interviews to "research therapy" appointments, and several even expressed the desire to have them more frequently. Finally, we had many informal interactions with them. Our student researchers went along on their group hikes and bike rides when invited, and sometimes ate lunch with them. We were all invited to holiday parties and to dissertation defense celebrations. Some students asked for career guidance or letters of recommendation.

1.4.2 Research Sites

I now provide a brief overview of the makeup and the kinds of problems addressed in each lab in order to aid the reader in understanding the different challenges of interdisciplinary problem-solving faced by the researchers as they attempt to integrate engineering and biology in these environments. All the labs we investigated were conducting research for which there was little or no precedent when they began. The members repeatedly told us about the pioneering nature of their research, using expressions such as "it had never been done before" or "no one has approached it this way before" or "no one was thinking this way" or similar such expressions.[21]

In the two BME labs we conducted intensive data collection over the first two years. For approximately five years, we further continued data collection on selected dissertation projects up until they were completed, including

Table 1.1
Data summary

Laboratory	Interviews	Meetings	Field observations (hours)
BME A	72	15	~350
BME D	75	40	~450
ISB G	44	7	~40
ISB C	62	22 (plus 2 joint C and G)	~250

additional interviews.[22] We began data collection in lab D approximately a year after we began collecting in lab A. In all, we worked on the BME project for ten years, before moving on to ISB. As noted before, both labs designed, built, and conducted experiments on hybrid living physical models, locally called "devices," which is the shorthand they use for "bioengineered modeling devices."[23] Given the distant nature of their respective research, there was little interaction between the directors of these labs, beyond the attention and the informal mentoring one would expect a quite senior member of a department (lab A) would provide to a quite junior member (lab D) of his department.

Lab A was a tissue engineering lab. Its overarching research problems were to understand mechanical dimensions of cell biology, such as the effects of the forces of blood flow on morphology, proliferation, and gene expression in cardiovascular endothelial cells. The researchers saw their research also as a contribution to the eventual medical application goal of creating a living substitute blood vessel to implant in the human cardiovascular system. Examples of intermediate problems that contributed to the daily work included constructing specific living tissue models ("constructs") that mimic properties of natural blood vessels; using biomechanical forces to create endothelial cells from adult stem cells and progenitor cells; designing environments for mechanically conditioning constructs; and designing means for testing their mechanical strength and functionality.

When we entered lab A, it had been in existence for thirteen years. It closed ten years after our study, when the director retired. During our study, the main members included a male director, a male laboratory manager, a female postdoctoral researcher, seven PhD students (two male, five female, two of whom, one male and one female, graduated early in our study, the other five after we concluded our formal data collection), two MS graduate students, and four long-term undergraduates. Additional undergraduates from around the country participated in summer internships, and international graduate students and postdocs visited for short periods. The laboratory director was a senior, highly renowned pioneer in the field of biomedical engineering, who had started his career as a mechanical engineer in aeronautical engineering. Near the end of his career, he liked to characterize its trajectory as "from astronauts to stem cells." All of the researchers had engineering backgrounds, mainly in mechanical or chemical engineering, and some were currently students in the BME program that was just starting. Some had spent time in industry before joining the lab. The lab manager had an MS in biochemistry. The researchers frequently consulted with a histologist located in the building, and some traveled to other institutions for various purposes, including to collect animal tissues and to run gene microarray analyses. Lab meetings were held irregularly, when the director, who traveled a significant amount of time, would be in town (approximately every three to four weeks).

Lab D was a neuroengineering lab. Its primary research problem was to understand the mechanisms through which neurons learn as networks in the brain. Here, again, the researchers had dual scientific and engineering goals. They aspired to use this knowledge to develop aids for neurological deficits or, more generally (as the director liked to say), "to make people smarter." Here are some examples of intermediate problems that contributed to the daily work. They developed ways to culture, stimulate, control, record, and image neuron arrays. They designed and constructed feedback environments (robotic and simulated) through which the main device (the model-system comprising a "dish" of cultured neurons) could learn. They used electrophysiology and optical imaging to study "plasticity." One researcher developed a computational model of the dish model-system that played an unanticipated pivotal role in the research. All the projects centered around the "dish," and, as the research unfolded, there developed significantly more interaction among research projects than we witnessed in lab A.

Lab D was just taking shape as we began our research. It closed when the director moved to another position, which was nine years after our study ended. During our study the main members included a male director, a male postdoctoral researcher, four PhD students in residence (one female, three male; one male left after two years to pursue neuroscience, and the remaining three graduated after we concluded formal data collection), one PhD student at another institution who occasionally visited the lab and was available via video link, one MS student, six undergraduates, and one volunteer for nearly two years, who was not pursuing a degree (already possessed a BS in engineering) but who helped out with breeding mice. Because the lab was new and had limited funding at the start, the director made more use of undergraduates, who usually had short-term research projects, which were supervised by the director for course credit.

When we began, the laboratory director was a new tenure-track assistant professor, fresh from a lengthy postdoc in a biophysics laboratory that develops techniques and technologies for studying cultures of neurons. He already had attained some recognition as a pioneer. His background was in chemistry and biochemistry, with his engineering knowledge largely self-taught, though highly sophisticated. The backgrounds of the researchers in lab D were more diverse than those in lab A and included mechanical engineering, electrical engineering, physics, life sciences, chemistry, and microbiology; some were currently students in the BME program, but also in electrical engineering and mechanical engineering. The wet lab was in a separate room. The main lab had the look of a computer lab, with copious wires connecting the incubator for its main object of study—the dish—to computers, and with small robotic devices, connected with a dish, scattered (or rolling) around the lab. They held lab meetings and a journal club (to discuss recently published research) weekly. Unlike the traditional configuration of a stand-alone lab, lab D was embedded in an open space designed to promote interdisciplinary collaboration among neuroengineering labs. It was shared by seven faculty members, their postdoctoral researchers, and graduate and undergraduate students.

The members of the two labs we studied in systems biology, lab G and lab C, preferred the name "integrative systems biology" (ISB) for the area in which they worked. They explained that "integrative" stressed both the integrating function of building a model of a biological system, as well as their research aim to integrate a range of resources from biosciences, engineering, and computational sciences in their investigations. In the ISB study we had less funding and fewer researchers for our project, so we conducted intensive data collection in both labs over the first year and followed selected dissertation projects through to completion for a total of five years. We started data collection in the two labs at the same time. In both labs the primary focus was building computational simulation models. There was significant interaction between the directors of these labs, though not much among their students. There were few ISB researchers in the department, and lab directors were hopeful they could build the area together, and possibly start an educational program aimed, specifically, at aspiring ISB researchers. They worked together with us to develop a graduate-level introduction to biosystems modeling course, which they co-taught. The lab G director provided some mentoring to the quite junior lab C director, even though, as we will see, they had quite different "philosophies" about how to conduct ISsB research and for what purposes.

Lab G is a purely computational systems biology lab, with the clever motto, "where life becomes numbers and numbers come to life." Its research problems focus on computational simulation modeling of biological systems at the genetic, metabolic, and cellular levels. The focus of the modeling is on the interactions among different components of biological systems (such as metabolic and signaling pathways), rather than on structural properties of specific components (such as DNA and ribosomes). The problems addressed are wide-ranging, and usually brought to the lab by biological researchers from universities and industry because of the outstanding reputation of the lab director as a pioneer in ISB. For instance, one of the problems tackled by the lab was to develop a model of the production and transport of dopamine and of how this system is affected in Parkinson's disease. In this research, the lab worked with experimental data provided by a medical research group specializing in neurodegenerative disorders. Another problem was to develop a model of ethanol production using algae, based on data provided by researchers at a biofuels company. In general, the domain-driven problems are provided by bioscience researchers of various kinds who approach the lab, asking the director to "model our data," usually with little understanding of what that means or entails. The overarching focus of the lab's own agenda is on methodological problems specific to computational modeling of biological systems, especially developing mathematical techniques and algorithms to improve the estimation of model parameters and the optimization of these parameters.

During our study the main lab members included a male director, four postdoctoral researchers (two female, two male), and four PhD students (one female, three male). The members of the lab varied widely in terms of educational background, although most were from engineering (mechanical, electrical, telecommunications, biomedical, computer). Other backgrounds included pharmacy, applied physics, bioinformatics, and information sciences. The main criteria for being accepted into the lab were applied mathematical and/or systems computational skills. A postdoctoral student who was from a collaborating experimental lab in Europe visited periodically for a month or so at a time; he had a PhD in biochemistry and was transitioning to modeling. A striking feature of the lab is that all members were from outside the United States. Eight were from Asia (China, Taiwan, Japan), two from Europe (including the director), and one from the Middle East. As a consequence of the sophisticated computational modeling skills the research requires, there were no undergraduates. The lab director is a senior pioneer in ISB, with an undergraduate degree in natural sciences and mathematics, two master's degrees (one in biology and the other in mathematics), a certification in philosophy and education, and a PhD in theoretical biology. The lab had been in existence for five years when we entered (the director's previous lab ran for fifteen years at another institution). The lab space consisted of desks with computers, and was quite often empty, since lab members could just as easily work at home on their laptops. For this reason, we stopped aiming for field observations after a few months. Research meetings were largely conducted one-on-one with the director, and they did not hold lab research meetings, though they did organize a few so we could get an overview of the research, and the researchers noted that these were beneficial to them as well. The lab researchers had a range of biosciences collaborators external to the lab (some of whom we interviewed).

Lab C is an ISB lab that conducts both computational modeling and biological experimentation in the service of model-building. Its research is guided by an overarching biological problem: to understand the impact of redox (reduction-oxidation) environments on proteins through systems modeling approaches. Under normal physiological conditions cells maintain a reduced internal environment. However, oxidizing molecules and free radicals that are produced in the cell as a part of physiological processes, or that enter the cell, can react with cellular components such as DNA, cell membranes, and proteins. Such reactions have physiological consequences and have been implicated in several diseases. Lab C's research focus is on the impact of alterations made by oxidants on proteins, which are part of signaling pathways, and on the dynamics and outcomes of these pathways. Based on her own training, the director has been training the graduate students, who have engineering backgrounds, to do biological experimentation in the service of building and testing their computational models. One student also engaged in engineering design through a collaboration to develop a microfluidic device ("lab-on-a-chip") to produce high-throughput single-cell and population data, which are the time-series data more amenable to quantitative investigation. The lab's overarching problem translates into specific research projects as varied as modeling chemotherapeutic drug resistance in acute lymphoblastic leukemia cells and modeling senescence in T cells. However, everyone in the lab was aware of what the others were working on, and provided feedback on the research projects of others in the weekly lab meetings and in weekly journal club meetings. We witnessed many instances of joint troubleshooting, in particular.

Lab C was just taking shape as we started our research. During our study the main lab members included a female lab director who was a new assistant professor, five PhD students, two of whom joined the lab after we started our observations (three male, two female), six undergraduates, and a female research technologist/lab manager with an MS in biology (who transitioned to a PhD student, while remaining manager, late in our study). A striking feature of this lab is that the researchers spanned four continents (North America, Europe, Africa, Asia). The lab director has an undergraduate degree in nuclear engineering (with a minor in biomedical engineering) and a doctoral degree in bioengineering, during which she first trained as a modeler and then as an experimentalist, followed by a postdoctoral period in a bioengineering lab that comprises both computational modelers and bioscientists. The graduate student backgrounds were predominantly engineering-related (electrical, biomedical, biotechnology, material science). A joint MD/PhD student had a background in chemistry and mathematics. The undergraduates mainly ran western blots and other experimental procedures for the graduate students. The lab's experimental biology research is conducted in-house, but they had a few external engineering and bioscience collaborators and bioscientists with whom they consulted during the period of our investigation, some of whom we interviewed.

1.4.3 Data Analysis

Numerous qualitative methods can be used singly or jointly in cognitive ethnographic data analysis. We have been using a variety of mutually complementary qualitative methods, specifically, qualitative data coding, case study analysis, thematic analysis, and cognitive-historical analysis. These are among a wide range of qualitative methods that have been developed and critiqued extensively over the last half century, especially in psychology and sociology (for an overview, see Patton 2002).[24] There are no formulas or recipes for how best to apply those qualitative methods in any specific case, so we have needed to tailor and innovate our data analysis with respect to our research goals and questions—as is standard in qualitative analysis—while adhering to accepted canons of what constitutes "trustworthy" (Lincoln and Guba 1985) and "validated" data collection and analysis procedures. Although "valid" is reserved in philosophy for logical argumentation, it is often used to signify credible qualitative research in that field. I prefer to use Guba's more neutral term, "trustworthy," when considering issues of warrant.

To establish trustworthiness, we have, in particular, taken into account the American Psychological Association standards (see, e.g., Eisner 2003, who argues standards need to take into account that qualitative analysis is both science and an art). We have followed three standard principles in particular: structural corroboration, referential adequacy, and consensual validation (Eisner 2003). Structural corroboration requires that a sufficient number of data points converge on a conclusion to support an interpretation. Referential adequacy addresses the richness and clarity of the description and interpretation, and how it aligns with member understanding. Consensual validation refers to the level of agreement that can be reached among two or more researchers in developing and using the coding schemes ("interrater reliability"). Adherence to these principles required that we would systematically collect the range and kinds of data sufficient to triangulate data from multiple sources in order to corroborate and determine the referential adequacy of interpretations. "Triangulation" in qualitative analysis refers to the processes of building warrant for an account through establishing consistency of findings across methods and sources of data collection. Our research conducted long-term studies that provided a variety of longitudinal data, which (as noted) consisted of persistent observations, of multiple interviews of each participant, and of the kinds of archival data previously mentioned.

Data collection in an ethnographic study always risks the dual charge of being not representative and/or subject to bias stemming from the researcher's own interests, values, and motivations. Ethnographic investigation demands continual self-scrutiny so as to mitigate researcher bias, which is an issue in all empirical research. Such self-scrutiny, for example, would control for asking leading questions in an interview or for importing favored notions into data analysis. In general, it is important to keep in mind that all ethnographic research is interpretive. As such, the researcher is the instrument of data collection and analysis, and, so, the researcher's interests, values, and motivations are always present, and it is a necessary part of good research to be explicit about and confront these (Osbeck and Nersessian 2015, Nersessian and MacLeod 2022). We were aided in our attention to potential researcher bias in data collection and analysis by the unusual approach we took to conducting ethnography. Unlike traditional practices, where the ethnographer is a single researcher, we decided to practice what we dubbed "team ethnography." In each given lab, more than one ethnographer was responsible for observations and interviews, and our more senior members worked across the labs. As project director, I oversaw that in all labs we collected comparable data to the extent possible. Our research group varied in size and composition over time (undergraduates through senior faculty) but remained highly interdisciplinary and thus provided multiple lenses through which we could examine the data.[25] Our weekly research group meetings provided the venue for scrutinizing and evaluating the ethnographic work together as it unfolded, and for reaching consensus on coding, theme development, and other forms of data interpretation. As data analysis progressed, we related our findings to the appropriate philosophical and cognitive science theoretical frameworks. We also formulated hypotheses with respect to issues within these frameworks, and together we evaluated, revised, or refined these in comparison to our empirical analyses.

Since coding is the first method through which one starts to make sense of the data, I next briefly describe a few of the procedures we used in our coding analysis of data. Although we used a variety of complementary methods of data analyses, as noted earlier, procedures that relate to systematic, fine-grained open coding and to grounded theory development (Corbin and Strauss 2008; Glaser and Strauss 1967; Strauss and Corbin 1998) provide the primary basis for our interpretations.

Coding is an interpretive procedure by which to partition the data by attaching descriptive categories to units of interview texts and field observation notes. Our approach to coding can be broadly characterized as "grounded" in the sense described by Corbin and Strauss (2008). We understood this to mean, in particular, that we remain open to seeing what categories/themes might emerge from the data. While, obviously, our coding was guided by our research questions and objectives, it was by no means restricted by them. We developed our coding procedures in several phases. We established coding procedures to mitigate, to the extent possible, issues of subjectivity—we even hired an external coding auditor midway in our investigations, by way of a check on our procedures.

We began with "open coding" directed toward identifying, categorizing, and describing what the text of the interview is about. During this process, coding pairs worked together on each transcript. We analyzed a subset of interviews progressively, line by line, with the aim to provide an initial description for as many textual passages or "meaning units" as seemed appropriate. In our research meetings, the entire group discussed the clarity, fit, and logic of the codes assigned. In early coding, we presented interpretations to the research lab members by way of checking whether their views aligned with our understanding. More than that, we used feedback from all pertinent sources to make adjustments.

We continued coding additional interviews, revisiting previous coding, and assessing descriptions for adequacy and for fit throughout the process, as is consistent with the goals of analytic induction (codes emerging from data and leading to hypotheses) and constant comparison (codes compared against possible alternative interpretations) (Lincoln and Guba 1985; Corbin and Strauss 2008). After about 20 percent (the standard) of the interviews were coded intensively in this manner, we coded the rest more selectively, focusing on categories of most relevance to our research questions and building out those categories. During research group meetings we reviewed all codes, and further grouped and arranged codes into superordinate categories and subcategories. We then related the codes to each other and developed the categories/concepts more directly with respect to our research questions as a start toward building "theory." In this context, we understood this process, broadly, as formulating "a set of well-developed categories (themes, concepts) that are systematically interrelated through statements of relationship to form a . . . framework that explains some phenomenon" (Corbin and Strauss 2008, 55) and allows forming hypotheses. Theory development in effect, then, took the form of developing increasingly refined conceptual models.

We coded separately for each lab, and then assessed the candidates for transfer across the labs in BME and ISB. Exemplars of lower-level codes from the BME study include analogy; model-based reasoning, understanding, or explanation; problem formulation; anthropomorphism; epistemic values; and pragmatic focus. Exemplars of superordinate categories include model-based cognition; seeking coherence; norms; and affect. In all, we developed seventy-three codes, and, with respect to codes that transferred across the BME practices, we organized these into thirteen superordinate categories. We did not use a coding software, preferring instead the traditional method of coding by hand. We developed our own coding database, using

MS Excel, which lists each category and code with an associated description and memo discussing it, and the codes have multiple exemplars from the interview texts attached to them from each lab. Codes can be easily organized and reorganized into categories with this software. We used codes and categories to create case analyses, which are finely detailed descriptions that follow practices of a specific researcher, or small group, as they worked toward solving a complex problem. We also developed longitudinal case studies specific to learning in the labs.

Code development, however, is more than mere description. It is an abstractive process, in which a code is both derived from and scrutinized in light of multiple exemplars across different interview texts within the study. Codes provide, also, the basis for cross-study comparison and for developing hypotheses to consider and assess for transfer, when detached from case-specific details. As our research progressed, we continued to assess transfer of selected major categories and themes across the labs. We were especially interested in what commonalities there might be in the general features of the model-building practices in these subfields of biological engineering and how these advance the epistemic goals of the subfields, in practices developed to support learning in the context of research, and in the challenges presented by the kind of interdisciplinarity. A major example of a cross-cutting category—or "theme"—that emerged from the BME labs and is developed in the chapters that follow is the multidimensional system notion of "interlocking models." This notion serves to articulate how multiple dimensions of these interdisciplinary research labs are built and fitted together as cognitive-cultural systems. Models interlock biological and engineering concepts, methods, and materials. They interlock in their design and construction and in experimental processes. Mental and material models interlock in model-based inference. In the latter instance, "interlocking models" is a specific kind of coupling between researcher mental models and artifact models as components of a distributed model-based reasoning system. Further, epistemic and sociocultural practices interlock in building models. We found the challenges of building systems of interlocking models to be central to research and learning for hybrid researchers. Table 1.2 provides a schematic overview of the major interlocking models with respect to the tissue engineering lab A, which is elaborated in chapters 2 and 4. The models interlock both within and across the categories.

Table 1.2
Interlocking models in lab A

Interlocking models			
Biological, engineering, medical models in the wider community (as detailed in journals, textbooks, etc.)			
cell biology	electrical engineering		
biochemistry	mechanical engineering		
fluid dynamics	disease processes		
Bioengineered in vitro artifact models			
flow loop	pulsatile bioreactor		
construct	baboon model-system		
Researcher mental models			
in vivo and in vitro phenomena			
devices qua in vitro models			
devices qua engineered models			
Sociocultural models			
Mentoring	Identity	History	Epistemic values

Finally, we never ceased using our codes and categories to analyze the data further and to examine them through various theoretical lenses in order to develop thick descriptions and analytical insights—in particular with a goal to extend, enrich, and critique philosophical concepts and theories in "productive interplay." As any ethnographer would agree, an ethnographic analysis is never complete. The greater the depth of the analysis, the more one sees what needs to be analyzed, as well as what additional data it would have been useful to have collected. This book presents yet a further analysis, which, hopefully, provides an exemplar of how to develop a conceptual model of the dynamics of cognitive-cultural integration in scientific problem-solving. Although undoubtedly there are other strategies to develop a cognitive ethnography, ours shows, in particular, that philosophers are well able, and well positioned, to work with ethnographic norms while pursuing philosophical targets of investigation.

1.5 Overview of the Chapters

Chapter 2: Building Hybrid Simulation Devices: Distributed Model-Based Reasoning. This chapter focuses on the BME in vitro simulation devices as cognitive-cultural artifacts that enable distributed model-based reasoning. The chapter provides an analysis of the iterative and incremental processes of *building*: designing, constructing, redesigning, evaluating, and experimenting with in vitro devices. The devices are physical simulation models comprising part-living, part-engineered materials. One tissue engineer called the practice of building these devices *"putting a thought into the bench top and seeing if it works,"* which the chapter interprets as building "distributed model-based reasoning" systems. The hybrid model-systems simultaneously provide simulations of biological processes, the researcher's current understanding of these, and the epistemic culture's norms and values. The chapter introduces the analytical theme of "interlocking models" and examines how devices provide hubs for interlocking many dimensions of practice. It further examines how these in vitro models are "built analogies" that are designed to simulate the behaviors or functions of selected in vivo biological processes. This form of model-building expands the epistemic practice of "building the source analogy" (Nersessian 2008), wherein analogies are designed for the purposes of scientific investigation by analogical displacement. It details several examples of how the researchers build epistemic warrant for in vitro devices, as well as for the methodological practice itself, in the processes through which they create the models, focusing, specifically, on the relationship between analogy and exemplification. (Lab A and lab D data)

Chapter 3: Engineering Concepts: Conceptual Innovation in a Neuroengineering Lab. This chapter focuses on the interplay between conceptual innovation and modeling practices. The pioneering nature of the labs leads to researchers investigating novel phenomena that have been conceptualized only partially or not at all. Thus, frontier problem-solving often requires conceptual innovation. The chapter follows the researchers in lab D in their quest to understand and control the behavior of a living network of neurons. At the outset of the research, they transferred concepts from engineering and single-neuron studies to get a grip on the model. These resources both facilitated and hindered their problem-solving. Ultimately, the research led them to develop fundamentally novel concepts. The analysis starts at the point where the researchers were failing to understand and control perplexing in vitro model behavior, which led one to introduce a novel practice for the lab: computational (in silico) modeling of the in vitro model. The epistemic affordances of the in silico model promoted concept formation and change as the researchers worked toward interpreting both the in silico and the in vitro behaviors, including the relations between the models. The chapter details how, over the course of two years, the cross-breeding of these two kinds of simulation models created a cluster of scientifically novel (and potentially highly significant) concepts, while also building a D-cog system comprising all of the researchers. Armed with these new representations of the behavior, this system was able to leverage the affordances of both models to productively control the behavior of the dish model-system, and, ultimately, attain the lab's goal to establish and demonstrate that the in vitro network of neurons could learn. (Lab D data)

Chapter 4: Interlude: Building "the Lab." This chapter focuses on the theme of how "the lab" builds itself as a cognitive-cultural system. It analyzes the dynamics of how, starting from broadly framed complex interdisciplinary problems, a research lab on the frontiers of science creates and develops the cognitive-cultural structures for productive research. It examines this building process in detail for the tissue engineering lab. The chapter examines how intersecting trajectories of problems, methods, and researcher-learners develop in relation to the practice of building in vitro devices, and details how the lab's *signature* devices (examined in chapter 2), in particular, provide structuring constraints for articulation of the lab as a distributed cognitive-cultural system in ongoing flux. It ends with a brief look at the educational infrastructure developed to foster the BME goal of interdisciplinary hybridization. (Lab A data)

Chapter 5: Managing Complexity: Modeling Biological Systems Computationally. This chapter focuses on the challenges of computationally modeling complex dynamical biological systems in the absence of domain theories that can provide significant resources for building models, such as for physics-based modeling. It examines how ISB researchers, engineers with limited biological knowledge, develop practices around computational modeling and simulation that enable them to manage the complexities of modeling biological systems. The analysis shows how a close examination of the processes of building models, rather than a focus on the finished products, is needed to fathom the epistemic achievements of this emerging approach to discovery in systems biology. The chapter details a case in which an engineer with little experience in biological systems modeling and little biological knowledge was able to make a fundamental discovery in biology by means of his "adaptive problem-solving" processes. It examines, in particular, the epistemic affordances of in silico simulation for building the model while also developing a close coupling between the modeler's mental modeling processes and the biological system model. The developing modeler-model D-cog system provides another instance of what Hutchins called creating "cognitive powers," which in ISB enables a modeler to make novel, verifiable inferences about the biological phenomena that outstrip his own understanding. (Lab G data)

Chapter 6: The Bimodal Model-Building Strategy. This chapter focuses on a novel method for managing the complexity of building models of biological systems. In this practice, modelers conduct their own biological experimentation in the service of building their models. The chapter highlights the methodological flexibility available to ISB as an emerging field, which affords researchers the opportunity to tailor methods to manage complexity. The chapter develops two case studies of modelers following this strategy. The first examines, briefly, how a modeler collaborated with engineers to design a microfluidic "lab-on-a-chip device" (LOC) to integrate complex activities, actions, processes, and operations in wet-lab experimentation that would usually be carried out in many steps, by many persons, and using a range of equipment. The modeler built the LOC to solve the difficult problem of collecting time-series data needed to develop her computational model of T-cell signaling. The second case examines, in detail, how one modeler built a tightly coupled methodological system that used computational model-building and simulation to direct and focus her wet-lab experimental investigation of a biological system, while also using the experimentation to further develop her model. The epistemic affordances of this D-cog system, in particular, helped her to triangulate uncertainties and missing elements in her models without having to deal with complex problem spaces of many open parameters. In both cases the novel methodological strategies enabled the modelers to manage a range of constraints—data, computational, cognitive, collaborative—prevalent in ISB research. The chapter illustrates and provides further insights into the possibilities for adaptive problem-solving in this emerging field and how they provide researchers with considerable flexibility to create different methodological strategies and lab organizations to manage the complexities of modeling biological systems. (Lab C data)

Chapter 7: Interdisciplinarities in Action. After providing a high-level summary of the major insights gleaned from the previous chapters, this chapter goes on to consider implications with respect to the epistemic situation of interdisciplinary science as such. The analysis offers insights gleaned from all our investigations of what we call the "adaptive problem spaces" of biological engineering: spaces where interdisciplinarity is enacted in research and learning. It examines challenges, differences, and similarities across fields, and assesses their implications for broader application to interdisciplinary practice. The chapter provides a nuanced account of two major kinds of interdisciplinary practices: hybridization and symbiosis ("epistemic interdependence"). It proposes specific characteristics, *interdisciplinary epistemic virtues*, that foster creativity and collaboration in twenty-first-century interdisciplinary science, at least of those varieties, and illustrates how these can be cultivated in different research communities with targeted interventions, such as those we developed for BME and ISB. Although the focus of the analysis is on the cases we have investigated, I hope the insights in this chapter, and the book as a whole, lay the ground for future research into interdisciplinary epistemic virtues in situations beyond those cases, and beyond science to interdisciplinary practice per se. (Data from all labs)

2 Building Hybrid Simulation Devices: Distributed Model-Based Reasoning

Research in biomedical engineering sciences (BME) has dual aims: to develop understanding of complex biological systems and to manipulate, control, or intervene in them. In this respect these fields aim at basic research while sharing the goal of application that Mieke Boon (2011, 2017) has claimed distinguishes the engineering sciences from the sciences. I concur that, unlike the sciences, which philosophy has traditionally understood to have the objective of creating knowledge, the primary objective of engineering sciences is what I call "getting a grip"—to understand sufficiently to manipulate, control, or alter in specific respect(s)—a pragmatic, engineering goal. However, in many fields of BME, as in the labs we studied, the application potentials are at most aspirational and in some instances do not even come into view until the research is significantly under way. This is because the basic biological phenomena are not yet sufficiently understood and, in some instances, have not been investigated at all.

Research in BME often confronts a problem that is not feasible, or would be unethical, to carry out experiments with animal or human subjects. Importantly, such studies, even if possible, would lack the requisite kinds of experimental control. Thus, in order for investigation to be possible, the biological system of interest must be reengineered in ways that *manage the complexity* of the biological phenomena. That is, researchers need to devise ways to emulate selected aspects of in vivo phenomena to a degree of accuracy sufficient to warrant their transferring outcomes of in vitro experimental simulation to in vivo phenomena in the form of provisional understandings and hypotheses. This is a complex interdisciplinary challenge. Our investigation examines how specific labs in the fields of tissue engineering and neural engineering address this challenge. The kinds of practices we have been examining, however, are not unique to the labs we have investigated. Rather, our studies provide insight into the basic epistemic landscape of biological engineering: the use of engineering concepts, methods, technologies, strategies, materials, and epistemic norms and values to reengineer biological phenomena so as to get a grip on complex dynamical biological systems. This chapter examines, specifically, the processes through which in vitro simulation models of complex biological phenomena gain their capacities and credibility. These models are pared-down representations of dynamical phenomena that have the capacity to enact selected biological processes under experimental conditions.

In our studies of two pioneering BME university research labs, we have found a common investigative practice is to create greatly, but appropriately, simplified living in vitro systems that parallel selective features of the in vivo biological systems of interest. The features chosen are those relevant to the goals of the research. These hybrid systems comprise artifacts composed of living cells and tissues and engineered materials. They perform simulations of biological processes and afford various possibilities for experimentation. They are epistemically and ontologically hybrid. The researchers refer to their individual in vitro models as "devices."[1] In my analysis, devices are built analogue models, designed to exemplify specific in vivo biological processes. That features of a model exemplify selected features of the in vivo system is meant in the sense advanced by Nelson Goodman (1968) and extended to scientific practices by Catherine Elgin (2018). As introduced by Goodman, a representation exemplifies a certain feature if it "both is and refers to" that feature; that is, "exemplification is possession plus reference" (1968, 53). BME researchers strive to design in vitro models that both instantiate and refer to features of the in vivo biological system germane to their epistemic goals. Specifically, a device exemplifies a selected feature of the in vivo system, such as the force with which the blood flows in a human artery, if it instantiates this feature, and it refers to this feature via its instantiation of it.[2] But, unlike the original notion, which addresses finished representations, here I will show exemplification to be a dynamic process in which *models are built toward exemplifying features of the target biological system*. Researchers design and perform in vitro simulation experiments with devices in processes they claim "*parallel*" or "*mimic*" salient aspects of in vivo situations, by which they mean, as we will see, the devices possess the features relevant to mimic in vivo behaviors. Here, it is clear that similarity (however one cashes out that notion) could not be the appropriate representational relation. The model needs to instantiate the feature. Exemplification, then, provides the criteria by which to evaluate whether and in what ways a built model can serve as an analogical source through which to investigate the target in vivo phenomena. That is, the building process not only establishes the reference of the model, but also enables the researcher to determine whether the model does or does not instantiate the target features relevant to the problem. These criteria assist researchers in their determination of what warrant they have to transfer inferences about the model to the in vivo system as provisional hypotheses.[3]

The in vitro models provide the primary cognitive-cultural artifacts of the D-cog systems of the lab's research; specifically, they are the part of the lab's material culture that has epistemic functions. As we will see, the main epistemic dimension of BME problem-solving practices is building in vitro simulation models, and so it is important to be aware from the outset that these devices also instantiate culture-specific epistemological assumptions. In a different context, Evelyn Fox Keller has noted such assumptions include at least those that "underlie the particular meanings they give to words like theory, knowledge, explanation, and understanding, and even the concept of practice itself" (Keller 2002, 4). Significantly, in the BME case, the design of a device embeds norms, values, and assumptions primarily associated with aims of control and intervention and the kind of quantitative analysis required by engineers rather than biologists. Many of the researchers in our labs characterized this epistemological difference in the following way: biologists focus on how *"everything interrelates to everything else,"* while engineers *"try to eliminate as many extraneous variables as possible."* In this epistemic context, a device, then, serves as a site of simulation not only of biological processes, but also of the researchers' epistemic assumptions, norms, and values.

I take the practice of in vitro model-building to be rooted in the engineering practice of building prototypes, but in the BME case the practice is dependent on what it is feasible to do with biological materials using engineering materials, methods, and technologies. *Building,* I remind the reader, is the technical term we use to encompass iterative and incremental processes of designing, constructing, redesigning, evaluating, and experimenting with a model. Building in vitro models is a bootstrapping process, which, to borrow a notion developed by Hasok Chang in a different context, entails "epistemic iteration": "a process in which successive stages of knowledge, each building on the preceding one, are created in order to enhance the achievement of certain epistemic goals" (Chang 2004, 45). In vitro model-building processes center on how the biology can be shaped by the engineering and vice versa to provide a legitimate source for displaced biological experimentation. As we have witnessed, obstacles that can lead to impasse or failure are ubiquitous, but researchers learn from each incremental step. Importantly, through these incremental and iterative processes, BME researchers develop reliable principles for building models of a specific type, as well as for the novel epistemic practice of in vitro modeling itself. The director of lab A was a pioneer in inventing the paradigm of in vitro modeling, and both he and the lab D director were pioneers in their respective modeling practices.

Although they are simplified biological systems, devices themselves are complex dynamical systems that are used in experimental processes either individually or in configurations with other devices. Researchers refer to the experimental systems, whether a single device or a configuration, as "*model-systems.*" As one researcher stated, "*when everything comes together, I would call it a model-system. . . . I think you would be safe to use that [notion] as the integrated nature, the biological aspect coming together with the engineering aspect. So, it's a multifaceted modeling system.*"[4] A specific model-system can be the locus of an experiment or just one step in a multi-model experimental process. From the perspective of cognitive-cultural integration, these hybrid in vitro model-systems are "multifaceted" in another respect: they constitute the material culture of these communities, they participate in epistemic goals, they give rise to interactive practices, and they perform as cognitive artifacts in their problem-solving processes. Developing skill at a specific building practice makes one part of the lab research cognitively and socioculturally. Further, as chapter 4 develops, building in vitro models not only creates understanding about biological phenomena; these processes also create "the lab" itself, both materially and as a way of doing science. Each generation of researcher (~five years in a lab) provides the methods, artifacts, and ways of thinking that serve as cognitive-cultural ratchets that scaffold the next generation of the lab's problem-solving activities. This scaffolding not only provides stability but also affords creativity in a research program.

In this chapter I introduce the primary devices of each lab. These figure in different analyses in chapters 3 and 4. Here, I describe the devices and unpack some of the epistemic affordances of these artifacts in the problem-solving practices of the BME lab, which we understand as a dynamic ecosystem of people, artifacts, and embodied skills (Hutchins 2011). Devices, as we have come to understand them, are sites of cognitive-cultural integration. They serve as hubs for interlocking biological and engineering concepts, methods, materials, values, and norms in mental and artifact representations, in design and lab history, and in research and learning. "Interlocking models," as noted in chapter 1, is a system-level interpretative theme we developed to capture intersecting aspects of research practices that emerged from our coding categories. We use "interlocking" in these interdisciplinary contexts, rather than the customary "integrative," to specify that integration in BME is a process of fitting things together. "Interlocking" signals, for instance, that researchers do not need to learn all the biology in an area of interest, just what fits their engineering and medical goals, or, for example, that in building a model-system, specific constraints of the biological and engineering components need to be fit together, as do the various component models in the model-system.

In section 2.1, I provide a brief look at the history of the development of the *signature* devices of each lab and lay out some of the reasoning underlying the researchers' choices of what to instantiate in a particular model. This reasoning is important for our objective to understand how such in vitro models can provide the basis for warranted inference. Section 2.2 discusses how the processes of building the models interlock mental and artifact representations into D-cog systems that can perform distributed model-based reasoning. Section 2.3 considers how analogy and exemplification work together to provide credibility for the researchers' claims that the practice of in vitro modeling provides understanding and predictions about in vivo systems.

Section 2.1 might be difficult for some readers, but it is important to at least skim the details of the model-systems because, first, to my knowledge our research provides the only account of the modeling practice of experimental simulation by means of building hybrid in vitro devices, which is central to major fields in biological engineering, and, second, the details are necessary to fathom how they function as analogue models that can support inferences about complex biological systems.

2.1 "An Experimental Model That Predicts": The Epistemic Practice of Building to Discover

Simulation by means of devices is an epistemic activity that forms the basis for understanding, explanation, and prediction in many areas of BME. Much of our research on the tissue and the neuroengineering labs has been directed toward trying to understand the nature of their in vitro models, their epistemic affordances and the sociocultural practices surrounding them, and, in general, how they relate to the epistemic and practical goals of the labs. In the BME labs we studied, devices are built mostly in-house. We refer to the primary devices as signature artifacts because references to the lab both internally and externally are often by means of that device, such as "the flow-loop lab" for lab A. Further, these devices often play a role in standardizing research in the area, for instance, flow-loop studies in tissue engineering or neuron dish studies in neuroengineering.

Devices are in vitro models that, in the words of lab members selectively *"mimic"* or *"parallel"* in vivo biological processes of interest, either normal or aberrant, under experimental conditions. That is, they are dynamic physical systems that simulate biological processes. Although we have not been able to determine how and when the idea for the practice of building in vitro simulation devices arose, the practice does appear to extend to biology the engineering practice of building facsimile models that mimic engineered phenomena, such as the wind tunnels that have been used to experiment with different aerodynamics for at least a hundred years (now largely replaced by computational simulation models). Of course, researchers cannot control living model-systems as fully as inanimate engineered models.

BME researchers aim to build models that allow them to transfer inferences that derive from experiments they conduct with in vitro models to in vivo phenomena as candidate understandings and hypotheses. As a researcher explained about her model-system: *"We typically use models to predict what is going to happen in a system [in vivo]. Like people use mathematical models to predict . . . what's going to happen in a mechanical system? Well, this is an experimental model that predicts what would happen—or you hope that it would predict—what would happen in real life."* Such prediction is a form of analogical transfer. The research in lab A, for instance, aims to create dynamic physical models that will enable inferences about disease processes related to normal and abnormal arterial blood flow processes. Thus, the primary epistemic function of the practice of in vitro model-building is to enable inference by means of model construction and manipulation, which I have elsewhere called *simulative model-based reasoning* (Nersessian 1992, 2002, 2008). We will see, as Helen Longino also has pointed out, that "treating reasoning as a practice reminds us that it is not a disembodied computation, but takes place in a particular context and is evaluated with respect to particular goals" (1990, 215). In sections 2.1.1 and 2.1.2, I unfold some of the context, goals, and considerations that have gone into building the primary model-systems of each lab.

On one level the researchers in the tissue engineering and the neuroengineering labs belong to different cognitive-cultural systems. On another level, however, and for the purposes of this chapter, there are important commonalities—ways of approaching research that we found to transfer robustly across both sites. Although the specifics in each case differ, our insights about the nature of the model-based reasoning that devices afford and the primacy of engineering assumptions, norms, and values in conducting this kind of in vitro research are two such interrelated dimensions. To understand how they interrelate, I begin by unpacking the processes through which the devices and model-systems are built to perform as cognitive artifacts and epistemic tools.

2.1.1 Lab A: The Flow-Loop Device and Model-Systems

From the outset, the assumption guiding lab A's research was that mechanical forces produced by blood flow in the cardiovascular system have biological effects and contribute to disease processes. The lab A director began this research program as a mechanical engineer who worked in aeronautics. NASA requested his help to understand the effects of the vibration forces of launch and reentry on the cardiovascular systems of the astronauts. In our first interview, he formulated the insight he had during that investigation that would transform his research into a bioengineering program as "*characteristics of blood flow [mechanical stress/strain forces] actually were influencing the biology of the wall of a blood vessel. And even more than that—the way the blood vessel is designed. The way the blood vessel is designed is—it has an inner lining, the endothelium. It's a monolayer—it's the cell layer in direct contact with flowing blood. So, it made sense to me that, if there was this influence of flow on the underlying biology of the vessel wall, that somehow that cell type had to be involved.*" His research, thus, started with an engineering framing of a biological problem and a goal to understand complex biological processes of the cardiovascular system in terms of mechanical engineering concepts and methods. The hypothesis that mechanical forces were *"influencing the biology"* was radical at a time when the nascent field of vascular biology was focused on biochemical processes, and biologists initially rejected it. His statement also reveals the design perspective on biology of an engineer, which came to pervade his investigative program. This engineering framing provides a means for managing the complex biological problem of the nature and effects of the dynamical processes within blood vessels by reducing it to understanding the effects of flow (mechanical forces) of blood on a specific cell type. The director proposed a novel hybrid "placeholder" concept (Carey 2009), called *"arterial shear,"* that is, the frictional force of blood on the endothelium as it flows in the parallel plane through the lumen (the inner space of the arterial tube), and the goal of articulating this concept was a driver of the research for more than forty years. In the course of developing and following out his research program, the director was also a pioneer in creating the "interdiscipline," (by which we mean a field that has coalesced into an "interdisciplinary discipline") of BME.

In the configuration in which we encountered lab A after nearly twenty years in existence (and more than thirty after the then director had begun the line of research) some of the complexity had been added back into their research. Further, an applied goal had been added to the lab's research agenda. Researchers now sought to understand how more components of the blood vessel wall respond to both mechanical stress forces (shear and strain) and to understand the requirements for designing a fully functional vascular tissue replacement for repairing the human cardiovascular system. As the director formulated their dual basic and applied research goals at that time, *"When it comes to a blood vessel substitute, what are the mechanical properties and how do we engineer those in—how do we fabricate something with the appropriate mechanical properties?"* Thus, the goal of creating new tissue engineering techniques had also now become a significant part of their problem situation. To achieve their applied goals, however, required basic research to understand what properties and processes give blood vessels the strength to withstand the in vivo forces of circulating blood. This understanding would enable the researchers to develop ways to design vascular tissue models (*"constructs"*) with sufficient strength, which could eventually lead to vascular implants. Researchers faced another problem, related to the functionality of any implant: to develop a ready source of immune-resistant endothelial cells. These cells are the most immune sensitive in the body, and implanted vascular grafts would need to be seeded with cells that would not be rejected by the host. The lab, again, formulated this problem in engineering terms and sought to find a way to create endothelial cells through mechanical manipulation of stem cells and progenitor cells. Whether this was even possible was an open question at the time.

Lab A had the look of a biology lab, with flasks, pipettes, a sterile workbench with a hood, incubators, petri dishes, and hazardous waste containers. Researchers could often be found at the workbench scraping cells from animal tissues they acquired from outside the lab. Indeed, the first thing the researchers learned upon entering the lab was to make cell cultures, which proved a daunting task for engineers and required considerable mentoring. However, unlike a biology lab, there were a variety of mechanical artifacts inside the incubators and on the bench tops. Most of these artifacts were lab-built in vitro simulation devices and instruments, with which new researchers familiarized themselves while learning to perform cell culturing. Here I focus on the two devices on which the research of the lab centered, the flow loop and the construct. Together these constituted the main experimental model-system of the lab. These models were designed to interlock in investigations of specific behaviors of the target blood vessel wall of human cardiovascular system.

In his research prior to founding lab A, the director had used animal models: cows in which vascular pathologies (stenosis) had been induced surgically. He then studied the changes in morphology (elongation and orientation) in the harvested endothelial cells of the sacrificed animals. Additionally, he studied velocity patterns in the extracted vessels, which were filled with liquid plastic and hardened ("replica model"), with doppler laser techniques. The results of the two lines of study were correlated to provide insights into the quantitative relations between variations in wall shear stress due to particular velocity patterns and the morphology of cells lining these vessels. After several years he abandoned the elaborate and cumbersome practices of the animal studies as "*too uncontrolled*" along several dimensions, but they had enabled him to understand factors and constraints that needed to be taken into account for launching a program to study the impact of shear stress flow on cultured endothelial cells with engineered devices. What he called the "*move in vitro*" opened the possibility of controlled experimental studies, amenable to qualitative and quantitative analysis, of both normal and pathological flow processes. It also solved problems related to the fact that it takes twenty-four hours to see the results of interventions, and in animals many confounding physiological changes take place during that period. The director set out to design an analogue model that instantiates the selected functions of blood flow in vivo such that endothelial cells behave in response to flow as they would in vivo, under those conditions. I unpack the latter sentence in the rest of this section.

At the outset, researchers need to understand what abstractions might be feasible from a biological perspective in designing an in vitro model, while yielding relevant and important information about the dynamical processes of interest. One lab member expressed the design process this way: *"As engineers we try to emulate that environment [in vivo], but we also try to eliminate as many extraneous variables as possible, so we can focus on the effect of one or perhaps two, so that our conclusions can be drawn from the change of only one variable."* In one major abstraction, the director decided, in line with his initial insight, to isolate and study only the endothelial cells and not include other components of the blood vessel. The researchers reasoned that this abstraction is warranted because these cells line the inner blood vessels, and thus are in direct contact with the blood flow forces and so bear the brunt of the frictional force. Further, as one researcher justified the choice, *"Cell culture is not a physiological model; however, it is a model where biologic responses can be observed under carefully designed and well-defined laboratory conditions."* This fact enables them to derive reliable quantitative measures. Another important abstraction was to begin with studying laminar flow, which is steady and uniform, in contrast to in vivo blood flow, which is turbulent and pulsatile along much of its pathway. The in vitro model-system is, thus, greatly simplified, but to investigate just the response of endothelial cells to laminar flow would at the very least provide baseline information on biological responses of cells to fluid forces.

Given the initial hypothesis of the centrality of the endothelial cells, an in vitro model of the target system requires at a minimum that it can replicate the shear forces of blood on the cells. The channel flow device (*"flow loop"*) is a functional model of that process that enables controlled experimentation directly on endothelial cell cultures, thus creating a model-system. The important modeling parts of the flow loop in use during our investigation comprise a peristaltic pump, a liquid, and a channel in which the liquid flows over cells. The speed at which the pump operates reflects a range of potential blood flow in vivo, and the pulse dampener allows control over the constancy of the flow; for instance, it can turn pulsating flow into laminar flow. Both normal and abnormal flows can in principle be studied. The channel through which an incompressible fluid flows over the endothelial cell cultures on slides is engineered to exact geometrical specifications in a physiologically meaningful range. The liquid medium has the viscosity of blood, a cell-friendly Ph, and other in vivo features. The current flow loop was the product of years of design and redesign, dating back twenty years.

The initial flow loop was designed in 1981 with the capacity only to produce laminar (steady, uniform) flow. The flow was redesigned in 1989 to allow *"studies in which fluid mechanic conditions can be systematically varied,"* which include pulsatile and oscillatory flows, in order *"to determine the extent of any such flow effects"* that can occur in vivo. Given the state of technology at the time, it was a large benchtop system that, as a former student remarked, *"had bulky tubes that looked like some time machine from the 1950s."* It had insulated heating coils to keep the cultures at the requisite temperatures and used hydrostatic pressure difference to derive the flow. Contamination was a constant problem because to keep cell cultures alive requires placing them in incubators that have appropriate CO_2 levels and a specific temperature range, which was impossible with the benchtop system. Over 50 percent of their experiments failed because of contamination.

New technology made significant redesign of the flow loop to address the contamination issue possible in 1995. In an interview, a recent graduate of the lab chronicled the process (see Kurz-Milcke et al. 2004). He had taken on the job of *"model-revising this design to go into the incubator,"* which made long-term (twenty-four hours or more) experiments possible. This was important because it takes twenty-four hours for the effects of flow on the cells to be seen, and contamination increases with time. The researcher claimed to have brought ideas for miniaturizing technologies from a lab he had worked in previously in Japan. *"Model-revising"* entailed a redesign of the model to replace the heating function of the coils with the incubator and to use a pump rather than pressure difference to derive flow. The revision also made the components sufficiently decomposable to allow for independent redesign if needed as the research program advanced. In fact, minor modifications continued to take place throughout our investigation. The redesigned flow loop (figure 2.1), in use when we entered, was

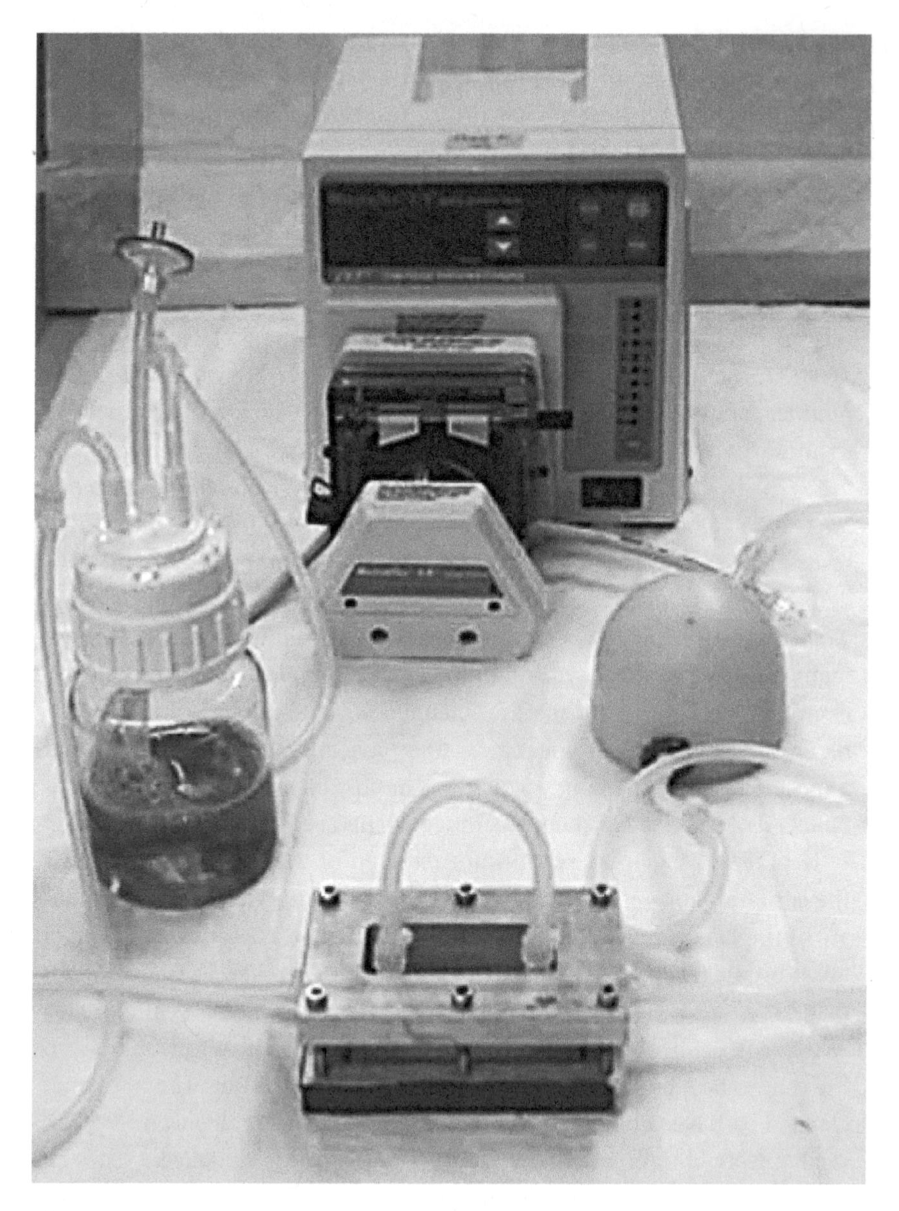

Figure 2.1
Flow-loop setup

assembled under a sterile hood, operated in an incubator, and had an integrated peristaltic pump. The geometry of the flow channel, where cells-in-culture interface with mechanical parts, was left unchanged. This redesign of the flow-loop device was central to its function, as in the model-system, because its viability as a model-system is totally dependent on the ability of the endothelial cell cultures to resist contamination. Experiments flowing cell cultures on slides continued to be conducted throughout the period of our investigation, but investigations with a more complex vascular wall model, the construct device (discussed below) were the focal point of lab research when we entered.

The flow loop, then, is a dynamical model that when in operation can simulate normal and pathological forces of blood flow through the lumen of an artery. In most experiments the process instantiates the shear forces of a steady (constant speed), laminar (straight stream lines) flow over a flat surface (cells on slides). The flow is two-dimensional and unidirectional. The researchers listed all of these features as contributing to their assessment that the model-system "*emulates*" in vivo shear to a "*first-order approximation . . . as blood flows over* [sic] *the lumen.*"[5] They argued that instantiating this process with only characteristics of first-order flow is justified because it provides a "*way to impose a very well-defined shear stress across a very large population of cells such that their aggregate response will be due to*" it and enables them to "*base . . . conclusions on the general response of the entire population.*" That response is determined by removing the cells from the chamber to examine them with various instruments, including the Coulter counter and confocal microscope, which provide information about proliferation, alignment, alive/dead status, morphology, migration, and so forth in the form of qualitative and quantitative representations, that is, numerical and visual (graphical, diagrammatic, color-coded) representations. The models and the instruments function as cognitive artifacts. To use Hutchins's characterization of information flow in a D-cog system: the forces generated by the flow loop represent shear stresses (to a first-order approximation) as it manipulates the endothelial cells, which, in turn, generates "conditioned" cells that researchers then manipulate with instruments that generate quantitative and qualitative information in various representational formats that propagates through the D-cog system.[6]

With this basic analysis of the flow-loop–cell culture model-system, we can begin to see how researchers build the warrant for the inferences they derive from it and transfer to the in vivo system ("*model that predicts*") as they build the models. What warrant there is depends both on their determination of whether the in vitro model does indeed instantiate the selected in vivo features and on the relevance of in vivo features that have been selected or eliminated in designing the model of the biological system of interest. The former is largely a technical issue about whether a desired process is achieved by the design: for example, does the flow loop achieve the desired kind of flow and shear forces on the endothelial cells? Assuming that the device can be made to instantiate the selected features, the main issues researchers need to consider are these, given the goals at hand: Have they selected the relevant in vivo features to instantiate in the in vitro model? Have they left out anything important? and Do any abstractions from the biology they have made for engineering purposes matter for their goals?

Addressing these kinds of questions is an ongoing process in which the researchers create a rationale for their decisions during the incremental and iterative processes of building a kind of model. In the design of the flow loop, for example, the researchers decided to instantiate a first-order approximation to in vivo flow. This means that the in vitro flow is laminar (steady, straight stream lines, no eddies, and non-pulsatile). As we saw, the researchers justified using flow with these characteristics because it provides significant experimental control and simplifies the mathematical analyses. However, as researchers noted, in vivo "*blood sloshes around in the blood vessel,*" that is, the in vivo process has turbulent flow as well. One of the reasons they gave in support of using laminar flow is that there is variation in the flow along the cardiovascular system, with quite high pulsatile flow near the heart but laminar flow near the extremities. So, the in vitro flow does instantiate a relevant feature of the in vivo process, just not all of its features. The issue is whether the features of "*sloshes,*" that is, higher-order effects, are relevant. The researchers were aware that if they found significant discrepancies in the behavior of the endothelial cells, for instance, in the event that, in vivo, "*there's a whole different pattern of genes that are up-regulated in pulsatile shear,*" they would need to instantiate higher-order features. In another feature selection, the model-system was designed to emulate the dynamic but not the diachronic nature of the in vivo environment. Blood flow in vivo changes, for instance, when eating and sleeping. Such changes were seen as experimental confounds with the animal studies, and the goal to control these features motivated the move in vitro. In general, the researchers recognized that flow-loop simulations are *"very abstract because there are many in vivo environments and many conditions within that environment."* Their objective was not *"to mimic the exact conditions found in vivo,"* but to build models good enough to provide reliable, but suitably qualified, understanding of phenomena that are inaccessible to in vivo investigation. The researchers, in fact, consider inferences made from the in vitro simulations to be more trustworthy than if the experiment had been performed in vivo because of the increased experimental controls. But, of course, as a researcher noted, the in vitro simulation experiments only *"indirectly answer [their] questions,"* which are about the in vivo system, so the issue of what qualifications inform the warrant for the transfer of the findings is always present. I discuss the issue in more depth in section 2.3.

Improving the devices and model-systems to instantiate more relevant features was an ongoing part of the research. For instance, for nearly twenty years the researchers used cells in culture on slides in the flow chamber. As we saw above, they justified using the cultures because the endothelial cells are the closest part of the vessel in contact with the blood forces and bear the brunt of shear forces. Simulations with the endothelial cells in isolation from other components of arterial tissue could enable them to get a grip on cell response to shear, but the researchers were always aware that *"cell culture is not a physiological model"* of the blood vessel wall. It leaves out many features of the blood vessel, and thus, the in vitro model-system provides limited understanding for their target problem of the effects of mechanical forces on the blood vessel wall, which has other components. In a first attempt to add relevant features, they created a *"co-culture"* of endothelial and smooth muscle cells, but the limitations remained much the same since it does not capture their structural relations in the tissue of a blood vessel. Specifically, as the director noted, *"putting cells in plastic and exposing them to flow is not a very good simulation of what is actually happening in the body. Endothelial cells, which have been my focus for thirty years, have a natural neighbor, smooth muscle cells. If you look within the vessel wall you have smooth muscle cells and then inside the lining is* [sic] *the endothelial cells, but these cell types communicate with one another. So, we had an idea: let's try to tissue-engineer a better model-system for using cell cultures."* Their aim became *"to use this concept of tissue engineering to develop better models to study cells in culture"*; that is, to build *"a more physiological model"*—one that would instantiate more features and would function as an in vivo vessel along mechanical, physical, and biochemical dimensions. With this more complex model they could study the effects of shear on more components of the blood vessel wall, as well as the interactions of different cell types. It also offered the possibility to investigate the forces of pressure on the wall and other effects (chapter 4). But the "*the big gamble*" the lab took to build a model that could instantiate all the features of a blood vessel wall was only possible because new engineering techniques and materials had been developed that enabled researchers to construct living tissue models. Part of their research focused on furthering the practices of tissue engineering. If successful, building the construct model would also open a novel application possibility: to turn the model into a vascular graft to repair diseased arteries in vivo. Within the lab, this tissue-engineered model was referred to, variously, as "*the construct*" device, the "*tissue-engineered blood vessel wall model,*" and, underscoring its application potential, the "*tissue-engineered vascular graft.*"

An in vivo blood vessel is tubular in shape and comprises several layers: the lumen where the blood flows; a first, monolayer of endothelial cells that sit on collagen; an internal elastic lamina; a second layer of smooth muscle cells, collagen, and elastin; external elastic lamina; and an additional layer of loosely connected fibroblasts. The construct device is first grown on a specially designed structure (for which they used the engineering term "*mandrel*") that comprises tiny silicon tubes that allow cells to attach and grow on them, and is then slipped off the structure (figures 2.2a and 2.2b). Although the primary motivation for the construct was to provide a better in vitro model, its application potential also figured into its design. Specifically, to achieve the goal of repairing systems in the human body, the constructs must replicate the functions of the tissues to be replaced. This means that the materials used to grow them must coalesce in a way that mimics the properties of native tissues. It also means that the cells that are embedded in the scaffolding material must replicate the capabilities and behaviors of native cells so that their higher-level tissue functions are achieved. For instance, in vivo the cells create an extracellular matrix, a network of proteins and other molecules, which provides growth factors and mechanical properties. So too, then, the in vitro culturing process needs to ensure that "*the cells, once they recognize they are in the construct will reorganize it and secrete a new matrix and kind of remodel the matrix into what they think is most appropriate,*" as they do in vivo.

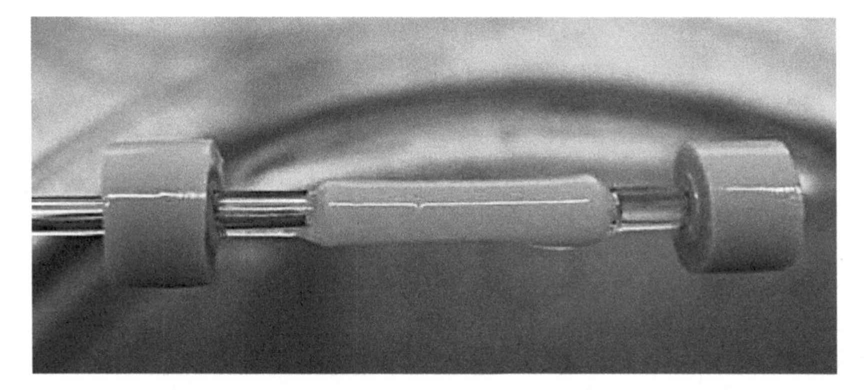

Figure 2.2a
Constructs seeded onto mandrels

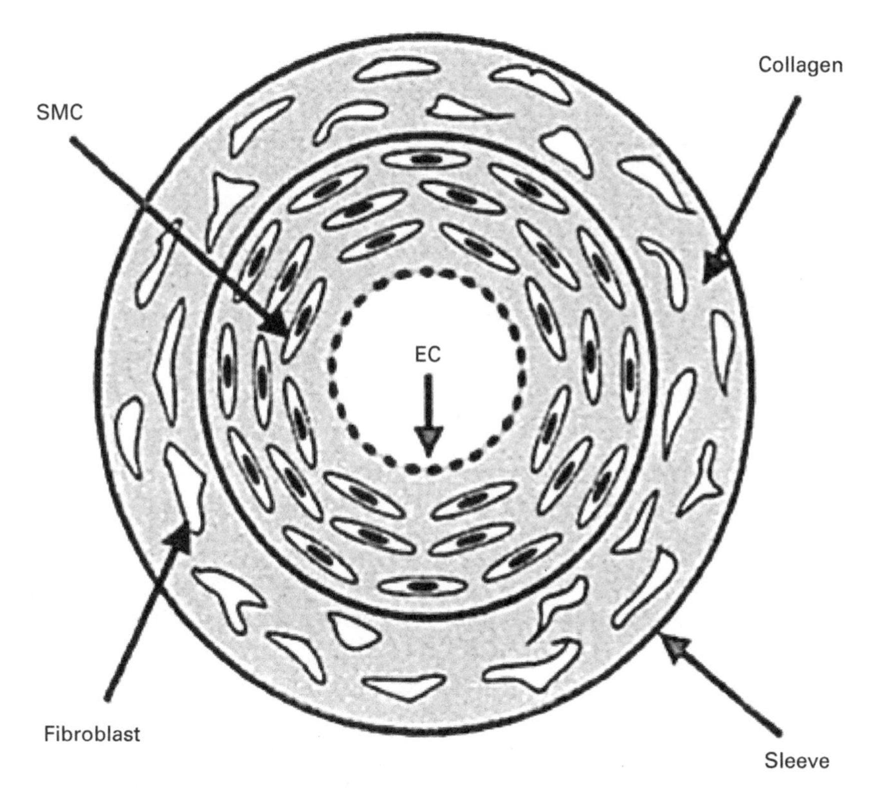

Figure 2.2b
Cross-section of a construct. In this case a Teflon sleeve has been added to strengthen it for the specific experiment.

The construct design is based on what was understood at the time of the biological environment of endothelial cells and vascular biology, on the kinds of materials available, and on the tissue engineering techniques developed in the lab and in the field thus far. The lab's ongoing research sought to advance all these aspects. So, with the move to tissue engineering, the lab's major research question became, as one researcher expressed it, "*The big, big question is how do our constructs act like a modeling tool, how do they respond to—or biological markers respond to—mechanical stimulation. So is there a certain correlation to the stress and strain and the distribution being applied to these constructs to certain biological markers. . . . Does it respond in the same manner? That's the big, big question.*" To "*respond in the same manner*" means, among other things, that it expresses the in vivo proteins and genetic markers, and possess the in vivo mechanical properties.

For a device to perform as a "*modeling tool*" requires that researchers recognize both how it represents in vivo phenomena (device qua model) and how it is an object in its own right (device qua device), an environment for biological experimentation with constraints and affordances due to the nature of the design, the materials, and the engineering challenges. All these factors need to be taken into account when researchers plan experiments, make inferences, and evaluate outcomes. The construct was a new model when we arrived, and so much of the research was directed toward understanding the behavior of the various features of the construct itself. Depending on the goals of the experiment, a construct can be built to instantiate some or all of the in vivo features. It is possible, for example, to use only collagen and not add elastin. Some experiments are conducted with a single layer of the blood vessel wall that has been seeded with either endothelial cells or smooth muscle cells. Often experiments do not require the third layer of fibroblasts to be developed. Thus, the construct forms a family of models, which can be designed for different experimental purposes. These models, in turn, can interlock with other models (devices) to form a variety of vascular construct model-systems.

The researchers had originally intended the construct to move the research from cells on slides to "*tubular studies . . . trying to understand how we can apply physical forces to these tubular constructs to stimulate the cells inside there . . . to recognize—just to appreciate their surroundings.*" The tubular shape of the construct did provide possibilities to experiment with respect to the effects of pressure and strength on the cells and tissues, and led the lab to develop a range of new devices for experimenting on it, some of which are discussed in chapter 4. However, the flow-loop–construct model-system was the locus of many investigations, and so the construct needed to interlock with the flow loop in order to carry out investigations of shear forces. The researchers would have needed to undertake a costly and major redesign of the flow chamber to accommodate their tubular shape. Instead, they decided to cut open the construct so it would lie flat in the existing chamber. This required only a slight redesign of the chamber to allow for a spacer to accommodate the thickness of the constructs. The researchers justified their use of the flat constructs by arguing that since the cells are so small with respect to the construct, the shear forces they experience are the same as if they were in a curved vessel. One researcher explained their reasoning this way: from the "*cell's perspective*" a cut-open construct is not an approximation because "*the cell sees basically a flat surface. You know, the curvature is maybe one over a centimeter, whereas the cell is like a micrometer—like 10 micrometers in diameter. It's like ten thousandth the size, so to the cell—it has no idea that there's actually a curve to it.*" That is, flowing the fluid over a flat construct instantiates the force the cell experiences in vivo because the cell is so small with respect to the arterial wall, the cell's in vivo experience of the wall is as though it lives in a flat world.[7] Thus, the in vitro construct topology instantiates the in vivo artery wall topology as experienced by the cells.

Lab A had considerable experience with, and understanding of, the flow-loop and cells-on-slides model-system when we entered. During our investigation, the lab's challenges centered largely on building an understanding of the construct and of the requirements to develop it into a fully functional blood vessel wall model and possible implant. Although the cognitive-cultural system was continuing to evolve, many practices were well-established and there were significant structures in place to support the participants in the research. The situation in lab D was quite different when we entered, since these researchers were just starting to build the cognitive-cultural system with a pioneering research program.

2.1.2 Lab D: The Dish Model-Systems

When we began our research, lab D was starting to set up. Broadly framed, the researchers were seeking to understand learning and memory in networks of living neurons. For more than thirty years, neuroscience research on living neurons had been focused on the electrophysiological properties of single neurons. The lab D director argued that to study learning there needed to be a way to study the properties of networks of neurons, because learning in the brain takes place through the communication of signals among neurons. Early investigations into the networks of neurons were conducted with brain slices, with neurons fixed in formaldehyde and their physical structure examined with a microscope. The director wanted to *"look at living things, while the interesting parts would happen."* He recounted that, in graduate school, while working on a completely different, and what he felt was uninteresting, problem in biochemistry, he began *"moonlighting as a cognitive scientist,"* reading, attending conferences, and taking courses on the psychobiology of learning and memory. He tried building computational neural networks, but he saw their *"relevance to the in vivo case [as] very tenuous."* He wanted to understand how real-world neurons learn in the brain, and for this purpose, living networks are needed. He recalled having had the idea, *"Perhaps you could make a cell culture system that could learn something. I thought to do that you would have to see the cells while they were doing the learning."* The lab D director's goal was to understand the neurobiological processes that take place in networks of neurons as they learn, but to carry out an investigation of the envisioned cell culture system required the resources of engineering.

For his postdoctoral research, he found a bioengineering lab that was developing the technology needed to study living networks of neurons, and over the next ten years in an extended postdoc, he transformed into a neuroengineer. During that time, he helped perfect the ability to culture a network of neurons that could live, be recorded from, and be imaged over an extended period (days, months, even years), and he developed novel technology for imaging it. This cultured network of neurons became the signature model-system of lab D, the hybrid device they called "the dish." The major goal of lab D's research was to understand how information is processed in the dish, and, especially, whether—and, if so, how—it could learn. If the dish neurons could learn, they would have a minimal criterion for network learning, which requires controlled feedback and memory.[8] The lab's application goals were largely aspirational and stated quite generally, such as *"to create aids for neurological impairments"* and *"to make some fundamental difference in human nature."* The lab director expressed the belief that the kind of research he was embarking on could create the latter, if *"we understand how learning works, how memory works, and how pattern recognition works—and if we come up with new kinds of [brain-style] computation."*

The neuroengineering department in which lab D was located had open lab spaces to facilitate interaction among labs, although lab D did have some walls defining it. The labs have benches on wheels, which can be moved around to accommodate changes in lab configuration. In lab D these benches were occupied by computers, not the usual biological equipment (flasks, pipettes, and so forth). Indeed, a glance around the lab provided no telltale signs of biology. Instead, one was struck by the copious wires that crisscrossed the space. These carried electrical signals back and forth between the neuron dishes and the computers. The wires indicated that the main activity of the lab was, as we will see, digital signal processing and analysis. The dishes were hidden from view in various kinds of insulating enclosures, wrapped in insulating foil—some with microscopes sticking out. All the biological work was done in an adjacent cell culture room where the researchers built the dishes. For a while the lab used an additional shared mouse/rat colony room in the basement to breed its own genetically engineered rodents that produce fluorescent proteins and a shared dissection room for harvesting embryonic cortical neurons, but they quickly realized it was less expensive, time consuming, and equally good to purchase these neurons from a commercial vendor. Fluorescent neurons help to differentiate the neurons from other elements (such as glia) in the cultures when they are imaged. In addition to regular microscopes, the lab used the latest advance in imaging, a type of scanning fluorescent process called two-photon imaging. Their initial two-photon microscope was built by the lab director.

All the graduate students during our observational period had arrived in lab D around the same time, and the director had taught them how to build the dish. Plating new dishes was usually a communal activity in which all the graduate students crowded into the room for "*plating parties*" when new cells were available. Everyone built their own dishes and gave them names. They created a "*dish log*" for each dish, to keep track of experiments run on it, since every experiment changes the properties of that dish, and researchers needed to be aware of these changes. Unlike the lab A constructs, which were created anew for each experiment, dishes were used in multiple experiments, so the researchers wanted to keep the dishes alive for as long as possible (two years was the longest we observed). Thus, keeping the dishes well-fed (removing the old medium and adding new nutrients) and "*happy*" was a constant concern. As with lab A, much discourse focused on the well-being and happiness of cells, and frustrations with their errant behavior impeding experiments. One researcher expressed her frustration at the dish's response: "*Pffft, you keep them happy by feeding them, by taking care of them, hopefully stimulating them and telling them to do something! I don't know what to do to make them happy. I don't know how to make them happy—that will make my neurons happy [pointing to her head].*" If the cells are not happy, neither is the researcher. There is an important dynamic interplay between "making the dish happy" and the researcher's cognitive goals ("making [her] neurons happy").

The construction of the dish interlocks concepts, methods, and materials from biology, chemistry, neuroscience, and electrical engineering. The dish comprises cortical neurons (15K–60K, depending on desired kind of network) and glia (support cells), which are harvested from embryonic rodents, dissociated (to remove any in utero connections), and plated in a single layer on a multielectrode array (MEA). The MEA is a small, glass petri-style dish with an 8x8 grid of microelectrodes, spaced around 200–300 micrometers across, embedded in the bottom (figures 2.3a and 2.3b). The researchers mainly used the MEA that was designed by the director of the lab in which the lab D director had been a postdoc, although they did experiment with design modifications. This technology had been a major advance, and the MEA was rapidly becoming the standard in in vitro neuron network investigations.[9] The electrodes poke into the neurons (without damaging them), which gives the researchers the capacity to record from, and inject, electrical activity ("*stimulation*") into the network. A sugary cocktail of biologically appropriate chemicals feeds the cells, and the lid of the dish is a thin nontoxic Teflon film that allows oxygen and carbon dioxide through while keeping out contaminants. The dish lives in a specially designed enclosure that provides the requisite external environment for the cells (temperature, humidity, carbon dioxide level). It takes around two weeks for connections to grow among the neurons and spontaneous activity to start. At that point experiments can begin.

The construction of the dish interlocks constraints of the device qua in vitro model and device qua engineered system. For instance, the in vitro model instantiates a monolayer network of neurons instead of the rich three-dimensional connections found in vivo, which reduces the complexity of the system by reducing the number of variables. Several interlocking constraints underlie this decision. The choice of monolayer cultures

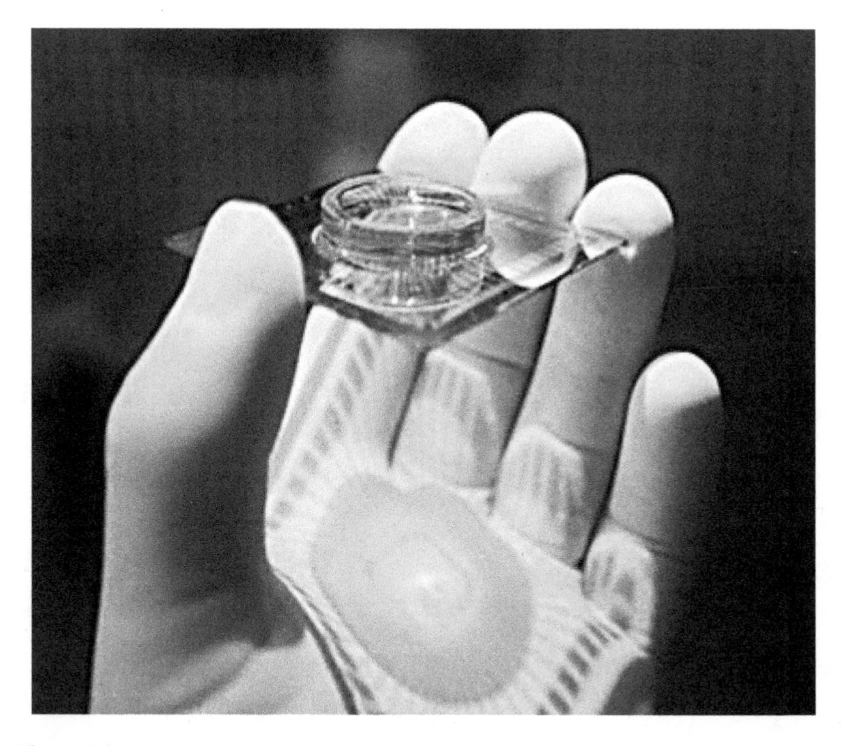

Figure 2.3a
A neuron dish device

provides *"one single layer of neurons . . . a simpler system to study."* But also, a monolayer is easier to feed. With three-dimensional cultures, *"the medium does not go into the inner layers. . . . [They] die off,"* which would impede their goal of studying the dishes *"over the long term. So, we want to keep them alive over months, years."* Additionally, if the model instantiated the more complex system of a brain slice, the researchers would have to understand their preexisting network connections before they could initiate learning experiments—a complex, if not impossible, undertaking. Finally, another design constraint figured into the choice of a monolayer. The recording technology is limited to the grid of electrodes (MEA) embedded in the bottom of the dish, so only neurons close to those electrodes can be recorded. In all, the monolayer fits the constraints of the dish technology and provides a reasonable reduction of information for data analysis, while (they hoped) still instantiating the salient features of interneuron communication.

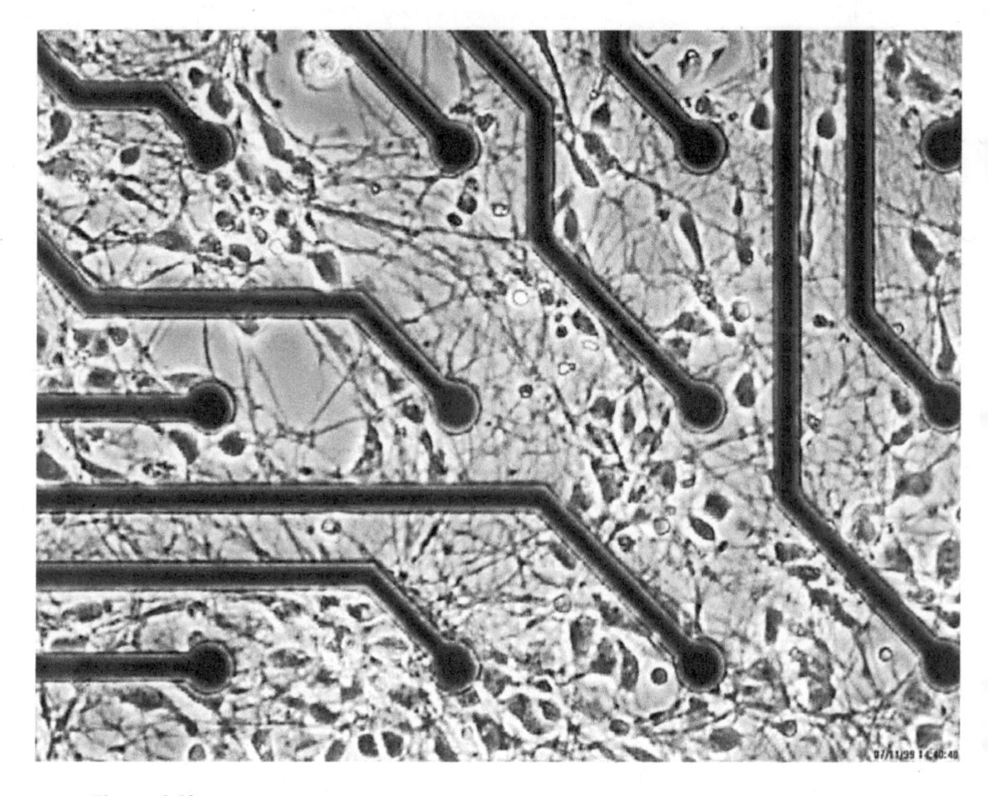

Figure 2.3b
A network of neurons plated on an MEA

The researchers chose cortical neurons because the cortex is thought to have the most adaptable ("plastic") neurons and to be where most general learning occurs in vivo. The dish model-system, then, provides both a means to investigate the basic features of interneuron communication and also whether learning can be induced in a system of neurons with just the network properties of the brain, abstracted from other brain structures. The lab director thought this abstraction was justified because the research focus was to understand the network properties of neuron communication and learning, and for this goal, he reasoned, *"it probably isn't necessary to include all of the details . . . but it may be. So that's part of our job to find out which details of the biology are important . . . and which are incidental."* So, determining the warrant for the dish as an analogue model was an ongoing part of their research. In comparison to lab A, there were many more open questions about the status of the dish as a basis for inference about in vivo neural networks.

Clearly no one saw the dish as realizing the philosopher's vision of "a brain in a vat." However, everyone agreed that the novel practice of in vitro modeling they were advancing would, at a minimum, yield understanding of the basic behavior and function of network-level cortical neurons. The director expressed the warrant for the research program: *"First of all, it's a simplified model; I say that because the model is not—it's artificial, it's not how it is in the brain. But I think that the model would answer some basic questions, because the way the neurons interact is the same whether it's inside or outside the brain. . . . I think the same rules apply."* Here we see understanding and control are linked once again in the goals of the bioengineering research. The goal to determine these *"rules"* for the simple network would require the lab D researchers to develop a control structure for supervised learning. If they could achieve that, then it might be possible for the research to move on to design a more complex dish model-system that would instantiate other relevant features of brains, such as *"cultures with different brain parts mixed together or specific three-dimensional pieces that are put together."* This move would be comparable to lab A's move from a model-system that uses endothelial cells on slides to one that uses tissue-engineered constructs—but recall that move took nearly thirty years of research.

When the field began to use MEA dishes to study neuron network cultures, researchers focused on recording and analyzing *"spontaneous activity"* of the network (open loop electrophysiology). Lab D aimed to determine whether networks can learn, which they operationalized as creating a *"lasting change in behavior resulting from experience."* In vivo neuron learning involves embodied interaction with the world. So, to create behavior, as the director specified, *"an in vitro learning system somehow has to be connected to the outside world. In order for learning to have any definition at all, there has to be some way for it to behave, and to see whether there was some change in behavior, there has to be sensory input to see whether you could even influence that. Those were my prerequisites for learning."* The researchers contended that to achieve learning requires the ability to evoke neural activity through controlled electrical stimulation and feedback (closed loop electrophysiology) through an *embodied* dish model-system. To carry out their investigation, the lab D researchers needed first to develop their own technologies to stimulate the dish and to study its responses—a process they called *"communicating with the dish"*—and then to create appropriate *"bodies"* for the dish. This work included that the researchers learn to interpret, control, and mathematically represent the dish's behavior under various conditions. Chapter 3 examines their electrophysiology research in my analysis of a process of concept formation and change that was instrumental in their solution to the problem of getting the dish to learn, so here I provide only a brief outline of the technologies the lab developed to carry out open-loop and closed-loop electrophysiology research.

Every experiment involves electrical stimulation and recording. Every piece of data collection and analysis involves multiple interlocking models. During our investigation, the stimulator board (transmits signals to the dish) and preamplifier (amplifies dish output signals) were placed inside the incubator with the dish, making up what they called the "*stimulation site*" (later these were moved outside). The researchers would program the board with a "*stimulation protocol*" formulated from a rough hypothesis about the "*character*" of a specific dish, based on observations of its spontaneous behavior or behavior derived from the experiments that had been conducted on it as recorded in the dish log. The MEA electrodes pick up and amplify the electrical signals from the neurons in response to a stimulus. These voltage changes (analogue) are transmitted to a data acquisition card, which samples the signals (25K per second) and transforms them into digital (numeric) signals. The researchers developed a suite of custom software tools to process the digital signals, MEAbench, after the software they purchased failed to have the various functions they required. The software captures signals and simultaneously saves them to the computer hard drive as a permanent record to provide the lab with a memory of the experiments performed on each dish, and sends them to their real-time visualization software tool, MEAscope. Just building all this software took nearly a year. In essence, the MEAbench software provides a series of filters to transform the raw numerical data into usable information. The researchers designed each of the filter algorithms on a model of selected electrical signals. Thus, even the ability of the researchers simply to "*see*" the neuron network activity is conditioned on a number of interlocking mental, physical, and algorithmic models. I say more about how they manipulated and interpreted data in chapter 3; here I introduce the main data of interest: spikes. The lab researchers conceptualized the notion of plasticity, a biological property, in terms of a quantitative measure based on recorded spikes of electrical activity in the neuron network.

Historically, the term "spike" designates the electrical trace left behind when a single neuron fires. In single-cell recording, neuron firing produces a steep jump in voltage potential as the neuron depolarizes, followed by a proportional drop in potential as the neuron recovers (action potential). Recorded visually, this process produces a spike figure. The lab D researchers estimated that the electrical activity recorded by an individual MEA electrode comes from three to five neurons, which possibly fire simultaneously. Thus, it is impossible to tell the difference between single and multiple neurons firing. Usually spikes are tagged manually, but the researchers replaced that tedious process with their own software that automated the process of tagging spikes. Their *"spike detector"* embodies the lab's model of a multineuron spike that includes *"height"* (difference from average voltage) relative to the noise on the electrode and the *"width"* (duration) of the change, along with a few more subtle characteristics. The spike detector checks for jumps in voltage that match the model, tags the spike, and keeps a snapshot of the electrical activity in the immediate surroundings of the spike.

Earlier, I said the copious wires indicated that the lab's main activity was signal processing. We see why here. The information provided by neural signal data provides the basis for understanding how the networks represent and transmit information. In all, the researchers built several pieces of such software, which they called *"filters,"* designed on various models of aspects of neural signal data. These filters transform the neural signals before they become the final filtered spike data on which their analysis began. The filters can miss actual neuron firings or provide false positives. The researchers performed data analysis in light of their understanding of all the transformation processes the neuron signals have undergone with all the software they have built. The simplest form of analysis uses spike data that have been transformed into a visualization by MEAbench and displayed by MEAscope on the computer screen in an 8x8 grid arranged topographically to match the layout of the electrodes of the dish (figure 2.4). When the spike detector is turned on, the visualization software places little red dots at the peaks for easy visualization of spikes. Because the researchers' primary interest was on network learning, they focused on spikes detected after electrical stimulation.

The simplest model-system was the dish itself, which, when connected to their stimulation and recording technologies, provided the site for in

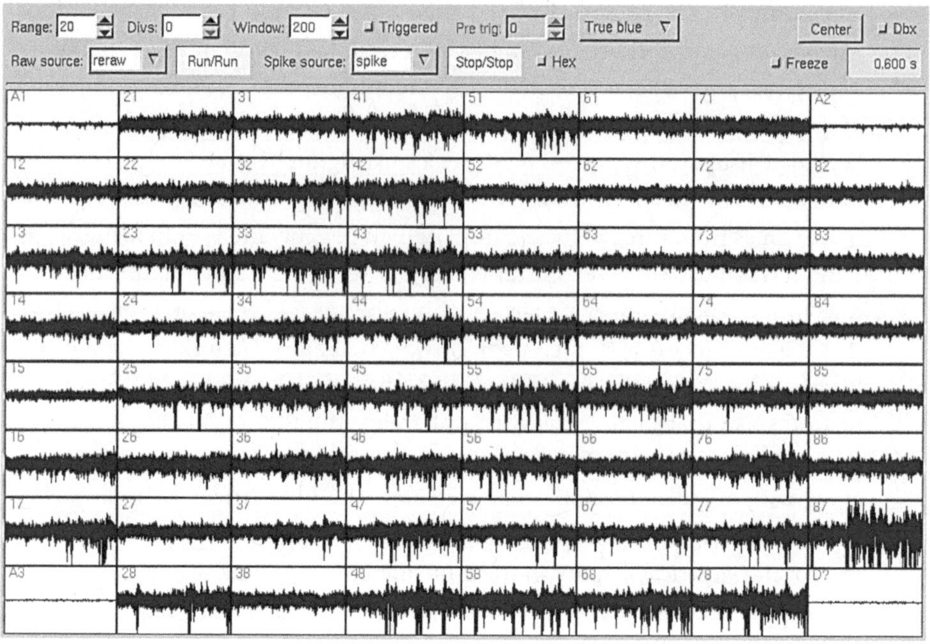

Figure 2.4
A screenshot of the lab's MEA visualization of output spike data, which topographically matches the layout of the electrodes on the MEA grid. This visualization is displayed on an oscilloscope.

vitro open-loop physiology (without feedback) experiments. But in vivo, the brain is embodied and learns in an environment, so, as discussed above, in vitro closed-loop physiology is needed to instantiate neuron learning behavior. As the director described, *"[in] the traditional way to do in vitro physiology . . . the closest thing to behavior is little waves on the oscilloscope screen. It has nothing to do with behavior. . . . And there is not any sensory input other than electrical pulses. . . . It's very disconnected."* To exemplify behavior requires that the in vitro system have features of a body relevant to neuron learning with which to interact with the environment. The director claimed to have gotten the idea of building embodiments for the dish from the proceedings of a conference on adaptive behavior in the 1990's: *"All of the people in that book are simulating animals or what they called 'animats.' . . . They were simulating these things on a computer or they were building robots that were animal simulations. They were continually emphasizing the importance of embodiment and being situated."* The goal to build embodied dish model-systems was, at the time, unique to this lab.

The lab borrowed the computational practice of modeling animals, but transformed it into a more realistic in vitro model that had a living "brain," the dish. The researchers created two different kinds of model-systems: "animats," simulated animals connected to the dish that move in a computational world; and "hybrots," hybrid robotic devices connected to the dish that perform in the real world. They also created animat model-systems of hybrot model-systems to investigate hybrot behavior in more detail computationally. Building these model-systems, in turn, required the researchers to create more software models to capture such features of embodiment as motor ability. A major epistemic affordance of an embodied dish model-system is that it provides the opportunity to investigate supervised learning in a controlled manner, *"'cause you define what it's going to learn based on the body you give it and the environment you allow it to work in."* As with the dish monolayer, in building these embodied model-systems researchers aimed *"just to try and simplify everything . . . to make the data easier to analyze"* and to be able to *"just say, 'okay this part of the simulation does this.'"* These abstractions, which I will not detail, enabled the researchers to define more clearly questions and experiments that, in turn, produced data that were less complex and more easily interpreted.

Just as there are many creatures in the real world, lab D created a variety of animats and hybrots. The simplest animat was a simulated *"moth,"* located in a circle (environment) and at the center was a dot (light). Neuron activity in the dish is translated into motor commands that determine how the moth moves, and that movement is translated into *"sensory"* information in the form of patterned electrical stimulation that feeds back into the dish. The neurons read the change in information from the dish electrodes and the behavioral loop continues. The hybrot model-systems consisted chiefly of commercially available robots on wheels to which researchers attached dishes in special containers. Their most complex model-system was a collaboration with artists, the hybrot mechanical drawing arm MEArt, so named because it was both a research project and a mechanical art exhibit (figure 2.5).[10] The robotic arm communicated via satellite with the dish in the lab and as it traveled the world, often accompanied by the lucky graduate student who was the primary lab member responsible for that research. Some of their research on learning using this model-system is discussed in chapter 3.

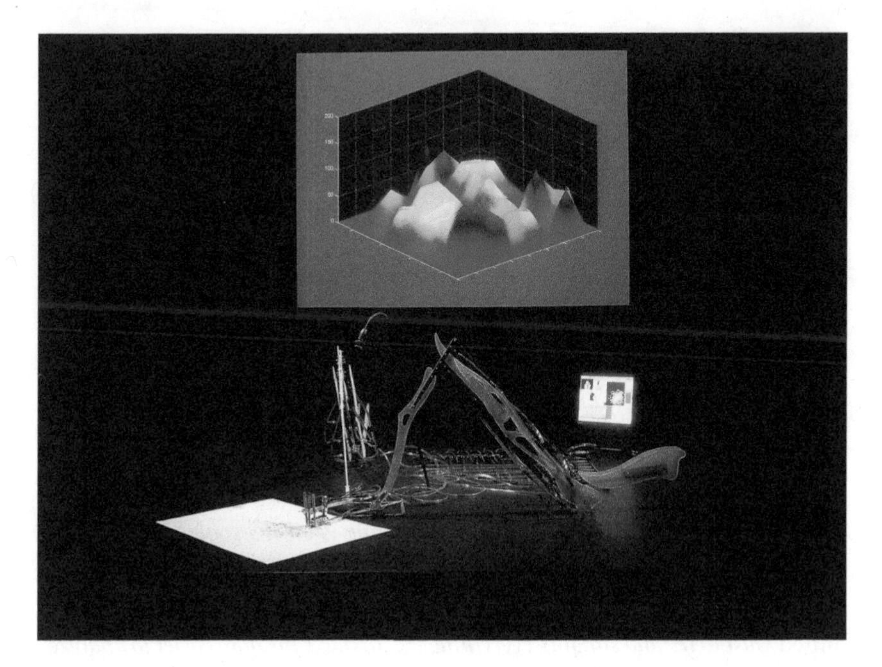

Figure 2.5a
A photograph of the MEArt hybrot robotic arm drawing in a feedback loop with the dish of neurons. Behind it is a projection of the activity of its neurons (the dish) in real time. The arm and the dish are usually in different countries so it communicates with the neurons via satellite.

Now that we have a basic understanding of each lab's signature model-systems in hand, we are in a position to examine how, in building these model-systems, the researchers create understanding through building D-cog systems that perform distributed model-based reasoning. In section 2.2, I show how we extend the D-cog framework to accommodate inferential practices as a process marked by ongoing interaction between models in the researcher's memory and artifact models, both under development. In section 2.3, I address how BME researchers build epistemic warrant for devices and model-systems as analogue representations that support understanding and inference about in vivo systems. I want to stress at the outset that in both of these theoretical analyses we see the "productive interplay" between case material that derives from an open empirical inquiry ("concrete") and concepts and theoretical notions from cognitive science and philosophy of science ("abstract"), including my own prior work on

Figure 2.5b
The photograph provides a sample of MEArt's output behavior. (Photographs courtesy of Guy Ben-Ary)

conceptual modeling practices. The new modeling practices we uncovered both here (in vitro) and in systems biology (in silico) provide empirical findings from which to bootstrap further theoretical analyses.

2.2 "Putting a Thought into the Bench Top": Distributing Model-Based Reasoning

Occasionally the researchers would surprise us by the simplicity with which they would inadvertently express the essence of a theoretical notion we, with difficulty, were attempting to articulate in our own field. When asked to characterize the in vitro modeling research, one researcher described it as *"putting a thought into the bench top and seeing whether it works or not."* This characterization provides a concise, plain-English formulation of a concept

I have been advancing, with doubtless less clarity, for some time: distributed model-based reasoning. From a D-cog perspective, in *"putting a thought,"* BME researchers distribute a process of reasoning about complex biological systems, which comprise imagination, visualization, simulation, analogical reasoning, and so forth, to the in vitro devices and model-systems they build. Building such models is a means through which these researchers create cognitive-cultural artifacts that extend their natural cognitive capacities into distributed problem-solving systems. Specifically, building model-systems is the means through which BME researchers distribute processes of model-based reasoning across systems of interlocking models, as illustrated in figure 2.6. Such distribution creates a coupled model-based reasoning system comprising researcher and artifact models. We contend that the notion of coupling captures the interactive and system-level nature of reasoning better than the customary D-cog notion of off-loading. Reasoning is not off-loaded to the cognitive artifact, but, rather, carried out in interaction with it, that is, coupled with it. I develop this notion further in subsequent chapters.

In this section I consider, briefly, an exemplar of *"putting a thought,"* in which a researcher built an in vitro model-system to investigate her thoughts about the behavior of specific entities in the cardiovascular system, and then explicate our analysis as depicted in figure 2.6. Then, I consider how the BME practice of in vitro modeling, understood as *"putting a thought"* provides an instance of the broader notion of distributed model-based reasoning introduced in chapter 1.

2.2.1 An "Experimental Model That Predicts": The Construct-Baboon Model-System

One goal of lab A was to find a source of endothelial cells that could be used in a vascular graft without rejection by the host. As noted in the discussion of lab A's model systems, the researchers were investigating whether it was possible to apply mechanical forces to generate mature cells, which would not be rejected by the host, from stem cells or other sources. The experiment I discuss here was designed to see if a host's own endothelial progenitor cells (EPCs), which circulate in the cardiovascular system, might provide a source. How these cells become mature endothelial cells was not known, but the "thought" of the researchers was that mechanical forces of blood flow provide a likely in vivo mechanism for the maturation process. The researchers hypothesized that harvesting the host EPCs and *"preconditioning"* them by subjecting them to shear forces in a flow loop would enable them to function like mature cells in the vascular system. In particular, to *"see if it works,"* they targeted the function of mature cells to express anticoagulant proteins—produce thrombomodulin—that prevent platelet formation. To explore how preconditioned EPCs would behave under various preconditioning forces, they needed not only to conduct flow-loop simulations, but also to build a model-system that would instantiate the features required to provide a simulation of their behavior as they function on the blood vessel wall under in vivo blood flow.

We followed a researcher, "A7," from start to finish as she articulated the lab's preliminary hypothesis into specific goals and problems and designed a model-system to investigate it. It took approximately three years for her to move from the hypothesis to the initial experiment. The experiment is significant because it constituted the lab's first move in the direction of in vivo research and their application goal. Both A7 and the lab director emphasized its importance and centrality in the lab's research program, as I discuss in chapter 4. However, this was not an experiment about the construct as an implant, since it was still far from being able to withstand blood flow forces and needed to be scaffolded with a silicon sleeve (figure 2.2b). Rather, the experiment was about the function of the cells within a *"more realistic"* model-system, which required they move into animal model studies, where more features of the human in vivo cardiovascular system would be instantiated. They settled on using a baboon because of its availability at a lab nearby, and that lab director's willingness to cooperate in creating the experimental setup. The baboon animal model-system, to which the construct model was to be connected, would provide a model of EPCs as they behave after circulation in the human cardiovascular system. To prepare the baboon to function as a model, with minimal intervention and discomfort, the researchers in that lab surgically connected its femoral vein and the femoral artery with a permanent exteriorized shunt. During A7's experiment, a long silicon tube was used to connect them so that a small amount of baboon blood flow could be diverted through a construct attached to the tube, and the tubing could be situated in a gamma camera at a sufficient distance from the baboon, so as not to frighten it. The baboon's blood was injected with iridium so that any platelet formation would be visible as the blood flowed through the part of the tubing that was situated in a gamma camera.

As I mentioned earlier, because the construct needed to remain in tubular form for this experiment, A7 had planned to do a major redesign of the flow loop to accommodate that shape. However, the baboon shunt gave A7 the idea to also create an external "shunt" to connect the construct to the flow components with tubing. That shunt went through several designs and tests to ensure that it would keep the requisite flow characteristics of the chamber. This analogical redesign saved her the considerable time and expense of redesigning the flow chamber itself. Before we turn to our analysis of her problem-solving process as an exemplar of distributed model-based reasoning, it is instructive to examine her own succinct summary statement made in her final postgraduation interview with us.

What is most notable from her summary account of her experiment is how she seamlessly interlocks biological, engineering, and device concepts and models in thought and expression as she describes how she built and assembled the components of her model-system, conducted experimental simulations, and made inferences through these:

> **We used the shunt to evaluate platelet deposition** *and that would be—in other words—were* **the cells, as a function of the treatment** *that they were given . . . able to prevent blood clotting? And so, we specifically measured the* **number of platelets** *that would sit down on* **the surface.** *More* **platelets equal a clot.** *So, it ended up being that we were able to look at the effects of* **shear stress preconditioning on the cells ability to prevent platelets** *and found that it was actually necessary to shear precondition these blood derived cells at an* **arterial shear rate,** *which I used* **15 dynes per square centimeter** *compared to a low shear rate, which in my case I used like 1 dyne per square centimeter, so, a pretty big difference. But I found that the* **arterial shear was necessary to enhance their expression of anti-coagulant proteins** *and therefore prevent clotting. So, in other words, the shear that they were exposed to before going into the shunt was critical in terms of magnitude, for sure.*

The bold text items mark reference both to interlocking interdisciplinary mental models as they function in her understanding and reasoning and to interlocking physical simulation models. To unpack a few of her expressions, "*the shunt*" refers to the baboon understood as a simulation device that is part of a model-system. "*The cells*" are the endothelial progenitor cells she had extracted from baboon blood and had given the "*treatment*" of "*shear stress preconditioning*" (flow-loop device). She measured the "*number of platelets*" (an indicator of coagulation) on "*the surface*" (construct device). The objective of her research was to determine whether, and at what level of force, the preconditioning ("*arterial shear*" simulation) of cells (construct device) would *"enhance their expression of anti-coagulant proteins"* or *"prevent platelets."* In carrying out several iterations of the experimental simulation over a period of many months, A7 adjusted the force of the fluid on the EPCs in the preconditioning process to determine at what force the platelet formation in the baboon model-system ceased. She found, in the course of these iterations with the entire experimental model-system configuration (*"used the shunt to evaluate platelet deposition"*), that the in vivo human arterial shear rate (*"15 dynes/cm^2,"* a mathematical model) was required for sufficient protein expression (*"was critical in terms of magnitude"*) in the in vitro model system.

By putting her "thought" about the pretreatment of the cells into building the construct-baboon model-system (the "benchtop" in this case), then, A7 was able to "see" that it worked, provided that the human in vivo arterial shear rate was used. Although her immediate inference from the simulation was specific to the effects of preconditioning the cells on the performance of the model-system, she intended this system to be what she called *"an experimental model that predicts what would happen—or you hope that it would predict—what would happen in real life."* That is, the inference she drew about preconditioning from the model-system simulation provided a hypothesis about how a vascular graft seeded with EPCs would behave if it were implanted in the human cardiovascular system.

2.2.2 Interlocking Models in Distributed Reasoning

We developed an analysis of the construct-baboon model-system as a distributed model-based reasoning system. The diagram in figure 2.6 is a greatly simplified schematic representation of that analysis. We constructed it based on our data analysis quite a while before our final interview with A7, but each component appears in her account, which I presented above so that the reader could better follow our diagram. On the extension of the D-cog framework I proposed in chapter 1, mental models and artifact models form coupled (interlocking) configurations in experimental simulations. To construct the diagram, we used Hutchins's notion that in performing a problem-solving task, a D-cog system generates, manipulates, and propagates representations, in this case, models. The diagram in figure 2.6 traces parts of the propagation of mental and artifact models within the D-cog system that constitutes A7's experimental simulation. In the figure, the models are highlighted by thick lines. The flow arrows represent

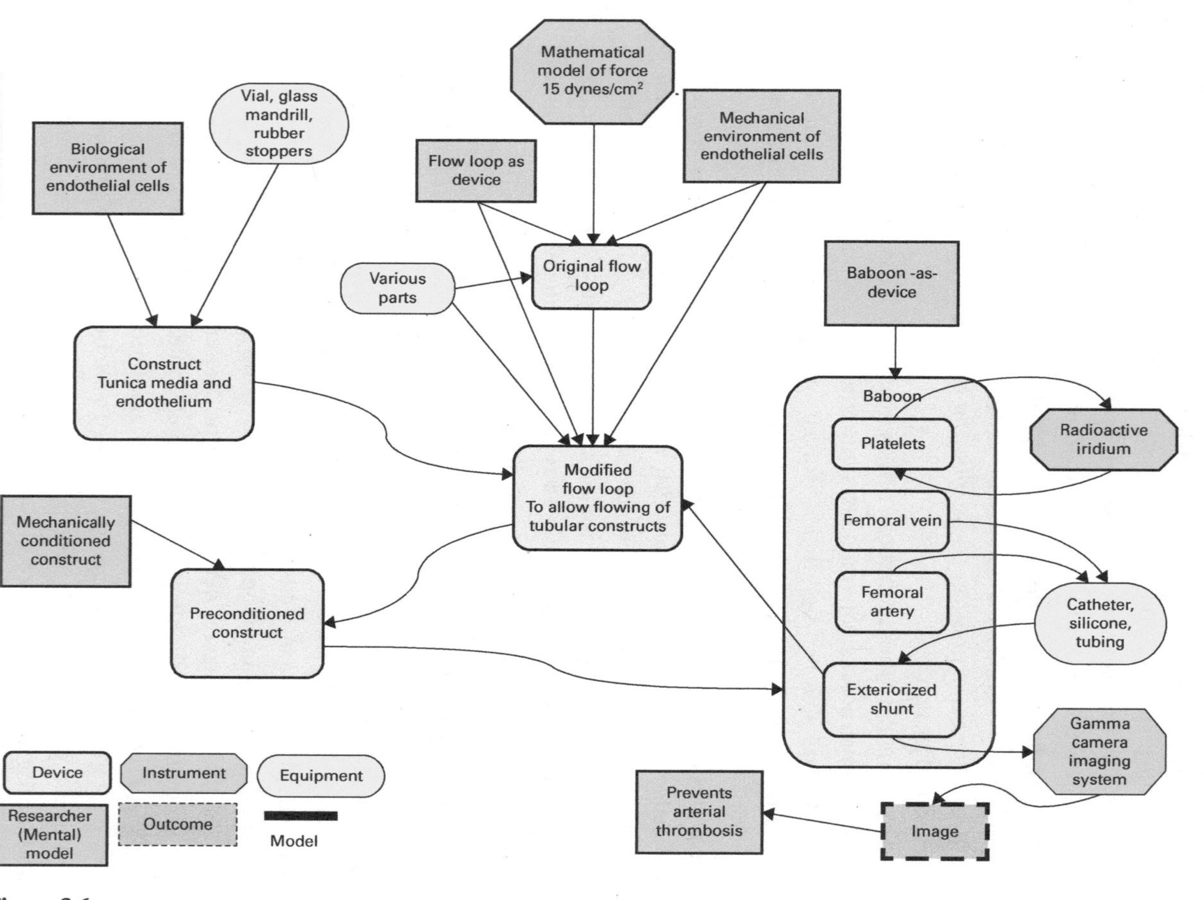

Figure 2.6

Our analysis of the construct-baboon vascular construct model-system (partial) as a D-cog system of interlocking models

the propagation of representations in the system as they are generated and manipulated. The categories of device, instrument, and equipment derive from a sorting task we conducted with the lab members of the artifacts used in their research (see chapter 4). To keep the diagram from becoming too complex, the numerous connections to other researchers and artifacts in the lab problem space are not included, but the construct, flow-loop, and baboon devices are to be understood as communal achievements, each representing years of research. In the process of preparing an experiment, including preliminary trial simulations, each artifact model and mental model can undergo numerous iterations.

Each researcher mental model is a part individual, part community representation. Each artifact model represents and performs as a selected aspect of the cardiovascular system, for example, a construct represents and performs as selected aspects of the biological environment of the blood vessel, the flow loop represents and performs as shear stresses on arterial walls, and the baboon model represents and performs as blood flow through a human cardiovascular system. Neither artifact nor mental models in an experimental set up are static; rather they are representations that can change *in interaction over time* as each component is developed and with each iteration and simulation of the experimental setup. In this way, these interlocking models provide an instance of a coupled system through which inferences are made.

There are three main components of this model system. I will call them "construct" (left third of diagram), "flow loop" (middle), and "baboon" (right third). Model-system configurations, though evolving artifacts, are long-term investments. It took more than three years of building the first two components before A7 could run the final experiments with the entire system over a further two-year period. Also, while she was preparing, the animal lab moved from a nearby institution to one that was in a distant state, which required her to fly there with the other artifact models. I provide a brief outline of what should be understood to consist of numerous iterations of building the models represented in each part of the diagram.

Considerable novel research went into building the construct (diagram left third) so that it could be used in the baboon model. Issues of mechanical integrity (strength) were significant, since it needed to be able to withstand nearly normal blood forces and also not leak. The desired final experimental configuration required that this construct model be what the researchers considered to be the most *"physiologic"* one the lab could create, in that it instantiated all but the outer layer (adventitia), which A7 thought was not necessary to the experiment since she would need to develop a scaffold to suture the construct to the baboon model. She also surmised it might even grow on its own. Endothelial progenitor cells (EPCs) were extracted from a baboon's peripheral blood. For a while A7 worked just with the EPCs on slides. She used the lab flow-loop model at various shear rates and then used several instruments to extract information about their protein and gene expression. She went through numerous iterations of manipulating EPCs and using various instruments to examine the cells for expression of molecules for thrombomodulin, which facilitated her building a mental model of their function and behavior in relation to shear stresses, and contributed to building the final construct device for these experiments. To build a construct that she could use with the baboon further required that she isolate an intact elastin scaffold, remove vascular smooth muscle cells from harvested carotid arteries, and determine the right collagen mix. To suture it to the tube connecting the baboon shunts, she needed to determine the right scaffold material for the sleeve, which would enhance its ability to withstand in vivo blood flow as well.

As I noted briefly earlier, A7 had originally proposed to redesign the chamber of the flow loop in order to condition the constructs in tubular shape (diagram, middle third). In the end, significant redesign turned out to be unnecessary because she had the insight that it should be possible to design a shunt for the flow loop to which she could attach the construct with tubing to precondition the cells. She needed to determine that using the tubing would keep the flow components the same as they are in the chamber. After I had hypothesized that A7 had made an analogy with the baboon shunt, I asked her how she had come up with the idea in an email and she confirmed this, while she also explained how the shunt works for the flow loop and why she was justified to use it. She wrote that, once the construct is sutured into the tubing that connects to the flow-loop components it *"fits in the place of the parallel flow chamber. . . . The shear stresses are the same and can be controlled like in the flow chamber. We consider the construct a rigid cylindrical tube of constant diameter and if we assume we have fully developed flow (just like the parallel plate flow chamber) going into the construct, we can predict the wall shear stresses on the cells based on fluid mechanical theory."* Here we again see the lab's engineering framing of the biological entities and processes. A7 ran many simulations to examine flow rates on the EPCs with just a construct–flow-loop model-system before advancing to the baboon system.

The baboon model (diagram right third) provided a functional evaluation of the preconditioned EPCs in an environment that parallels the human cardiovascular system. The model had been prepared to her specifications by the animal lab. Before running the experiment, she sutured the construct to a long tube attached to the shunt so that it could be placed directly on the gamma camera (a commercially available instrument the size of a small table) to capture and observe platelet formation during real-time blood flow. The baboon sat at a significant distance in a specially designed restraining chair, blindfolded to keep it calm, and with a suit over its body to prevent it from pulling on the shunt. As blood, which had been injected with iridium, flowed naturally through the baboon, A7 took pictures with the gamma camera so she could see any platelets that formed. The construct was then disconnected from the baboon, and she used other instruments and software programs to analyze the EPCs for information about expression of thrombomodulin, optical density, and electrical resistance.

However, although the baboon model exemplifies the in vivo human cardiovascular system better than the construct–flow-loop model-system, it is less reliable as a model than a fully in vitro engineered system. As A7 noted, *"In the lab we can control their [endothelial cells] environment completely. . . . We can control exactly what flow is like and we can monitor by visually seeing it. But when we move to an animal model, it's more physiologic—the challenge then is that it's a much more complex system."* To advance the research, then, some control and precision needed to be sacrificed. A7 made several trips over two years to the out-of-state lab where the baboon model resided to run experiments, modified her experiments on the basis of the simulation results, and, in the end, she was able to establish that, in preconditioning the EPCs, the normal human blood flow rate was required to prevent platelet formation. This understanding was an important contribution to the lab's goal to find an endothelial cell source for vascular grafts. If the hypothesis was transferred and successfully verified for the human system, it would mean that the host's own EPCs could be harvested for the graft. On our D-cog analysis, in this case, the *"experimental model that predicts"* encompasses the entire system of mental-artifact-animal models through which A7 could make that analogical inference.

2.2.3 Representational Coupling in Distributed Reasoning

"Putting a thought into the bench top" in a D-cog analysis is a process that turns a preliminary conceptual (mental) model into a physical model that affords a more detailed representation, a wider range of manipulations—including possible and counterfactual situations—and more control than would be possible in thought alone. In the processes of co-constructing mental and physical models, correspondences develop between them, which are updated through interaction between them. The artifact model, then, provides a site of simulation not just of some biological or mechanical process, but also of the researcher's understanding, both of which can alter in interaction. To capture this *interdependence* as a D-cog system, we understand the researcher mental model and the artifact model to form a *coupled system*. In the example, what this means is that in building a construct to precondition in the flow loop, A7 also builds a mental model that selectively interlocks what she understands from cell biology, vascular biology, and fluid dynamics. This mental model represents biological aspects of endothelial cells with respect to mechanical forces in terms of the hybrid concept of arterial shear (force of blood as it flows over these cells, causes elongation, proliferation, and so forth), which can alter in interaction with building and simulating in vitro models. In this co-building process, the researcher develops mental models both of the in vitro model as it exemplifies an in vivo system (model qua model) and as it is an engineered artifact (model qua device).

In general, this representational coupling is an interactive bidirectional process different from what is customarily understood as off-loading a cognitive function or process to an external artifact. The notion of coupling captures the idea that, as researchers gain understanding and make inferences by means of manipulating the artifact models, features of their mental models can change, which in turn can lead them to alter the artifact models. Such mental-artifact representation-building processes can take place over short periods of time during hands-on building or over longer spans. These coordinated dynamic processes couple the researcher mental model and the artifact model into a distributed model-based reasoning system. Cast in these terms, then, what the A7 referred to as the *"experimental model that predicts"* comprises the entire model-system: researcher mental models and artifact models interlocked in various experimental simulation configurations from which inferences derive. These inferences are, in the first place, about the model system, but the goal of the research is to derive inferences that are sufficiently warranted to transfer to *"what happens in real life."*

However, in this case and in general, researchers in BME are often far away from being able to test inferences from in vitro simulations in real life. In the circumstances of A7's research, the value of the model-system is that it provided a means of getting a grip on the dynamical behavior of the EPCs circulating in the blood stream, a biological system about which little was currently understood in the field. This research was also a step in moving incrementally toward achieving the lab's applied goals. The construct-baboon model-system was the lab's first step into the real world. A further step would be to implant a vascular graft directly into an animal rather than using the bypass shunt, but when and whether they would be able to do this was dependent on another line of the lab's research that aimed to strengthen the construct so as to be able to withstand the forces of blood flow in vivo without scaffolding (see chapter 4). For the lab's then-current state of research, the model-system enabled A7 to infer additional features of the behavior of EPCs in response to mechanical forces, from what the lab currently understood. This new understanding derives from a complex process of distributed model-based reasoning, but what enables the researcher to have some assurance that she is on a productive path with an in vitro device or model-system—that it has the potential to provide understanding of an in vivo biological process?

As discussed in chapter 1, scientific practice, unlike other practices examined in the D-cog framework, has epistemic goals. Thus, we need to examine not only how inferences derive from in vitro model-systems, but also to account for how these inferences could provide hypotheses about the in vivo system. In general, the warrant for such transfer depends on the kinds of considerations BME researchers advanced in section 2.1 in the assessments they make of the fit between the devices and model-systems and the target in vivo systems during the building processes, as well as in their evaluations of outcomes from experimental simulations. We saw, for instance, researchers query: Is a first-order approximation sufficient for representing blood flow? Do the effects of shear stress on the cells in a flat construct differ in relevant ways from the effects in tubular arteries? Can a network that uses only cortical neurons provide a model for learning? And so forth. These are questions about the relevance of the features they have selected to instantiate—or not—in the model. *To attain the epistemic goals of distributed model-based reasoning requires that selected model features match to the target phenomena in ways relevant to the problem and goals at hand, and that nothing essential to the problem is left out.*

2.3 Predicting "What You Hope Would Happen in Real Life": Building Epistemic Warrant

In vitro models are the primary means through which researchers in numerous fields of BME gain epistemic access to complex biological phenomena. The researchers develop epistemic warrant for a model through the *principled* decisions and rationalizations they make in the processes of building it. Building epistemic warrant for a model is a significant piece of the process of building to discover. The warrant for using these kinds of models as *epistemic* tools, then, is connected to how the models function as dynamic representations; that is, how they are designed to instantiate and simulate in vitro features. It does not matter to the airplane pilot how the system of the speed bug was built to be able to use it to solve the problem of adjusting the plane's speed during landing. But a biomedical engineer needs to know how a model is built and in what ways it does or does not exemplify features relevant to the research in order to claim to have gained insight into and, have hypotheses about, an in vivo target system from experiments conducted with in vitro model-systems. We saw ample evidence in our research that the researchers do have this knowledge of their models, as I briefly detailed in section 2.1. What I propose, and consider in this section, is that for philosophers to understand how the practice can achieve its epistemic goals through such model-based reasoning, we need to understand the epistemic affordances of the *models as built analogies*.

In the words of our researchers, models are designed to "*parallel*" or "*mimic*" features of the in vivo phenomena. I interpret their expressions to mean that in vitro physical simulation models are built to provide structural, behavioral, or functional analogue representations of selected dimensions of complex in vivo biological systems. They provide a way to get a grip on the behavior of a biological system by creating a virtual world through which to conceptualize, control, and experimentally probe aspects of a complex dynamic system. They are, to some extent, in the words of lab D researchers, "*building science fiction.*" But, the models can function as epistemic tools only if they have been designed with an appropriate representation of biological facts. Importantly, unlike computational virtual worlds, in vitro models are composed in part of biological materials, so the cells and tissues have biological functionality that needs to be maintained as they interface with engineered materials and perform under greatly simplified conditions, all of which figure into how they function epistemically. And, to add a level of complexity, most model-systems are *nested* analogies, that is analogies within an analogy (Nersessian and Chandrasekharan 2009). For example, the flow loop provides an analogy to hemodynamics, the construct provides an analogy to the blood vessel wall, and the model-system they constitute provides an analogy to blood flow in an artery. So, the considerations in play are not only about each model, but also about how the model-system fits together (interlocks).

So, what enables the researchers to have some assurance they are on a productive path with a device or model-system design? Despite their complexity, in vitro models are missing much of the in vivo target system. Determining what can and cannot be left out is a multidimensional problem that requires the researcher to track a number of constraints at the same time. What we saw in our data is that researchers were continually asking the question I phrase generically as, "Is the model of *the same kind as* the in vitro system along the dimensions relevant to the problem?" That is, are the features instantiated such that the researcher is warranted to infer that the behaviors of the model belong, along specified dimensions, to the same class of phenomena as those of the in vivo biological system? Answering that question requires an assessment both of the relevance of the features that are instantiated in the model to its behavior and of those that have been left out. As I have argued in my (2008) analysis of conceptual analogue models, the best way to interpret that question is as asking whether the built analogy *exemplifies* the features relevant to the research.

In the sense advanced by Goodman (1968) and Elgin (2009, 2018), "X exemplifies Y" means "X instantiates relevant features of Y and refers to Y by means of that instantiation." A paint chip, for instance, instantiates, and so refers to, a selected color that is relevant to the goal of the painter to reproduce that color on a wall. The notion of exemplification captures the representational relation the researchers aim for as they build models to parallel and mimic in vivo phenomena. The dish not only refers to neuronal network processing, it is doing such processing in a neuron network simplified to a monolayer. The flow loop, in performing, not only refers to shear stress forces in a process of blood flow through the endothelial cells in a blood vessel, it also produces those shear stress forces. The liquid has what the researchers judge to be relevant fluid-dynamic features of blood as it flows over the endothelial cell cultures or the construct device that has been designed to have relevant features of the blood vessel wall. The in vitro models, then, are successful as exemplifications if, indeed, they possess the features of the in vivo phenomena germane to the problem at hand, and much of the research is directed toward determining if this is the case. Such determination requires the researcher to consider the relevance of both what is and what is not instantiated to the behavior of the system. For instance, lab A researchers could examine and make inferences about cell proliferation and morphology when they used the flow-loop–cells-on-slides model-system, but not the functional relations between them and smooth muscle cells, which led them to develop the construct that instantiates both types of cells and supporting tissue. Importantly, then, what is not instantiated (either negative or neutral analogy) at a specific point in a research program provides a potential resource for further development. Building in vivo models toward exemplifying features is an iterative and incremental process of epistemic iteration. Models that are satisfactory exemplifications provide the researchers with warrant for analogical transfer of experimental outcomes. So, *analogy and exemplification work together in model-based reasoning,* as I consider next.

2.3.1 Analogy and Exemplification

The BME epistemic practice of building devices and model-systems is, fundamentally, an analogical practice. The researchers aim to design models to provide analogical sources that have the potential to provide understanding and control of complex biological systems. This analogical practice is quite unlike any considered in the customary philosophical and cognitive science literatures. Usually, analogy is cast as a process of making sense of what we do not understand (target) in terms of what we do (source). In the case at hand, little is understood about either source or target at the outset. In the usual case of analogical problem-solving, the reasoner retrieves a previous problem solution—or, more broadly, an existing representation—that provides a source analogy, determines a mapping between source and target, transfers features from source to target, and evaluates inferences with respect to the target domain. Mary Hesse (1963), whose account has been most influential in both philosophy and cognitive science, called the features that form the mapping "positive," if they match the target, "negative," if they do not match, and "neutral," if their status is unknown.[11] In her account, the neutral features provide a resource for further development. Although negative features as a possible resource are not addressed by Hesse, Tarja Knuuttila and Andrea Loettgers (Knuuttila and Loettgers 2014) have shown in their recent analysis of interdisciplinary analogy use (retrieved analogies) in synthetic biology that negative features also can lead to further development. I have argued this as well with respect to built analogies, where features not exemplified in a model can provide negative analogies that can be evaluated as opportunities for development (see Nersessian 2008).

Although models have pride of place in contemporary philosophy of science, scant attention has been directed toward the analogical dimension of models besides my own (Bailer-Jones 2009; Harré 1970; Black 1962; and Hesse 1963 provide exceptions). I venture this lack of attention stems from the fact that the literature focuses on models as derived, at least partially, from theories, which has brought to the fore traditional representational issues associated with realism and especially the problem of how "false models" can support predictions or provide explanations. I have been arguing that starting from the other direction—that of building models "from the ground up" in the absence of a theory of the phenomena under investigation—underscores how models and analogies are tightly bound (see, e.g., Nersessian 1992a,b, 2008). I contend, too, the analogical relationship between model and world has importance, thus far not addressed in the literature, for the customary, reverse direction (models from theories), since it provides the means to transfer prediction, explanation, and understanding from model to world. In addition, as Hesse (1963) pointed out in her groundbreaking analysis of models as analogies, the source model is always a false representation in that it cannot accurately or adequately represent all the features of the target phenomena. With her, I contend that "true" and "false" are not the appropriate categories for thinking about models. But neither is Hesse's, and the customary, notion of similarity, as an extensive philosophical literature has been arguing. It is not necessary to get into the intricacies of these discussions for our purposes, however, because in vitro simulation models must instantiate relevant biological features in order to function properly, and so "similarity" is also not an appropriate category for the representational relation between these kinds of models and the in vivo phenomena.

The vast cognitive science literature on analogical reasoning, too, does not attend to the creative work of building the source representation that is central in the kinds of analogical problem-solving I have been considering.[12] Cognitive science has made significant contributions to analogical reasoning that are useful to epistemic considerations, such as the overwhelming experimental evidence that, by and large, productive analogies rely on mapping relational structures (see, especially, Gentner 1983; Gentner et al. 1993; Holyoke and Koh, 1987; Holyoke and Thagard, 1989) rather than similarities between properties, and that multiple constraints, including goals, direct the mapping and transfer processes (see, especially, Holyoak and Thagard 1989). I have reviewed, analyzed, and critiqued the cognitive science literatures on analogy in my 2008 book, and I refer the interested reader to the discussion there in chapters 5 and 6, rather than reprise that analysis in full. Instead, here I focus only on the aspect of my account developed there of analogical reasoning in creative problem-solving in science that is addressed by neither the philosophical nor cognitive literatures: building the analogical source/base. If we attend to this aspect, it becomes clear there is a significant linkage between models and analogical inference. My prior analysis concerned building conceptual models. The in vitro model-building practices discovered in our empirical investigation provides an opportunity to advance that account.

The standard analyses assume that in problem-solving, source analogies are prior (though not necessarily easy to retrieve) problem solutions the reasoner has encountered. However, there is a significant *representation-building* aspect of analogy in science for which several sources of data provide evidence, including historical, think-aloud protocol, and ethnographic data (Nersessian 2008). Although what we customarily understand as analogy occurs in science, that is, a comparison to what is ready-to-hand, for frontier research problems there is often no such analogical source. Rather, the source analogue itself needs to be created in interaction with the goals and constraints of the target problem—a process that furthers the articulation of the problem itself. My original analysis of building analogue sources concentrated on conceptual models in physics, in interaction with diagrammatic representations. Here I list the following takeaways from that analysis that are relevant to BME in vitro models as built analogies.[13]

The main features of building analogue source models are these:

- Building processes are goal-directed.
- Building processes are iterative and incremental.
- Interaction between source and target is ongoing in the building process.
- Elements used in building analogies can derive from more than one domain (hybrid analogies).
- Various abstractive processes are used to select features and to merge target, source, and model constraints.[14]
- Mappings are established during the building processes, so in most cases mappings develop over time.
- Models are built toward instantiating features germane to the epistemic goals.
- Models are evaluated on the basis of whether they in fact exemplify relevant features.
- Features not exemplified can provide a resource for further development.
- Analogical transfer requires that a model instantiate relevant features.

As we saw, BME researchers aim to build physical simulation models to the degree of specificity they believe sufficient to examine an aspect of the in vivo phenomena in a cognitively tractable manner. This goal is informed by an assessment of both the current state of understanding of the phenomena and the degree to which the current state of the available materials and technologies constrains and enables design possibilities. With respect to the latter, for instance, the lab A researchers were aware that the cells-on-slides model did not instantiate some clearly relevant features of the blood vessel wall, but the development of the construct model had to await the advent of tissue engineering. Given the frontier nature of the research, all of these factors change over time; thus, the building process is incremental, as the representation is developed over an extended period. Further, to underscore the engineering dimension, models are hybrid constructions, and there is tension between the constraints on the design and functionality of a device that derive from biology and those from engineering. Some selections are made in order to merge these constraints, and these need to be considered in assessing the warrant for any inferences they transfer. For instance, part of lab D's use of a monolayer of neurons had to do with the construction of the MEA and its recording capabilities; part, with the ability to feed the neurons given the overall construction of the dish; and part, with the need to begin with a "*simpler system to study*" than if three-dimensional layers were to be used. These and other considerations added a level of uncertainty as to whether the dish model could actually achieve learning and what it would mean for learning in the brain if it did, that is, how and what it exemplifies (see note 5).

During our investigation, in both labs the researchers' concerns about a model's relation to the in vivo system informed decisions about design and redesign, as well as how they evaluated experimental simulation outcomes. Importantly, in vitro models are dynamic systems, and a model needs to instantiate those features that enable the cells and tissues to behave ("mimic") in an experimental simulation as they would in the in vivo phenomena under those conditions. A major epistemic task, therefore, is to determine what those features might be and whether or how any abstractions that have been made can impact behavior. Take, for instance, a flow-loop simulation that instantiates first-order (laminar) blood flow. This is a counterfactual situation because there are always higher-order effects in vivo, but for their initial epistemic goal to simply understand in what ways forces can affect the morphology and proliferation of endothelial cells, the researchers argued that there is no need to capture the full complexity of the in vivo blood flow at the outset. The reasons researchers gave for this choice included such considerations as these: there are places of laminar flow in the circulatory system as the flow gets further away from the heart, laminar flow enables them to impose a well-defined shear on a population of cells, and if indeed the cells functioned differently in significant ways in vivo (e.g., gene expression), the device design affords (or can be redesigned to) the possibility to simulate higher-order effects. These reasons, in order, are of the following sort: the model instantiates a germane feature of a part of the in vivo system of interest, the model achieves an important engineering goal that reduces the complexity of the analysis, and the model can be made to instantiate other features of the system if in vivo biological function is importantly different. They did use the flow loop's capacity to simulate higher-order effects in later research when it became technologically possible to examine gene expression, which made it worthwhile to investigate these effects (initial negative analogy).

Importantly, redesign is an overarching agenda in in vitro modeling in BME. Some redesigns have to do with improving the engineering and others are made for practical purposes, such as enhancing the viability of cells. The most important redesigns, however, are to improve the nature of the parallelism to the biological phenomena of interest, if only in minor ways, as they are made to provide better or different exemplifications. This process is often motivated by known negative analogies. Redesign can be driven by a change in understanding of the phenomena or of the problem or by a change in technological and material capabilities as the research progresses. At any point in time, in vitro models are in different stages of development. During the period of our investigation, the flow loop was quite stable, and the construct was still undergoing design changes, aimed mostly to improve its mechanical strength toward that of the in vivo artery. Although the dish design was stable, experience with it was quite limited, and the embodied systems were newly under design. Thus, exemplification, at least in this context, needs to be understood as a historical process. There are usually quite numerous obstacles along the way, but determining how to overcome or get around these provide opportunities for learning about the model and the in vivo system. Once the kinks have been worked out of a design and the researchers assess that it has met their current epistemic goals, change is largely incremental. In vitro systems are meant to be sites of long-term investment so as to enable systematic experimentation.

The programmatic agenda of redesign makes lab history a hands-on resource, since the current design needs to be understood as conditioned on the problem situation as it existed for the lab at a prior time. It is thus important in moving the research forward that researchers know what, how, and why design choices and changes have been made. Although the flow loop was a stable design at that time, the researchers were able to recount its redesign history, as well as to envision potential changes down the road. Notably, we saw few instances in which a kind of model was abandoned completely. In those cases, the reason was either that line of research had come to an end or researchers found a better, often simpler, way to achieve their goals. Because of the incremental nature of design, the analogical mappings between built source and target are not fixed as they are with retrieved sources, but rather are what we called "creeping," as models are built toward providing better analogical sources (Nersessian and Chandrasekharan 2009; Chandrasekharan and Nersessian 2017). A better, more satisfactory analogue model is one that improves or enhances the relevant features of the target system that the model exemplifies. For instance, some important information could be gleaned from the dish by stimulating it with electrical signals and recording its response (open-loop physiology) but a closed-loop model-system in which the dish is connected to a robotic or computational model is a more satisfactory analogue because it instantiates "sensory" feedback as occurs in the in vivo brain.

Now that I have laid out my idea of how analogy and exemplification work together to build analogical mappings and provide warrant for transfer, it is useful to take another look at the processes through which the lab A researchers designed and redesigned various flow-loop model-systems as a concrete illustration, which itself exemplifies the BME practice of in vitro modeling. When we entered this lab, in vitro research had been under way for more than twenty years, so much more was understood about the in vivo phenomena than would be the case in more preliminary research, such as in lab D. Scientific understanding of the biological phenomena was informed now by decades of this (and other) lab's in vitro research and animal research, as well as vascular and cell biology research by bioscientists.

2.3.2 Building Analogue In Vitro Model-Systems

At the start of his research into the problem of the effects of mechanical forces of blood flow through arteries on the cardiovascular system, the lab A director decided to greatly simplify the vascular system by focusing only on the endothelial cells, which form a monolayer that provides the inner lining (endothelium) of the blood vessel. The reason he gave is that "*it's the cell layer in direct contact with flowing blood*" and so "*it made sense to [him] that, if there was this influence of flow on the underlying biology of the vessel wall, that somehow that cell type had to be involved.*" Thus, the initial in vitro model system was designed to exemplify the shear forces in the process of blood flowing (flow loop) over the endothelium (cell on slides). In vivo hemodynamics is a complex process that has areas of laminar flow (smooth path without interference) and areas of turbulent flow (whirlpools) as the heart pushes the blood through the system. Further, the process varies over the course of twenty-four hours. Things change constantly in human bodies over the day and over lifetimes, including physiological flow rates. These changes had been a significant problem in the director's earlier animal studies, and motivated his move in vitro. So, researchers saw flow-loop simulations as "*something very abstract because there are many in vivo environments and many in vivo conditions within that environment.*" The lab used mainly unvarying laminar flow in its simulations: it instantiates the shear stresses during blood flow in an artery to a "*1st order approximation of a blood vessel environment . . . as blood flows over the lumen.*" The researchers maintained that this reduction was a warranted selection because it enables "*a way to impose a very well-defined shear stress across a very large population of cells such that their aggregate response will be due to [it] and we can base our conclusions on the general response of the entire population.*" That is, this abstraction enables them to determine changes in cell morphology and proliferation across a population, characteristics that can be easily determined visually by means of a confocal microscope and Coulter counter, respectively, and can be related directly to the controlled shear stresses and quantified.

These studies provided provisional understanding sufficient for the researchers' goal of getting a grip on these effects of blood flow about which nothing was known at the outset. The researchers, were, however, aware of several negative analogies from the outset: cells on slides, laminar flow, and lack of diurnal blood flow variation. These were sources of further development of the lab's in vitro model systems. The flow loop, as designed, did have the capability to produce a range of flow rates. Flow-loop simulations could also be made to instantiate higher-order effects if there were reasons to do so, such as "*if there's a whole different pattern of genes that are upregulated in pulsatile shear.*" In this circumstance, however, for many years there was no way to investigate possible salient differences in gene regulation. That potential came quite late in the research program, when gene array technology was developed, at which time they made an agreement to use the new technology at a nearby medical school. The prior basis for partial comparison of their results was provided by studies of morphology and proliferation in vascular biology and whatever biological markers were available from biochemical studies. The possibilities for comparison from biological research were always fluid and not fully adequate for lab A's purposes.

Two other negative analogies were important to furthering the lab's research program. First, the flow-loop model exemplifies only one of the in vivo mechanical forces: shear stress. This is the force with the greatest impact on the endothelial cells. Blood vessels are also subject to strain forces from the blood pressing on the vessel wall, but to instantiate this force requires a model that instantiates the topology of the vessel. As we will see in chapter 4, new simulation devices were built to investigate pressure or strain, once the construct model was introduced. The other negative analogy, as we have seen, concerned the use of slides with endothelial cells in culture in flow-loop simulations. The researchers recognized that this model-system does not provide *"a physiological model,"* that is, *"putting cells in plastic and exposing them to flow is not a very good simulation of what is actually happening in the body."* What this means is that this simulation does not instantiate some of what they knew to be relevant mechanical and biochemical features of blood flow through the lumen of an artery, and thus limits the understanding obtained from it. For one thing, endothelial cells have a *"natural neighbor,"* smooth muscle cells. The researchers first tried a co-culture of both kinds of cells to address this issue, but it, too, is not a satisfactory physiological model. Not until the technologies to engineer complex tissues started to develop in the 1990s did it become feasible for the lab to take the *"big gamble"* and attempt to build a blood vessel wall model that could also instantiate smooth muscle cells and other in vivo components in the model—that is, build the construct family of models.

As the director stated, the research when we entered had the goal to *"use this concept of tissue engineering to develop better models to study cells in culture. . . . So, we had the idea: let's try to tissue engineer a better model-system [construct] using cell cultures."* Unlike endothelial cell cultures on slides, a construct has a three-dimensional tubular surface in which both kinds of cells can be embedded, along with various native tissue components, such as collagen. It exemplifies a blood vessel, because with those components instantiated, it *"behaves like a native artery because that's one step closer to being functional."* This redesign of the cells-in-culture model from slides to constructs provided a different kind of simulation—one that more closely mimics the in vivo system. Simulations with this model-system could also be used to determine whether there were relevant differences in the behavior of the endothelial cells between the simulations with the slides and with the constructs. And, finally, the tubular construct afforded the possibility for the lab to develop new in vitro devices through which to simulate and investigate the strain forces (the other negative analogy), some of which I discuss in chapter 4.

How the construct was used with the flow loop demonstrates that engineering and pragmatic constraints interlock in the assessment of what is relevant to instantiate of the in vivo phenomena. They had intended to redesign the flow-loop chamber to accommodate the tubular construct design, which they assumed would be a time-consuming and costly process.

In the end, the researchers made the decision to cut the construct open so it could lay flat in the flow-loop chamber. The researchers reasoned that the reduction in dimensions would not matter (neutral analogy) since the endothelial cells that line the vessel are so small with respect to the surface "*the cell has no idea that there's actually a curve to it [vessel].*" Because cells embedded in flat tissue experience the same shear forces as when the fluid flows over them in vitro as in vivo, they concluded there is no need to instantiate their tubular shape to study the effects of these forces. The thickness of the construct required a small modification of the design of the flow chamber to accommodate them. However, once there was the possibility of doing an implantation experiment with an animal, the constructs needed to be kept in tubular form. When it came time to redesign the flow loop, as we saw, the solution was quite simple and inexpensive, once A7 saw the possibility of an analogy (ready-to-hand) with the animal shunt, which, indeed, worked.

As noted previously, the development of the construct in tubular form made possible investigations into strain forces for which the lab developed model-systems that instantiated different kinds of strain forces. Additionally, the tubular form opened a line of research that was directed toward figuring out what creates the significant mechanical strength of tissue of the in vivo vessel. This understanding was a prerequisite for them to redesign the construct to instantiate the in vivo mechanical strength, and thus exemplify a feature germane to the applied goal of a vascular graft.

Finally, to cast the representational nature of devices and model-systems in terms of exemplification underscores the sociocultural nature of representation that Goodman called to our attention. A tailor's swatch or a paint chip is understood to represent a fabric or a color only within specific sociocultural practices. In an epistemic practice, the norms, values, and assumptions of the culture contribute to the warrant of its practices. In unraveling some of these in the epistemic practice of building hybrid in vitro simulation models, we have been able to discern how an engineering culture is embodied in these technologies. The nature of the research questions in BME dictates the hybrid nature of the physical simulation models and also that the researchers attain a degree of hybridization as biomedical engineering scientists. Indeed, the interdisciplinary culture within which these labs reside self-consciously refers to itself as an "integrative interdiscipline." But, as I consider in more detail in chapters 4 and 7, the interdiscipline largely comprises engineers who become hybrid biomedical engineers through their education and research, and not an integration of, say, tissue engineers and vascular biologists. As we have seen, in the epistemic commitments of the researchers, engineering assumptions, norms, and values predominate. Within the engineering framing of the problem, the biological phenomena are to be explained and understood largely through engineering concepts (e.g., shear forces on endothelial cells; electrical noise in neuron cultures) and the application of engineering methods to screen, isolate, and control the "*messiness*" and complexity of biological systems. Experimental setups are designed to provide outcomes that can be turned into mathematical form. Although cell cultures that need to be created, sustained, and cared for are at the heart of the research, the ways in which they are thought about are as opportunities for design into in vitro simulation models. Our interviews provided substantial evidence that the design of the simulation models is dominated by norms and values associated with engineering, such as abstraction, approximation, control, quantification, constraint satisfaction, simplicity, and a good measure of pragmatism, especially, in the form of compromises with respect to what it is feasible to do vis à vis engineering. I and my research group have worked for many years in an engineering institution and recognize these norms and values as widely in use across the various engineering fields and as inculcated in engineering courses. Many of these share what Keller (2002) portrays as the values and norms of mathematical physics, which she contrasts with those of experimental biology. Within the context of our investigation, these epistemic norms and values did not seem to be undergoing the hybridization, as were the models and the researchers—at least we have no evidence to support such a claim. Perhaps epistemic hybridization takes place more slowly, and this kind of BME community is still young.

2.4 Summary: Getting a Grip with/on In Vitro Simulation Modeling

In our preliminary ethnographic investigations of BME research labs, we discovered the practice of in vitro simulation modeling, which is the primary means of research in many fields of BME, but is an investigative method novel to the philosophy and history of science. The practice involves building hybrid artifacts that merge living cells and tissues with engineering materials to provide sites for simulation of biological processes under experimental conditions. Such simulations are the primary means through which researchers attempt to get a grip on complex biological systems, otherwise inaccessible to experimentation. With our discovery, we decided to concentrate our research on understanding these "*devices*" and "*model-systems*" and the practices surrounding them within our framing of the research lab as a distributed cognitive-cultural system with epistemic aims.[15] Our research goals required that we determine how the models are created, how they function as cognitive-cultural artifacts, what are their epistemic affordances, and what provides the epistemic warrant for believing the models can provide information relevant to understanding and controlling the complex biological target system. These dimensions are interwoven in the lab's actual research practices, but in this chapter, I have disentangled some aspects for the purposes of analysis. In particular, we have seen that building these in vitro models requires researchers abstract and integrate constraints from engineering and biology. Our cognitive ethnographic investigation enabled us to follow out the reasoning of the researchers in detail over an extended period as they determined and justified their selection of what features to instantiate and the various abstractions and engineering and biological trade-offs.

Our pilot studies drew our attention to the devices as the most epistemically, cognitively, and culturally salient artifacts in each lab. As I noted previously, in one of our earliest interviews, a researcher characterized their practice of in vitro simulation modeling as "*putting a thought into the bench top to see if it works or not.*" From the perspective of our analytic framework, "*putting a thought*" designates building in vitro models as the means through which researchers actively distribute cognition in the environment, thus building the distributed cognitive-cultural that is the lab, and the problem-solving subsystems it comprises. The in vitro models have a dual nature. They are cognitive artifacts because they perform cognitive functions that enable analogical and other model-based reasoning processes, a major epistemic aim. They are also the material culture around which researchers develop the sociocultural practices pertinent to the lab's epistemic goals and in which researchers embed the epistemic assumptions, norms, and values of BME and the further subdivisions into tissue and neural engineering. Furthermore, individual and lab identity and history are bound up with the models researchers build and use in their epistemic practices.

To "*see whether it works or not*" refers to the iterative and incremental processes of designing, constructing, redesigning, evaluating, and experimenting with models—"building models to discover." I have characterized the practice of in vitro modeling as one of building analogue source models so as to exemplify features of complex biological systems germane to the researchers' goal to determine an aspect of the system's behavior. Examining how such models are built toward serving as source analogies provides important insights into a largely overlooked creative analogical practice central to this and other frontier scientific research. The BME methodological innovation of building in vitro models demonstrates the more widespread nature of the analogical practice of "building the source" that I identified with respect to conceptual models, as well as reinforces the generality of my earlier analysis of this practice (Nersessian 2008).

Research in both labs began with isolating cells for simplicity and to achieve control. Since the endothelial cells are the ones immediately in contact with the blood flow, lab A research began with studying their behavior in response to flow, independent of the rest of the blood vessel components and environment. In lab D, the embryonic rat neuron connections were broken apart, so all learning would be de novo, and plated on the MEAs as a monolayer, which leaves out other parts of the brain usually thought to be implicated in learning. These moves are warranted because even the greatly simplified systems instantiate features relevant to the goal of understanding aspects of the system's behavior under specific conditions. The flow-loop–endothelial cell model-system, for instance, exemplifies features of hemodynamics germane to the investigation of the effects of shear forces on endothelial cells. The dish model-system, as another instance, exemplifies features of interactions among neurons germane to investigating network activity in response to stimulation. Improving in vitro models toward providing better analogue sources was an ongoing process, especially as researchers aimed to instantiate additional features thought relevant to the behavior of an in vivo system (negative analogies). Much of the research was directed toward figuring out what features need to be instantiated in a model, what constitutes a legitimate abstraction from the in vivo system, and in what ways the current engineering capabilities afford and constrain model design. When we examine ongoing research in situ, we see that and how discovery and justification are interwoven processes in scientific research.

The tendency to isolate and control is, however, counterbalanced by the systems thinking also common to engineering. After a period of investigation with relatively simple models, researchers were willing to take the risk (*"gamble"*) to build back in some of the complexity that was considered to matter. The lab A researchers understood that endothelial cells on slides do not provide an adequate physiological analogue to the blood vessel wall, since at the very least there are smooth muscle cells with which the endothelial cells communicate in vivo. When tissue engineering technologies became available, they began to build the construct device, which could instantiate both endothelial and smooth muscles cells, as well as supporting tissues. The flow-loop–construct model-system exemplifies more of the in vivo system, which, in principle, should yield greater understanding of the behaviors of that system. One form that building complexity took in lab D was to design systems for *"embodied learning"* in the dish. An in vitro model with a feedback loop, that is, one that inputs sensory stimulation to the dish and outputs the dish response in the form of behavior, exemplifies learning in an animal better than the disembodied dish model.

In general, a satisfactory in vitro simulation model is one that exemplifies features relevant to the epistemic goals of the problem-solver(s), which are, in the BME case, ultimately, to understand and control the behavior of the in vivo system. I have argued that exemplification provides the evaluation criteria for warranted analogical inferences. The researchers' assessments of whether a model instantiates those features relevant to the problem at hand provides the basis for their inferences from the source model to the target system. Warranted predictive inferences (epistemic goal) about the behavior of the in vivo system affords understanding as well as possibilities to control and intervene (pragmatic/application goal) on the target system. However, the application goals require a high degree of confidence in the outcome of a simulation, as well as ethical and practical considerations, so in frontier research the realization of those goals is often quite far off. Instead, the objective of the labs we investigated, each during a five-year period, was by and large to understand and control the models themselves so as to provide better epistemic tools.

Given the pioneering nature of the research, in building the warrant for a specific in vitro model (construct, dish), researchers are also building the warrant for a specific domain practice (vascular and tissue engineering, neuroengineering), as well as the practice of in vitro physical simulation modeling across the spectrum of that kind of BME research. The warrant for the modeling practice rests, ultimately, on the ability of the model-building methods to create models that provide a reliable and successful basis for understanding and inference. Further, a successful practice of in vitro simulation modeling provides warrant for the broader epistemic project of biological engineering it exemplifies: to use the material, conceptual, methodological, and technological resources, together with the norms and values, of engineering to get a grip on complex biological systems.

3 Engineering Concepts: Conceptual Innovation in a Neuroengineering Lab

In chapter 2 we saw how building hybrid devices and model-systems provides the means for BME researchers to think about real-world biological systems. In this chapter I examine how the interplay of models of different kinds led to conceptual innovation in the neuroengineering lab. Solutions to the problems the BME researchers in our investigations posed required creating conceptual resources through which to understand both the in vitro models and the novel behaviors that emerged from experimental simulations. For instance, as noted in the chapter 2, the lab A researchers introduced and articulated the concept of arterial shear to mark their novel investigation of the effects of arterial blood flow on the endothelial cells by means of mechanical forces, in contrast to the customary biochemical investigations. The initial placeholder concept provided the means to characterize the various flow-loop model-systems and, eventually, to understand the behaviors of the cells in response to stimulation with mechanical forces. This chapter provides a detailed analysis of how building modeling environments facilitates processes of conceptual innovation. Conceptual innovation includes conceptual transfer and modification, as well as novel concept formation.

Frontier research areas are excellent candidates in which to investigate conceptual innovation in the development of an epistemic culture. Concepts are significant cognitive-cultural resources. They provide the means through which the researchers understand, reason, learn, and communicate. Often the phenomena under investigation, such as mechanical forces on cells or network-level neuron learning, are entirely novel, as are the means through which they are investigated. Research in BME requires, in addition, interdisciplinary synthesis, in which researchers tap conceptual resources from engineering to represent phenomena in biological systems. So, it was not surprising to find that solutions to research problems in these labs required conceptual innovation. What *was* surprising was the nature of the modeling practices through which the innovations actually took place. In the lab D case I consider here, this included building computational models as well as in vitro simulation models. As we will see, building the in silico model of the in vitro model played a pivotal role in the research and in building the D-cog system.

The analysis I develop here builds on the account I have advanced of conceptual innovation as a problem-solving process in which model-based reasoning plays a central role (most recently, Nersessian 2008).[1] In particular, it once again underscores the central role of analogy, visualization, and simulation in creating or adapting concepts. Thus, this chapter provides another exemplar of the "productive interplay" through which philosophical notions and frameworks are examined, modified, and extended in interaction with novel empirical data on scientific practices. In the earlier work I examined how problem-solving processes using conceptual models, which are developed in the scientist's imagination and explored in conjunction with pen and paper visual (especially diagrammatic) and mathematical representations, are instrumental in conceptual innovation (e.g., Nersessian 1984, 2002, 2008). In this chapter I focus on the relation between conceptual innovation and the problem-solving practices of building in vitro and in silico (computational) simulation models that we discovered in the bioengineering sciences.

During the period of our investigation, both BME labs provided novel data on the centrality of model-based reasoning for conceptual innovation, as well as on how the need to understand novel phenomena can promote new investigative practices that configure and reconfigure "the lab" as a distributed problem-solving system (chapter 4). In this chapter, I develop an extended case of conceptual innovation "in the wild" from lab D, the neuroengineering lab, where the cross-breeding of two models—one in vitro and one in silico—led the researchers to create a cluster of concepts to get a grip on puzzling dish phenomena and, ultimately, on their problem of neuronal network learning. In these conceptual innovation processes, the researchers both transferred and modified concepts from single-neuron studies and engineering to the network, as well as constructed novel concepts specific to network behavior.

Our analysis of the case centers on three graduate students whose research projects were brought together through the development of a computational model of the in vitro dish model-system discussed in chapter 2. The in silico model, which might be considered a second-order model, was constructed initially by one researcher in an attempt to understand the spontaneous, dish-wide firing of the neurons ("*burst*" phenomena) that was occurring in the in vitro model and that they assumed was an impediment to progress in the lab's research project of getting the dish to learn. Thus, D11 built the in silico dish to provide an analogical source for getting a grip on the artifact model, the in vitro dish itself. For the in vitro dish to function as an epistemic tool and help them gain understanding of neuronal network behavior in vivo, they first had to understand its puzzling behavior. This kind of second-order modeling of built prototypes (which we consider the in vitro dish to be) is a common engineering investigative practice. For instance, engineers develop computational models to provide wind-tunnel simulations of the behavior of airplane prototypes under various conditions. In this chapter, as well as in chapter 5 where I discuss systems biology cases, our investigations show specific ways in which this kind of practice is taking root in the bioengineering sciences. In the case examined here, once the in silico dish model was established through numerous iterations, it served as a platform from which the researchers derived novel insights and problem solutions that they, eventually, transferred for investigation to the target in vitro dish. The results of this investigation also opened up novel application possibilities with respect to the in vivo world. The interaction between the two kinds of models led the researchers to formulate a cluster of novel concepts by which to understand neuron network activity and, ultimately, to solve the problem of getting the dish to learn. Importantly, only once the researchers had developed a control structure for supervised learning for the in silico dish did it prove possible for them to exploit the built analogy to develop a control structure for the embodied in vitro dish, MEArt.

For the purposes of clarity of the exposition, I divide the research into three phases (sections 3.1.1–3.1.3), although in the actual research some of the activities in these phases overlapped. From the perspective of our ethnographic research, it was not until well into phase 2 that we realized that something highly significant for both their research and our own appeared to be taking place. Fortunately, we had been collecting detailed data on these research projects as we worked to make sense of these data ourselves. So, we both were able to collect additional data as it was happening and had the possibility to go back and reconstruct earlier phases as the research had unfolded. We continued to conduct follow-up interviews with the lab D members as their major publications and dissertations on this research were being written. After presenting our case analysis in section 3.1, I examine in section 3.2 the specific epistemic affordances of building the lab D computational model for this case of conceptual innovation, and consider, broadly, those affordances of computational simulation modeling as part of D-cog systems.

3.1 Concept Formation and Change in a Neural Engineering Lab

3.1.1 Phase 1: "Playing with the Dish"

As discussed in chapter 2, lab D was founded to pursue the general hypothesis that advances could be made in understanding the mechanisms of learning in the brain by investigating the functional properties of in vitro models of neuronal networks.[2] The current neuroscience paradigm for studying the fundamental properties of living neurons used single-cell recordings. The lab D director's postdoctoral bioengineering research lab had developed a completely new kind of model-system for studying network interactions and emergent properties that might arise, the MEA dish. Lab D was one of the first to investigate its properties and behaviors and, as far as we have been able to determine, the only one at that time to investigate embodied dish model-systems. As a reminder, to construct the in vitro dish model researchers dissociate the connections among cortical neurons extracted from embryonic rats and plate them (15K–60K, depending on the desired kind of network) on a specially designed set of sixty-four electrodes called a multielectrode array (MEA) whereon the neurons generate new connections to become a network. The design of the dish model-system incorporates constraints from neurobiology, chemistry, and electrical engineering.

For much of the first year, the researchers constructed and cared for dishes, established lab protocols, and created the technologies and software needed to interface with the neuron culture by stimulating it and recording its activity—what researchers called *"talking to the dish."* The technologies and software were developed together with a postdoctoral researcher, briefly in the lab, and a PhD student at the director's former institution.

During that period the researchers put significant effort into figuring out *"what we should be looking at,"* with respect to dish behavior and how to do so. Initially, the three researchers decided to focus on simply stimulating the dish with a *"probe"* electrical stimulus on a few electrodes to initiate activity and see what effects this caused in the network behavior. Their ultimate goal, however, was to figure out how to create and control learning in the in vitro network. Specifically, they aimed to discover how to train the in vitro dish to control an embodiment (animat, hybrot) such that it exhibited goal-directed behavior using feedback. From the start, their research posed significant problems with many facets, the solutions to which would involve conceptual innovation.

The researchers began to conceptualize learning in the neuronal network by transferring concepts from single-neuron studies, specifically those associated with what is known in neuroscience as Hebbian learning. They operationalized learning in terms of *synaptic plasticity* (basically, the ability of the synaptic connections, the structures through which neurons pass signals from one to another, to change in response to experience) and *memory*, which is the ability to retain and retrieve experiences. Plasticity is thought to provide the basis for learning, which they characterized as a *"lasting change in behavior resulting from experience."* They also transferred the mathematical formulation, known as the Hebbian rule, for learning (basically, "neurons that fire together wire together"). As researcher D4 recounted later, *"from Hebb's postulate—which talks about learning between two neurons—we thought our data will show that something has to be added to the known equation in order for it to manifest in a population of neurons. Our idea was to figure out that manifestation. . . . So, it has gone from what Hebb said for two neurons to what would translate into a network."*

Figure 3.1 is our schema for the activities of the distributed cognitive-cultural system of lab D for this part of its research as it evolved during the period of the episode, which began in year two. The system comprised the three primary researchers involved in the day-to-day activities of building simulation models and other technologies. Of course, ongoing interactions with the director were an integral part of the system. The three graduate students involved in the case were all recruited within a few months of one another. D2 had a background in mechanical engineering and cognitive science, D4 in electrical engineering, and D11 a joint degree in life sciences and chemistry, with an aspiration *"to make movies,"* which proved relevant

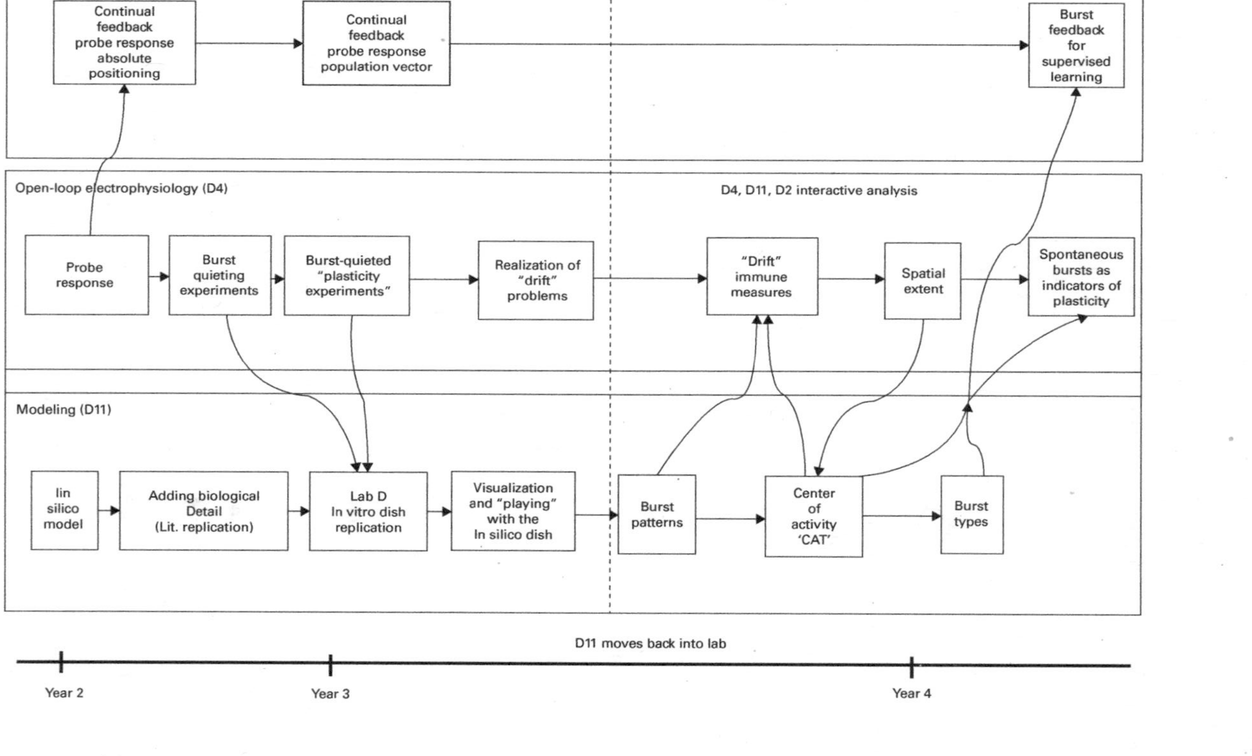

Figure 3.1

Our representation of the approximate timeline of the research leading to the conceptual innovations and the development of the control structure for supervised learning with the robotic and computational embodiments of the in vitro dish (years two through four of the existence of lab D). The dashed line indicates the period after D11 moved back into the main part of physical space of the lab and all three researchers began to collaborate actively to exploit the findings stemming from the in silico dish.

to the research. In general, it was an important part of the culture of the lab to create visualizations—what they called *"making pretty pictures"*—in their attempts to grasp dish phenomena, as well as convey their research to others in publications, presentations, and grant proposals.

During the first year, after the initial period of getting the lab up and running (discussed in chapter 2), the researchers all spent considerable time together *"playing with the dish,"* which consisted of exploring the space of possibilities through stimulating the neuronal network using different electrical signals and tracking the output. In addition, D2 and D11 were engaged in expanding the model-system by building the computational and robotic embodiments that could be connected to the dish and would enable real-time feedback experiments. Although there were a handful of research groups investigating in vitro dishes, these embodied dish model-systems were entirely novel to the research field.

Our analysis begins in year two, when the researchers had encountered a major problem. The middle section of figure 3.1 provides a schema of D4's research. She had attempted to use a probe stimulus to replicate a plasticity result reported by another research group, but was unable to do so, largely because the dish was exhibiting spontaneous synchronous network-wide electrical activity. This activity occurred in all dishes, but was greater in high-density cultures. To interpret this behavior, they had borrowed two concepts: the notion of *burst*, transferred from single-neuron studies, where it meant the spontaneous electrical activity of one neuron, and the engineering notion of *noise*, which is considered a random disturbance that carries no useful information and needs to be eliminated from a signal. They called the dish-wide spontaneous phenomena *"bursting,"* extending its meaning to the *population* of dish neurons. D4 had tried to introduce the term *"barrage"* into the community to focus attention on the network-wide nature of the phenomenon, but soon reverted to *"burst"* when her term did not catch on. This dish-wide phenomenon is visualized in figure 3.2 as the spike activity for each electrode per electrode recording channel, across all channels. Once busting began in a dish it was persistent throughout its life. Bursting phenomena do occur in the brain, during development, in epileptic seizures, and in some stages of sleep, but it is not a lasting phenomenon in the healthy adult brain. The researchers reasoned that a quieted dish would provide a better analogy to the in vivo brain because it exhibits its *"natural"* adult state—that is, bursts are not a feature of a healthy mature

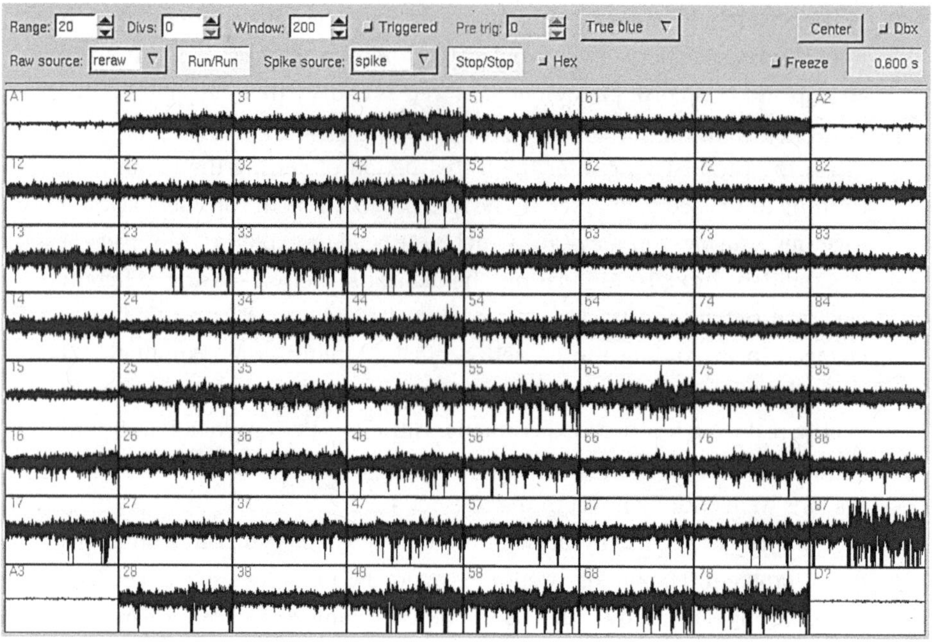

Figure 3.2
The MEAscope per channel visualization of in vivo dish activity shows spontaneous bursting across the channels of the dish. Bursting activity is represented by the spikes appearing in the channels. A relatively "*quiet*" dish would have few spikes across all the channels, which D4 eventually managed to achieve.

brain: "*So if you have an adult, mature awake brain then you never see such bursting, that's except in epileptic seizures. So, the idea is to move away from that kind of unnatural behavior to more natural behavior, where you always have some stimulation coming in . . . and on that you have your learning stimulus. So maybe, then, it's a more natural mode of activity for the dish and you'd be able to induce better learning—that's kind of the model.*"

There was much discussion at lab meetings about how to understand and, hopefully, eliminate bursting. They formulated several hypotheses about it that directed the research. First, they hypothesized, on analogy with an animal brain, which continually receives sensory inputs, that the cause of bursting is the lack of sensory inputs. Next, they hypothesized that "*spontaneous bursts erase the effects of plasticity-inducing experiments.*" That is, bursting creates a problem because it prevents the detection of any systematic change that might arise because of controlled stimulation. They concurred that *"bursts are bad"* and interpreted bursts as *"noise in the data—noise, interference. . . . So it's clouding the effects of learning that we want to induce."* The concept of noise provides, as we will see, an instance in which resource transferred from engineering to biology proved largely to be an impediment to problem-solving.

Given this understanding of bursts as noise, the group decided that they needed to get rid of them. D4 began working on developing techniques for *"quieting"* bursts in the dish by providing the network with artificial sensory input in the form of sustained electrical stimulation. She tried a range of stimulation patterns to lower the bursting activity in the networks. *"Probe stimulation"* is a one-time low-level electrical stimulus to a few electrodes that provides the researcher with an indication of the state of the network, without changing it much. *"Tetanus stimulation,"* is a fast, high-frequency, long-duration (~15 minutes) stimulation aimed to induce plasticity changes. D4 developed tetanus stimulation to simulate the stimulation the thalamus, which is essential to memory and learning, imposes on the cortex in vivo. *"Background stimulation"* is a series of low-frequency stimulations applied at random to about a third of the electrodes.

Without stimulation, a neuron's threshold for firing decreases, which D4 thought could contribute to bursting. She reasoned that constant background stimulation should keep the thresholds from dropping, and thus quiet the dish. After about a year, D4 achieved a breakthrough, managing to stop the bursting entirely. However, despite the meanwhile quieted network, for the next six months her attempts to induce plasticity in the network failed. The activity pattern evoked by a stimulus in the quieted dish did not stay constant across trials, but *"drifted"* away to another pattern. This drift prevented tracking the effect of a stimulus, because the network never responded the same to a constant stimulus pattern. The impasse was so frustrating that D4 told us she would likely leave the research project if no solution to the drift problem was found in the next six months; after all, she had just spent a year getting it to quiet. Instead, as we will see, this failure provided the group with an important opportunity for learning about the dish.

During the same period when D4 was trying to quiet the network and induce plasticity, the other researchers were engaged in largely separate, though interrelated, research activities. D2 (top section figure 3.1) worked on embodiment software modules that would transform the electrical signals passed between the neuron network and the motor commands used to control animats and hybrots. He initially used information from D4's probe experiments to begin closed-loop controlled-motion experiments with the embodiments. D2 began with his knowledge of control engineering to develop movement commands to a specific position with respect to a fixed zero point ("*absolute positioning*"). Not having much success with numerous trials of these commands, he then developed commands based on the neuroscience notion of a population vector (roughly, the sum of the preferred directions of activation of a population of neurons) to direct motor commands. Much of this laborious work centered on simply getting the embodiment to move, which required him to determine how to map the neural activity into motor commands and how to map the embodiment behavior into sensory feedback to the dish. D2 created numerous animats and worked with the small robots the lab had purchased, but the primary hybrot he worked was MEArt, the mechanical drawing arm, discussed in chapter 2 (figure 2.5). The MEArt project was perhaps the most exciting of all the lab projects for the researchers because it required that D2 spend considerable time traveling the world with the art exhibit of hybrid mechanical-biological art projects developed by their collaborators. As a biomechanical art exhibit, MEArt's creativity required only that it draw, but as a neuroengineering research project it needed to "draw within the lines," that is, it needed to exhibit goal-directed controlled behavior.

Since MEArt was the hybrot on which the researchers would focus their attention in the experiments to create a control structure for supervised learning, it will be helpful to have a general understanding of the model-system. The mechanical arm was created by a laboratory of artistic research, dedicated to using science and technology to explore the possibilities of "*wet art*," that is, art that fuses biological entities and technology. The arm itself was designed to "*resemble organic forms in function and aesthetics*." Its shape was modeled on bone structure, and the pneumatic muscles were developed based on physiological models so as to contract individually, as can the biceps and triceps. The arm drew in a circular motion about a fixed point by flexing and extending within a specified range, and additional small muscles pressed the pen to the paper. It communicated with the in vitro dish through satellite or Internet transmissions of electric signals from and to the neurons. These signals went through numerous processing programs that "*translated*" between the "*language*" of the neurons and that of the mechanics of the arm. The sensory feedback to the dish was derived from the images of the drawing in progress that were captured by a digital camera located above the movement space. To close the movement feedback loop, D2 had to develop numerous pieces of software to process the signals that directed MEArt and fed back to the dish (*"signal transform"* programs). Again, building this model-system was achieved thorough incremental and iterative processes. As with the vascular construct model-system discussed in chapter 2, we also analyzed the MEArt model-system as a D-cog system that generates, manipulates, and propagates representations through interlocking artifact models and researcher mental models, as represented in figure 3.3.

The movement software modules D2 developed were to be used in training the embodied dish once they had the sought-after control structure. The control structure, however, had an unexpected origin: it was derived through the group's interaction with a computational simulation of the in vitro dish model-system built by D11.

3.1.2 Phase 2: "Seeing into the Dish"

Early in the burst-quieting period, when the research seemed at an impasse, D11 decided to branch away from working with the in vitro system entirely and develop a computational model that would simulate dish phenomena. As he put it, *"The advantage of modeling [computational] is that you can measure everything, every detail of the network. . . . I felt that [computational] modeling could give us some information about the problem [bursting and control] we could not solve at the time [using the in vitro dish model-system]."* D11 felt that to understand the phenomena of bursting he needed to be able to *"see"* the dish activity at the level of individual neurons, to make precise measurements of variables such as synaptic strength, and to run more controlled experiments than could be conducted with the physical dish. Computational modeling was not part of the investigative practices of the lab, largely because the director was skeptical it would be of any benefit since he had rejected computational neural network modeling as a means by which to understand neuron activity in the brain. Fortunately, he did allow his students considerable freedom in their research, and he supported D11's proposal. D11 moved to a different physical space where he could work on a computer, and moved back into the physical space of the lab (indicated by the dashed line on figure 3.1) only after he had successfully

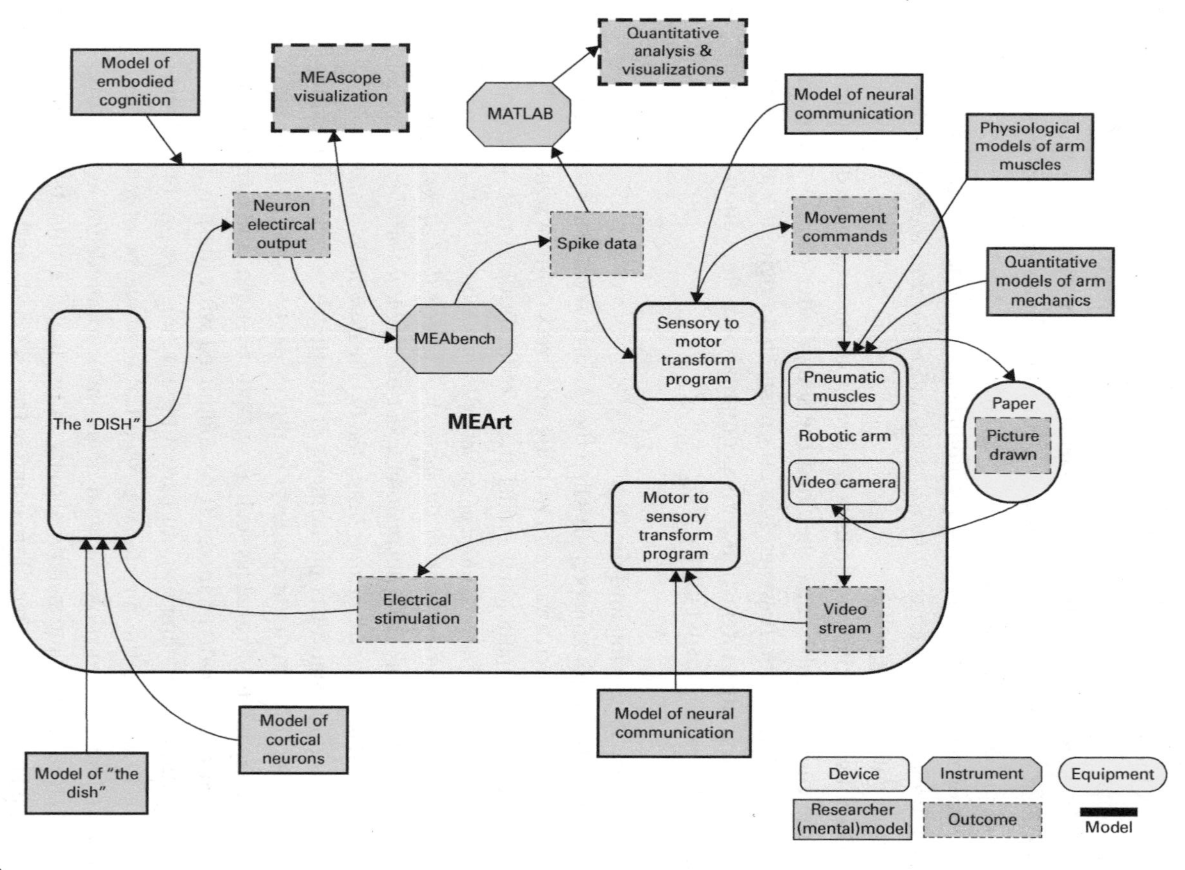

Figure 3.3
Our partial analysis of the MEArt model-system as a D-cog system of interlocking models. The arrows represent the direction of propagation of representations as they are generated and manipulated in the system.

replicated the behavior of their in vitro dish with the in silico dish. From that time, the researchers worked together to exploit the computational model's results in interaction with the in vitro dish and the animat and hybrot model-systems.

From the outset, D11 built the in silico dish model to serve as an analogical source for their in vitro dish model-system. That is, D11's goal was, eventually, to map and transfer insights derived from it to the target problem to understand and control the behavior of the in vitro model-system. However, he began with building a generic in silico dish, which he later instantiated with specific features of the behavior of their dish. Even more clearly than in the case of in vitro models, the in silico case underscores the *emergent nature* of built analogies, since a computational model can take scores of iterations of building and thousands of runs before it replicates the phenomena sufficiently to serve as an analogical source. As it turned, out, the in silico model provided a source analogue from which the researchers would both develop novel concepts to represent in vitro dish behavior and develop the desired control structure for supervised learning in the embodied dish model-system.

D11 developed and optimized the in silico dish model through an incremental bootstrapping process that comprised many building cycles. These cycles included integrating constraints derived from the design of the in vitro model and data from a wide range of neuroscience literature, as well as constraints of the in silico model. Among the latter constraints were those from the computational modeling platform and those that arose from the model as it gained in complexity. Figure 3.4 provides our schema of this complex process.

Many constraints contributed to the iterative and incremental building process, and I highlight only the most significant. First, D11 was not an expert modeler, so for a modeling platform he chose what he called *"the simplest neuron model out there—leaky-integrate-fire"* (CSIM modeling platform) to see whether he could replicate network phenomena without going in to too much detail, such as including synaptic models. The only constraints D11 took from their in vitro dish at the outset were structural: 8x8 grid, sixty electrodes, and random location of neurons (*"I don't know whether this is true, though, looking under the microscope they look pretty random locations"*). In the model he used only 1K neurons, 70 percent excitatory and 30 percent inhibitory, which he believed would produce sufficiently complex behaviors. Each neuron was connected directly to a number of

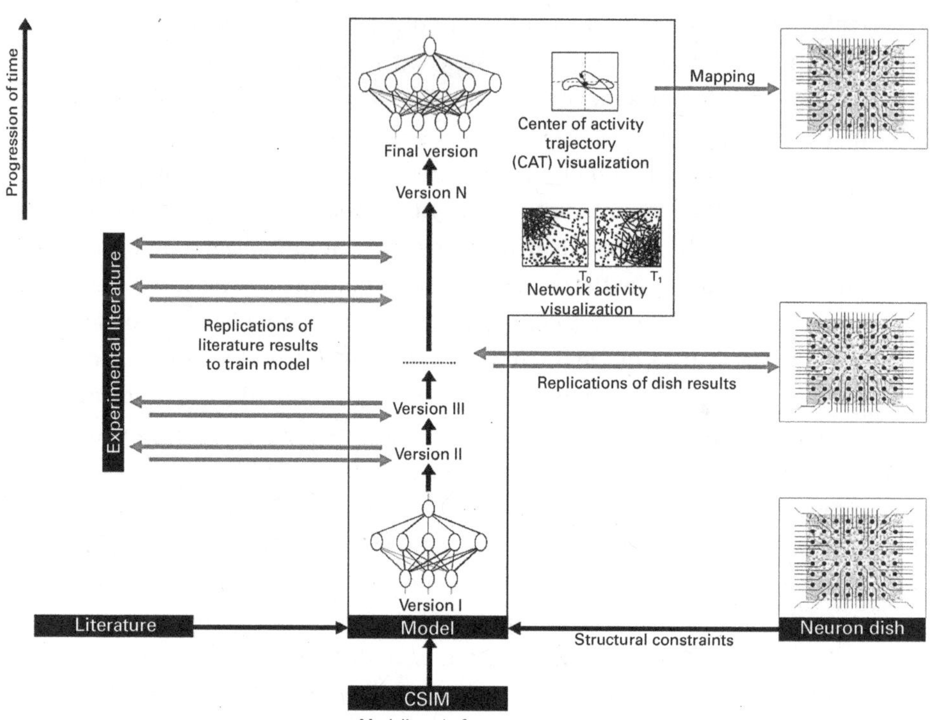

Figure 3.4
Our schema of the bootstrapping processes involved in building the in silico dish through numerous iterations. Once the in silico dish was able to replicate the in vitro dish behavior, the researchers worked to understand and control its behavior. They mapped and transferred those analyses of the in silico dish to the in vitro dish and evaluated their applicability.

other neurons in its surroundings through a statistical distribution that had been developed in neuroscience. In all there were 50K synaptic connections. He modeled each neuron as a simple unit whose activation slowly decreases over time ("leaks"). Through its connections to other neurons, each neuron constantly adds—or integrates—the activation of those neurons to its own activation. When the neuron's activation reaches a certain threshold, it fires and sends signals to all the neurons with which it is connected. D11 took all the parameters of the model—such as types of synapses, synaptic connections, synaptic connection distance, percentage of excitatory and inhibitory neurons, conduction velocity and delay, noise levels, action potential effects—from the neuroscience literature on single-neuron studies, brain slices, and experiments with in vitro dishes other than their own. His justification for using all of these data was that the behavior of the neurons should be pretty much the same as these instances. D11 then just let the model run for a while to see what would emerge, which is typically what modelers do before they conduct simulation experiments by changing parameters.

To *"validate"* the model (in this case, establish that it instantiated the relevant features of an in vitro dish), he first followed the same experimental protocol used with in vitro dishes other than their own to see whether he could replicate those behavioral data. For in silico simulation modelers, a "valid" model is one that has sufficient warrant to support the claim that predictive inferences about the target system are credible and worthy of pursuit. Once D11 had succeeded with the replications of the dishes in the literature (an outcome he called *"striking"* given the simplicity of the in silico model), D11 used data from the behavior of their lab's in vitro dish and was able to replicate these as well. The in silico model D11 had developed was, then, a model of a *generic* in vitro dish, because it incorporated replications not only of the lab's dish behavior, but also of disparate behavioral data from the wider neuroscience/engineering literature. By early year three, he had developed the model network sufficiently to begin *"playing with the [in silico] dish,"* by which he meant seeing how the computational network behaves under different conditions. He had started to get what he called *"some feeling about what happens actually in the [in silico] network."* Sometime during this playing period, he moved back into the physical space of lab D, and all three researchers began to work together.

For exploring bursts, the computational model offered many advantages over the in vitro dish. For instance, the in silico network could be stopped at any point to examine its states. Significant variables, such as synaptic weight, could be measured and manipulated. Synaptic weight was a variable of particular interest in the research, because it is a measure of how strong the connection is between the neurons, and is thus related to learning. The strength of this connection cannot be measured in the in vitro network, but it can be in the in silico network. Further, a large number of experiments, including counterfactual, could be run easily and at no cost, in contrast to the laborious and expensive processes involved in setting up and maintaining an in vitro dish. Most importantly, with the computational model, D11 was able to create a dynamic visualization that allowed him to track activity across the network as it occurred. These epistemic affordances (common to computational simulations, generally) proved to be a powerful combination that gave D11 immediate access to a range of configurations and data that the in vitro dish could not provide. He could design experimental simulations with different configurations and data, run these instantly, and examine and reexamine them at will.

The visualization, in particular, proved to be highly significant for the research in that it facilitated the group's articulation of a cluster of novel concepts by which to represent and understand dish activity and, ultimately, to achieve their goal of a control structure for supervised learning. D11 noted that he built the visualization as part of getting a feel for the dish: *"I am sort of like a visual guy—I really need to look at the figure to see what is going on."* It is important at this point to realize that computational visualizations are largely arbitrary. D11 could have visualized the in silico dish in any number of ways, including using the visualization the group was accustomed to seeing on MEAscope: a per-channel spike representation (figure 3.2). Instead, D11 built a visualization to capture the dish as he imagined it—as a network (figure 3.5). With the network visual format, he could *"visualize these 50K*

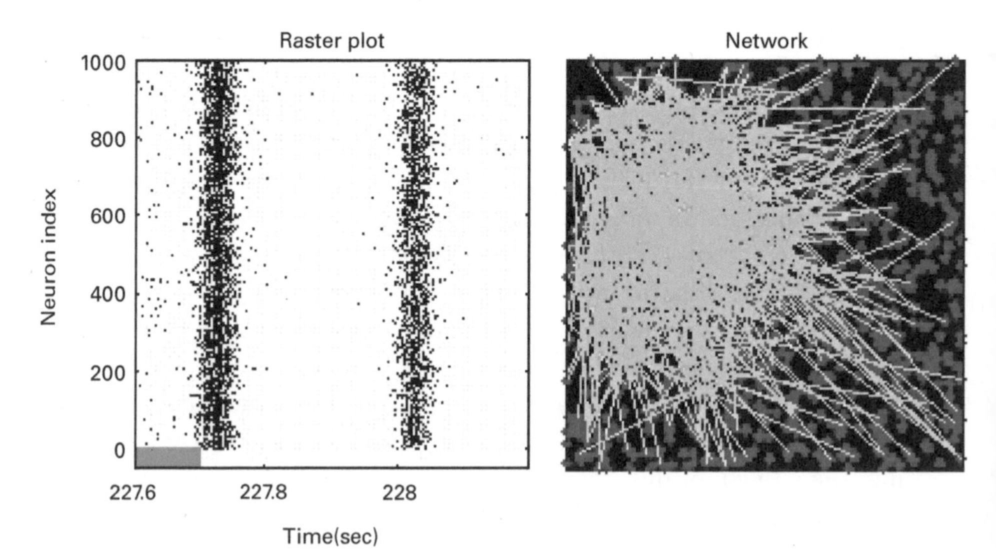

Figure 3.5
A screenshot of the network computational visualization of bursting activity across the in silico dish.

synapses and so you can see—after you deliver a certain stimulation you can see those distributions of synaptic weight change—or synaptic state change."

A major contribution of this visualization is that it enabled *the group* to notice—literally to see—interesting patterns in the way the in silico dish responded to different stimuli. These behavioral patterns were novel and distinct from anything they had thus far understood about in vitro dishes. In the MEAscope visual display of the in vitro model, the activity of *individual* neurons is hidden, as is the *propagation of activity* across the dish. Specifically, one can see activity across each separate channel (clusters of neurons) in figure 3.2, but this display does not exemplify the network configuration of the in vitro dish and it does not capture burst movement across the network. Thus, it was not possible to see that there were *patterns of movement across the network*. Further, with the computational visualization, D11 began to notice what appeared to be structurally similar looking bursts. To follow out this hunch, D11 made movies of the visualization as the simulation ran, and these showed the movement of activity patterns across the network over time. While running numerous simulations and repeatedly reviewing the movies, he began to notice something interesting: there were a small number of repeated *spatial patterns* in the activity, both when bursts arose spontaneously in the in silico network and when it responded to stimuli. As he expressed it, he found that there were "*similar looking bursts*" that propagated across the network, and only a limited number of these, which he called "*burst types.*" He stated that through the process of watching the movies "*you get some feeling about what happens in the network—and what I feel is that . . . the spontaneous activity or spontaneous bursts are very stable.*"

D11 showed his movies to the other researchers (and to us) so they too could "*come away with the same thing.*" They all agreed that, by analogy, it was possible that there were stable bursts in the in vitro dish as well, and began to work together to develop a means to track and mathematically represent the activity "*stable bursts*" across the in vitro and in silico networks. The radical implication of stable bursts was that the researchers' understanding of bursts transformed from "bursts as noise" to "bursts as pattern"—signals that might be exploited to develop a control structure for training the network. It is important to underscore that this change in understanding arose from the group running many visualized simulations with different conditions, a process that led to group consensus. This consensus was made possible by the manifest nature of the network visual display of in silico dish activity.

3.1.3 Phase 3: "He Was Thinking Like a Wave, while We Were Thinking of a Pattern"

From this point, things developed rapidly in the lab as the researchers worked together to develop a range of ways to quantify the spatial properties of moving bursts using clustering algorithms, statistical analyses, and experimentation to develop *"drift immune"* measures for the in silico network and equivalents (determined by a suitable analogical mapping) to transfer to the in vitro dish. With the drift problem solved, meaning that now that the activity pattern stayed constant across trials, they were in a position to determine whether the *"burst feedback"* in the in vitro dish could be used for supervised learning with the embodiments. This phase of research began with the idea that *"bursts don't seem as evil as they once did"* (D4)—or as the director like to say, *"bursts can be both good and bad."* That is, the researchers had modified their initial understanding of bursts, from the engineering notion of noise to be quieted completely to the notion that some bursting in the dish is useful because it can be tapped as a signal that might be used to control the embodiments. This shift in understanding led to their articulating several interconnected novel concepts:

- *burst type*: one of limited number of burst patterns (approximately ten)
- *burst occurrence*: when a type appears
- *spatial extent*: an estimation of burst size and specific channel location
- *CAT* (center of activity trajectory): a vector capturing the flow of activity at the population scale

D4 developed the concept of spatial extent directly for the in vitro dish. But the other three were developed first for the in silico network, and then mapped, with suitable modifications, to the in vitro dish. Each of these concepts is important, but they are complex conceptually and also mathematically. Therefore, in what follows, I focus only on the, for my purposes, most significant details of the development of one of them, CAT. This is an entirely novel concept for understanding neural activity, which the researchers claimed might well prove to be of major importance to neuroscience. Its importance for our analysis is that it emerged from the visualization of the in silico dish model, which instantiates the network behavior of the neurons, and would likely have been impossible to conceptualize and formalize without it. D2 recounted during the final stages of analysis, *"The whole reason we began looking at the center of activity and the center of activity trajectory is because we are completely overwhelmed by all this data being recorded on the 60 electrodes—and we just can't comprehend it all. The big motivation to develop this is to actually have something—a visualization we can understand."* CAT enabled them to understand how burst activity moves across the network and, thus, opened the possibility to control it.

The researchers formulated the mathematical representation of the in silico CAT concept by making an analogy to the physics notion of center of mass and by drawing from three resources within the group: (1) D11's deep knowledge of statistical analyses from the earlier period in which he had tried to create sensory-motor mappings between the dish and the embodiments; (2) an earlier idea of the graduate student at the director's old institution (who had worked remotely with the group) that it might be possible to capture *"the overall activity shift"* in the in vivo dish by dividing the MEA grid into four quadrants and *"using some kind of subtraction method"*; and (3) the idea that bursts seem to be initiated at specific sites as shown in a new graphical representation (figure 3.6) for the in vivo dish (*"spatial extent of a burst"*) that D4 developed after the in silico model had replicated her in vitro dish results.

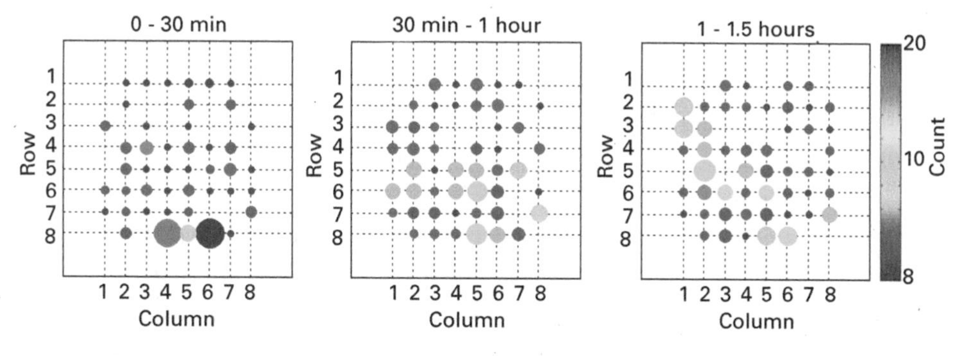

Figure 3.6

A screenshot of the display of information captured by the spatial extent representation. Spatial extent captures the location and frequency of bursts over time in the in vivo dish per channel. The figures show the burst initiation sites and the number of times (count) any neuron near an electrode initiated a burst during 30-minute segments of 1.5 hours of continuous recording. The color and size of the circles represents the number of times any electrode initiated a burst in the 30-minute segments. D4 kept the per channel grid of the MEAscope visualization (figure 3.2) but analyzed and displayed much different information in these new representations.

D4 had been trying to see whether she could get at some of the information the in silico visualization provided by graphing more specific spatial information about in vitro bursts, in particular, their location and frequency over time. She introduced a novel concept to represent this information: *"spatial extent"*: *"the number of times any neuron near an electrode initiated a burst in 30 minute segments of a 1.5 hour spontaneous recording."* The spatial extent of a specific burst is represented by the color and size of the circle in figure 3.6. These graphs clearly represent information different from the MEAscope representation of bursts as spikes per channel across the channels (figure 3.2). However, this representational format does not instantiate the *flow of the burst activity* as it propagates across the network. This is what the CAT representation was developed to capture.

D11 found the spatial extent representation useful but not sufficient: *"I not only care about how the channel's involved in the burst, I also care about the spatial information in there and the temporal information in there—how they propagate."* CAT tracks the spatial properties of activity as it moves through the network; that is, it tracks *the flow of activity at the population scale*, as displayed in the third visualization in the sequence of screenshots (figure 3.7), taken at the point when the CAT was moving from the center. D11 and D2 worked together to formulate the mathematical representation of CAT to include temporal as well as spatial dimensions. For our purposes it is not necessary to understand the details of the mathematics. Basically, the CAT is an averaging notion similar to that of a population vector. A population vector captures how the firing rates of a group of neurons that are only broadly tuned to a stimulus, when taken together, provide an accurate representation of the activity in response to the stimulus. CAT is more complex than a population vector because it tracks the spatial properties of activity as it moves through the network. That is, if the network is firing homogeneously or is quiet, the CAT will stay at the center of the dish, but if the network fires mainly in a corner, the CAT will move in that direction. The third screenshot in figure 3.7 shows the CAT as the activity moved from the center (shot one) to a new position (shot two). Thus, CAT tracks *the flow of activity* (not just activity) at the population scale. It is a novel conceptualization of neuronal activity. What their complete CAT analysis shows is that, if the simulation is allowed to run for a long time (tetanus probe stimulation), only a limited number of burst types (classified by shape, size, and propagation pattern) will occur—approximately

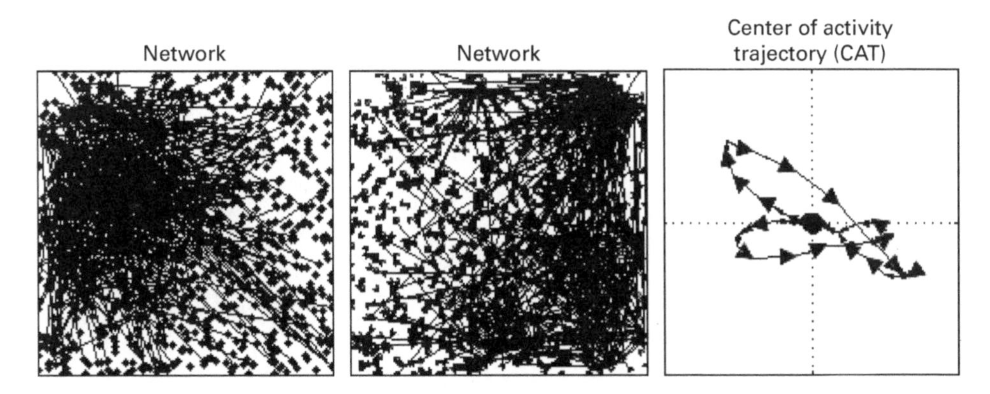

Figure 3.7
The first two screenshots of the computational visualization of the network show the flow of burst activity in simulated dish at burst time T1(first) and burst time T2 (second). The third screenshot shows a corresponding CAT from burst time T1 to burst time T2. The CAT tracks the spatial properties of activity of the population of neurons as the activity moves across the network.

ten. Further, if the probe stimulus is given in the same channel, *"the patterns are pretty similar."* Thus, the CAT *provides a signature for burst types.*

They developed the CAT first for the in silico dish, and D11 was unsure whether it would be possible to transfer the representation to the in vitro dish because of the potential role of the negative aspects of the analogy between the in silico and in vitro dishes: *"The problem is that I don't think it is exactly the same as in the living network—when our experiment worked in the living network, I am surprised—I was surprised."* One difference between them is that the in silico CAT tracks individual neurons, while the in vitro dish has a cluster of neurons at each electrode. For the in vitro dish, the researchers decided to try a mapping that replaced individual neurons in the in silico CAT representation with individual electrodes and began a range of experiments with the in vitro dish alone (open loop) and with the dish connected (closed loop) to various animats, including an animat version of a robotic drawing arm and, finally, to the real-world robotic drawing arm. The different representations in figure 3.8a and b show how CAT is conceptualized for each kind of model, that is, the mapping from CAT–in silico to CAT–in vitro. To visualize the dynamic nature of the simulation without seeing the movies, the reader should imagine the CAT as pictured in the "C" representations in the screenshots in figures 3.8a and b moving in the direction of

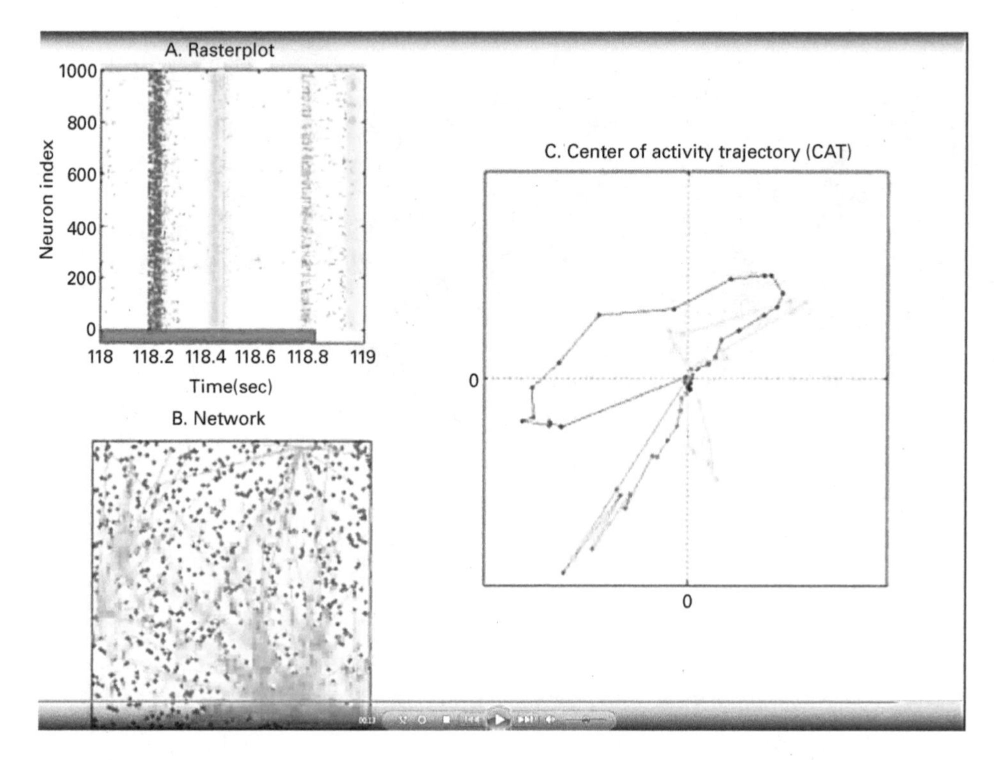

Figure 3.8a
(a) in silico CAT

the flow of activity, as pictured in the "B" (Network [3.8a] or MEA [3.8b]) representations in the screenshots.

D4 summed up the difference between the understanding of the in vitro dish behavior they had with CAT conceptualization and the way they had been thinking of its behavior prior to D11's simulation: *"[CAT] describes a trajectory. . . . We weren't thinking [before] of vectors with direction . . . so think of it as a wave of activity that proceeds through the network. So, he (D11) was thinking like a wave, while we were thinking of a pattern."* Even the spatial extent analysis she had developed to capture some of the information in the in silico visualization tracks a pattern of bursting across the channels (figure 3.6), but the CAT analysis tracks a wave of busting activity across the network (figure 3.8). CAT thus exemplifies the in vitro behavioral dynamics and, potentially by further analogy, the in vivo dynamics. Notably, she continued, *"We had the information always. . . . The information was always*

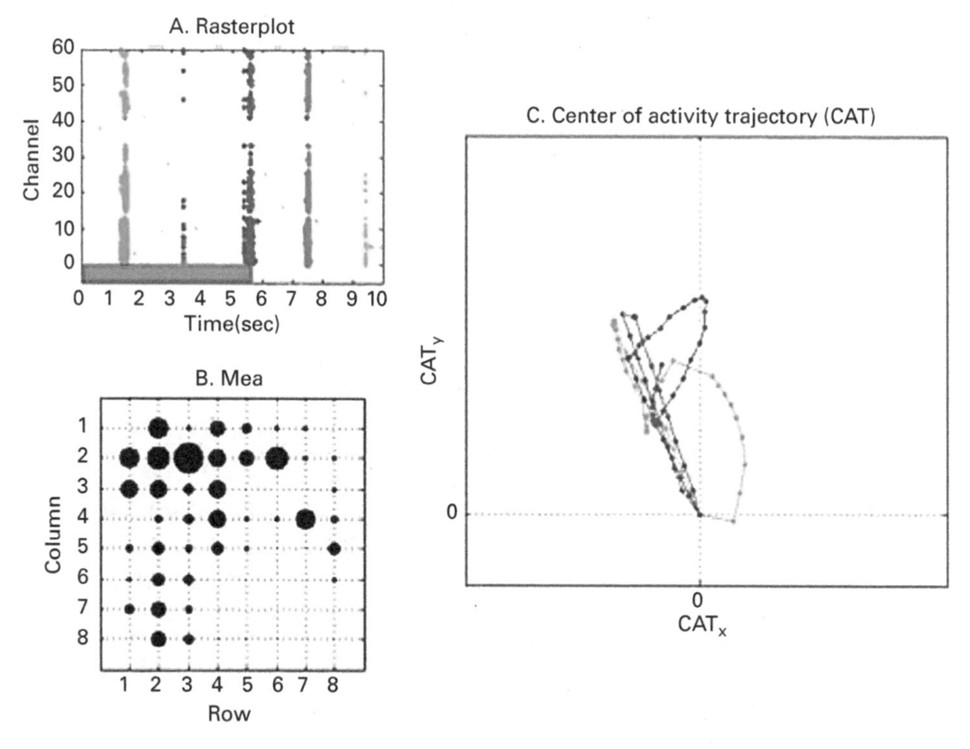

Figure 3.8b
(b) in vitro CAT
These screenshots show parallel CAT representations for the computational (3.8a) and in vitro (3.8b) models. The Raster plots of activity (A) on each shot show the activity at the level of the individual neuron for the in silico dish (3.8a, B Network) and for the electrode (cluster of neurons) in the in vitro dish (3.8b, B MEA), respectively. The CAT (C) tracks how that activity propagates through each kind of dish.

there." She is correct that the information was always there, at least in the raw data. However, the per-channel visualization of MEAscope hid that information from them. The computational network visualization enabled them to "*look inside the dish*" and see the dynamical behavior of the activity as it propagated across the network, which facilitated their ability to make perceptual inferences about its behavior and, then, to develop a mathematical representation that captures what they were seeing. What CAT enabled them to do was tap into the *network behavior* of the system and, eventually, exploit bursts as signals to control the system's learning, even though they did not understand the behavior of either dish sufficiently to explain it. The kind of understanding the in silico model facilitated is an exemplar, in this epistemic practice, of what Johannes Lenhard (2006) has called "pragmatic understanding" in the context of in silico simulation modeling—a mode of understanding that provides the potential for intervention, manipulation, or control, but is not explanatory. He and others see this as the worst-case scenario for simulation models. In practice, however, the allegedly *worst* case is often the *standard* case with respect to computational simulation modeling in the fields of bioengineering we have been investigating, and is often the case with in vitro simulation modeling as well; yet it is highly productive in enabling researchers to achieve their goals to control their systems.

To wrap up the story, D4 kept working with open-loop experiments to investigate network properties of the in vitro dish, while now using CAT to track poststimulation changes. She also added a new wet-lab investigation into the cellular properties of the neurons under electrical stimulation. She conducted this research together with a medical school researcher as part of preliminary investigation into whether the lab's new understanding of bursts as signals could be transferred to epilepsy (which they had speculated could be caused by bursting) and used to control seizures.

D2 and D11 stayed for an additional year after D4 had graduated and worked on combining CAT and techniques D4 had developed for burst quieting to develop a range of stimulation patterns for the in vitro dish that led to supervised learning for the embodied dish. To control the network required that it learns (through training), and that it retains and is able to access (have a memory of) what it has learned. Developing a control structure for embodiments was a difficult problem because of the sheer number of possible patterns. So, they began with animats, since numerous in silico trials could be run quickly. After thousands of trials, they first developed algorithms for control structures that produced adaptive goal-directed behavior for the "moth" animat D11 had created (a dot with the goal to move toward a light), and then for a MEArt animat (with the goal to draw in a constrained space). Finally, they made suitable mappings of the productive algorithms to direct the hybrot MEArt to draw within the lines through adaptive goal-directed behavior (figure 3.9). The goal was to make a sufficiently greater number of marks inside the lines than outside, using visual feedback from the camera to determine whether or not to tetanize the dish (provide it rapid electrical stimulation to contract the "muscle"). Through the in silico simulation experiments and the in vitro ones with MEArt, the

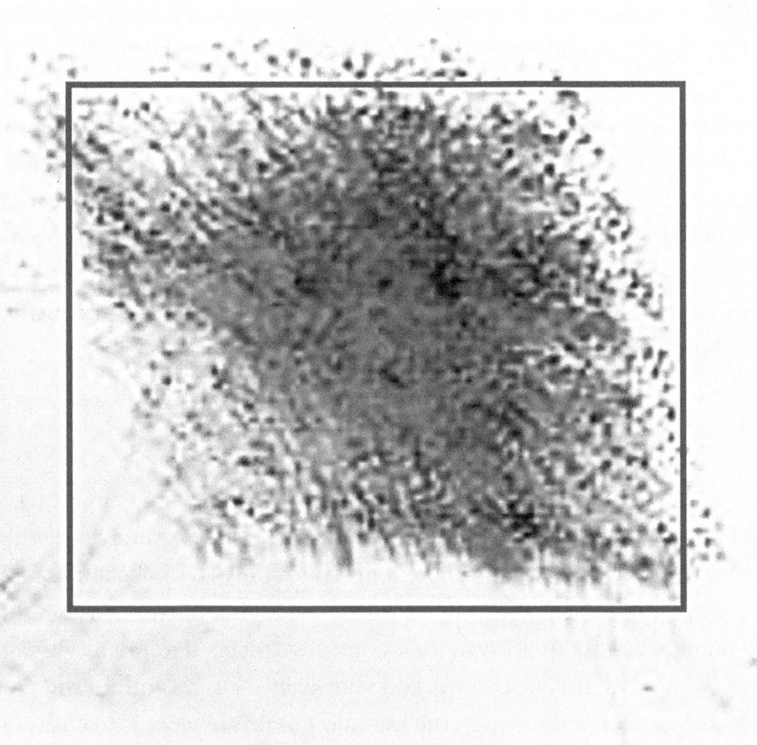

Figure 3.9
MEArt learns to consistently draw between the lines. (Photograph courtesy of Guy Ben-Ary)

researchers were able to establish the first demonstration of adaptive goal-directed learning in an embodied dish of cultured neurons. They cautiously reported that their results "*suggested*" that even though a cultured network lacks the three-dimensional structure of the brain, it can be "*functionally shaped and show meaningful behavior.*" D2 and D11 each wrote and successfully defended a dissertation on different aspects of this work.

Notably, their control structure differed in important respects from the customary structure for reinforcement learning, where the same stimulation is repeated continually. Their control structure consisted of providing the network with a patterned stimulation (a number of these worked) to induce plasticity, followed by providing a random background stimulation to stabilize the response of the synaptic weights to the patterned stimulation.

This method is counterintuitive with respect to existing notions of reinforcement learning, and it emerged only in the context of the researchers' attempts to gain sufficient understanding to control the in vitro dish model by creating a new representation of "the dish" with different affordances: a computational dish model built toward serving as an analogical source. The successful work to control the embodiments opened new application possibilities for the lab. After we finished our data collection, the lab director wrote grant proposals and received funding to conduct research that aimed to use the dish to control a limb prosthesis and to control the distribution of energy in the US national power grid.

3.2 Creating Scientific Concepts

A wide range of epistemic aims can lead to conceptual innovation and change in science (see, e.g., Feest and Steinle 2012). As I have been discussing, in BME, chief among those aims is not only to understand but also to control or intervene on complex in vivo biological phenomena in artifact or natural systems. To achieve their aims, researchers develop in vitro simulation models to isolate, control, and selectively focus on entities and processes of interest. From the outset, the intention of this practice is to create an analogical source from which to develop candidate hypotheses to transfer to the target system. As we saw in chapter 2, the specific nature of that analogy is usually determined incrementally, as the model is developed over time.

Often, to build an analogy requires configurations of more than one model, as we saw in the analyses of in vitro model-systems in chapter 2. These models are themselves complex dynamical systems with emergent properties and behaviors that, too, need to be understood and, possibly, controlled. In the case examined in this chapter, a researcher introduced a novel kind of model-building into the lab's epistemic practice—a computational model of an in vitro model—in the attempt to understand its "errant" behavior. The interaction of the different kinds of models in this configuration provides a rich and productive case of not only methodological novelty, but also conceptual innovation. In earlier work, I investigated how building conceptual models can promote conceptual innovation; in particular, the model-based reasoning processes of the analogical, the visual, and the thought-experimental kind (see, e.g., Nersessian 2008). The present chapter adds building in vitro models and in silico models, as well as the interplay among models of different kinds, to such innovation processes.

There is nothing particularly innovative about how D11 built a neural network (computational) simulation. What is innovative is that he built it to provide an analogy to an analogy—a second-order analogy to an in vitro model—and to exploit it in an attempt to overcome the impasse the research had reached with the in vitro model. At the outset there was no guarantee that this strategy would work. D11 thought the affordances of computational modeling would give him the ability to *"measure everything,"* which could not be done for activity in the in vitro dish, and, importantly, to visualize the dynamic behavior of the network. But it was an open question whether either those measures or the network visualization would provide any insight into the bursting behavior of the in vitro model.

The in vitro simulation model and the in silico simulation model are each complex dynamic systems with novel emergent behavioral possibilities. As with the in vitro models, D11 built the in silico model to provide a surrogate through which to reason about another system, but in this case to investigate and get a grip on emergent behavior (bursting) in the in vitro model. As we saw, the in silico model produced its own emergent behaviors, which, when rendered visible, indicated that bursts occurred in patterns, and those patterns were limited in number. Eventually the researchers came to understand the in silico behavior sufficiently to gain control of it and—by analogy—of the in vitro model. The primary objective of the research we have examined here was to develop further the in vitro model as an epistemic tool to investigate whether neuronal networks could learn, which in turn might provide insight into learning in the brain. First bursting behavior and then burst-quieting impeded D4's attempts to induce plasticity in the in vitro dish. D11 saw building a computational simulation as a potential resource for examining the perplexing behavior of the in vitro dish.

The physical dish model is a hybrid construction, merging constraints, methods, materials, and epistemic values from biology, neuroscience, and engineering. In chapter 2 we saw that the researchers transferred several concepts from the neuroscience area of single-neuron studies to provide a provisional understanding of behavior exhibited by the in vitro dish neural network, namely, spike, burst, and the Hebbian notion of learning in relation to plasticity, along with the associated rule for learning, to wit "neurons that fire together wire together." The engineering representation of electrical signals as spikes on an oscilloscope provided a manageable format in which to observe the behavior of the dish as their software picked up signals from the grid of electrodes. The engineering concept of noise as interference to be eliminated provided their initial understanding of spontaneous bursting behavior. The researchers were aware that the network nature of a population of neurons would likely require modification of the neuroscience concepts, but appeared not to have anticipated a problem with "noise." Nor did they anticipate that the per-channel activity grid visual representation would hinder their progress. In practice, these representations both facilitated and impeded the research. The concept of spike, for instance, facilitated their development of stimulation and recording methods and interpretations of the output of clusters of neurons surrounding an electrode. However, the grid visual representation hindered their thinking about neuronal activity as it propagated through the network. The single-neuron concept of burst, when extended to spontaneous dish-wide electrical activity, and understood as interference to be quieted, impeded the research for an extended period. But, in the end, they were right to focus on bursting behavior. The transfer proved fruitful when they realized that bursts could also be understood as "signals" (another engineering concept).

To deal with the impasse, D11 introduced an investigative practice with epistemic affordances different from those of the in vitro dish: an in silico simulation of a generic in vitro dish. Only after the in silico model gained sufficient complexity to replicate, and thus exemplify, the known behaviors of in vitro dishes, including their own, could the researchers consider transferring their findings to the in vitro model. As we have seen, to build that model D11 needed to draw on and integrate information from several resources (see figure 3.4): structural constraints of their dish, constraints of the CSIM modeling platform, data from single-neuron and brain slice experimental literature, and data from the literature on in vitro dish experiments in other labs. He built the model through numerous iterations, which infused data into it. With each iteration the model gained complexity. He evaluated each iteration by means of the criterion of how well it replicated the test data on established results from the experimental neuroscience and neuroengineering literatures. D11 used their dish data for testing only after the in silico dish had replicated these other results. The building process took nearly a year, at which time the in silico model was established sufficiently to serve as an analogical source to their dish model-systems. D11, together with the group, exploited the epistemic affordances of the computational model to develop a new understanding of burst behavior, which they articulated in novel concepts with which to get a grip on the in vitro dish behavior. D4 investigated the transfer of a hypothesis about bursting to the in vivo brain through collaborative research on animals. D11 and D2 developed the control structure for goal-directed learning, first for the in silico dish, and then for the embodied dish model-system, MEArt.

3.2.1 Epistemic Affordances of In Silico Simulation Modeling

The lab D case of computational simulation modeling provides an opportunity for us to begin to examine the affordances of the practice with a model that was relatively simple to build. As epistemic tools, in silico models have cognitive, manipulative, and experimental affordances, which I take to constitute epistemic affordances, not available with physical models. Although the systems biology computational model-building practices we will examine in chapters 5 and 6 are significantly more complex, the basic epistemic affordances of the practice are much the same. For a D-cog system in which this investigative practice is an epistemic tool, the major affordances are these:

- A computational model synthesizes a vast amount of data into a complex representation that can enact the behavioral dynamics of the target system.
- A computational model enables researchers to run an unlimited number of experiments and do so quickly. Although these experiments do not create new empirical data, experimental outcomes enable the modeler to examine implications of the data as synthesized in the model, as well as investigate potential experimental scenarios.
- The model can be used to run counterfactual scenarios. These scenarios enable the modeler to gain understanding by interrogating different possible variations of the target system and their existence conditions (parameter settings).
- It is possible to stop and start a simulation at any state and to track the system variables that generate the behaviors. This enables the modeler to understand the dynamic interplay among different model components.

- The model provides detailed measures of significant variables, often not accessible in vitro or in vivo (e.g., synaptic strength, in our case).
- The dynamic behavior of the model can be visualized, for instance, by graphs against which data can easily be compared or by visualizations that are themselves dynamic, as in the case at hand. A dynamic visualization, specifically, supports the modeler in thinking about three-dimensional phenomena, and processes taking place across time. The visualizations can be recorded and viewed, and compared repeatedly.

Taken together, these affordances help the modeler to form a global perspective on the phenomena—a perspective that cannot be obtained from the more limited in vitro and real-world experimental possibilities of the target system. This global perspective is what informs the *"feeling for the model,"* that D11 expressed, and that is ubiquitous among modelers (more about this in chapter 5).

These epistemic affordances of the model enable the researcher, among other possibilities, to determine dependencies among variables, to discover information about a model's potential behaviors, to develop new experiments, and to use perceptual inferences to discover patterns hidden in existing empirical data. How the model's behavior is visualized is largely arbitrary, but with judicious choices of how to exemplify behavior with respect to the problem goals, the visual representation can provide significant novel information about the system being modeled. One example is the choice D11 made to visualize the system behavior as it propagates across the network.

Finally, by the time a model successfully replicates a target system, the modeler has experienced thousands of simulations providing thousands of views of the system dynamics, both real and counterfactual. These experiences help the modeler build an understanding of the system in component terms and of how these components interact dynamically to produce behavior. They can also serve to alter the conceptual landscape of the researcher(s). In the case at hand, D11's experiences of in silico model bursting behavior led to his hypothesis that the behavior could be stable, not random, which required him to develop new ways to conceptualize the behavior and to represent it mathematically. Repeated experiences of the movies of the model behavior and the other researchers own interaction with these made conceptual innovation a group project. This aspect of our case shows that the visualization capacities of computational models offer the additional affordance of building collaboration ecospaces. Together, the piecemeal building of the computational model and the affordances of the model generate the task environment, collaboration space, and shared representations of the D-cog system.

Of course, computational models have their limitations. Two major ones are that the model relies on the quality of the experimental data and the nature and fit of the parameter space, and that, of course, a computational simulation cannot create new data. But, as with logical inference, in which inferential processes can reveal novel implications hidden in the premises of an argument, so too, model simulation processes can lead, as we have seen in abundance, to novel insights and to verifiable experimental predictions, which provide part of the warrant for the model.

I delve into these issues in detail in chapter 5, but, briefly, in addition to predictions, the warrant for the model rests on a pragmatic evaluation of the credibility of well-established modeling methods (see also Winsberg 2010), and on each increment of the building process being evaluated by means of a success criterion—that is, how well it fits the data. A computational model is built iteratively and incrementally through cycles of construction, simulation, evaluation, experimentation, and adaptation, and each cycle infuses data into the model, fitting the model closer to the target behavior. Each replication of the literature data changes the model's parameters in the direction of fitting the target system. As the model gains complexity through infusing data, fitting processes, and numerous runs under various conditions, it comes to instantiate known and hypothetical system-level behaviors of the target system. The researchers anticipate, though still need to evaluate through predictions, that it also can accurately enact novel system-level behaviors of the target. The "target system," is not a specific instance but a class of systems of that kind, which is what enables the possibility to transfer predictive inferences, for instance, to any in vitro dish model, as well as to in vivo neuronal networks, belonging to the class.

3.2.2 Distributed Model-Based Reasoning

Building computational simulation models provides another means of distributing cognition in the cognitive-cultural system of the research lab. We have been proposing a comprehensive account that places thought experimenting, simulation with physical models, and computational simulation on a spectrum of simulative model-based reasoning.[3] Specifically, all of these types of modeling build coupled mental–artifact systems that extend the researcher's natural ability to create and test real-world and counterfactual situations imaginatively. On our account of computational model-based reasoning, the artifact model gradually becomes coupled with the modeler's imagination (mental model simulation) during the course of numerous iterations of model-building and simulation. This coupling is built as the researcher performs systematic actions on the model and receives the feedback from those actions. The repeated two-way flow of information enables the researcher to build correspondences in her mental model as she is building the computational one. The developing correspondences enable her to make the requisite inferences to build, test, and draw implications from the model, as well as make changes to the model. The ability of the computational model to enact system-level behaviors makes manifest many details of the system's behavior the modeler could not have imagined before in view of the fine grain and complexity of those details (see also Kirsh 2010). In particular, the coupling of the mental model with the artifact model expands the researcher's imagination space so as to enable many possible scenarios (real and counterfactual) to be explored at a level of detail not possible in the mind alone. The enhanced ability to explore "what if?" scenarios furnishes resources beyond the natural human ability to conduct thought experiments, providing an exemplar of how scientists "create cognitive powers" through creating modeling environments. The computational model allows researchers to interrogate different variations—real and possible—of the target systems. Both thought experiments and computational models support reasoning about counterfactual scenarios, but only computational models enable the researcher to probe, in principle, all possible variations. An important difference is that thought experiments provide particular scenarios, while computational models are built using variables, which support examining a range of possibilities in a parameter space and allow the modeler to drill down into the behavior for details, and, potentially, to develop insight into why a specific one is instantiated in the target. The affordance of visualization that computational modeling offers is of particular value in building coupling in a D-cog system, as we have seen in the lab D case. I explore in more detail the epistemic affordances of computational coupling in chapter 5. Here I focus mainly on those of dynamic visual representation.

First, I consider the researcher's comment that they needed *"a visualization we could understand,"* which took the form of a dynamic computational visualization. As we saw in this case, the epistemic affordances of visualization can be especially productive because they make it possible for the researcher to see significant system behaviors and thus to support what comes relatively easy for humans, namely, making perceptual inferences. The computational visualization of dish behavior could have taken many forms, including the MEAscope graph with which the researchers had interacted throughout the research. However, D11 chose to visualize the activity as he envisioned it, as a network in which neural signals are propagated across the system of neurons (dish, brain). Of course, because the lab was investigating learning as a network phenomenon, everyone assumed that the activity in the dish was network activity. But the per-channel representation of the MEAscope graph, based on the oscilloscope visualization used in electrical engineering, displays the dynamical activity (spikes) of a cluster of neurons at an electrode in separate channels. As such, that visualization does not exemplify the network features of the in vitro dish or the in vivo phenomena, and so the researchers appeared not to have been thinking in those terms as they wrestled with the dish behavior. As we have seen D4 express, they had been *"thinking of a pattern"* (structure) instead of a *"wave"* (propagation behavior).

None of them had seen the network activity or a representation of it. D11 built the computational visualization on a counterfactual scenario—a thought experiment: "if we were able to see into the dish"—which altered his capabilities and those of the other researchers for simulative model-based reasoning. The network visualization provided a significant representational change in both the artifact model and the researcher mental model from a structure of burst patterns to propagation of burst activity (wave). The computational visualization was dynamic and captured the in silico network's structure and behavior as a whole, as relationships changed and moved across it. The processes of "playing" with the in silico model in numerous configurations and under various conditions, and of "seeing" the resulting behaviors as these were occurring served to develop a close coupling between the researcher(s) mental model and the artifact model as a D-cog system and enabled the group to develop a better grasp of the *system-level behavior* of the network. Even the spatial extent graph, which

D4 developed to capture some of the spatial behavior they were seeing in the computational simulation, displays only a pattern of structural information, whereas the CAT visualization captures behavioral information as it unfolds over time.

The researchers articulated their interpretation of what they saw in the visualization as indicating that bursts could be signals into a cluster of concepts:

- burst type: a limited number of bursting patterns (~10)
- burst occurrence: when a type occurs
- spatial extent: an estimation of burst size and specific channel location
- center of activity trajectory (CAT): a vector tracking the spatial properties of bursting activity—the flow of activity at the population scale—as it moves through the network

They modified and elaborated their initial concept of burst. However, the CAT is a completely novel conceptualization of a behavior heretofore unrepresented for both the lab and the field. It emerged initially from the computational model and the visualization of the movement of patterns of network activity and would not have been formulated, including represented mathematically, without a dynamic network visualization. The information might have been *"always there,"* but it was hidden in the MEAscope graph, which does not exemplify network behavior. The computational simulation and network visualization made it accessible. The manifest nature of the visualization served to align the researcher mental models and to facilitate the group in exploiting the possibilities of bursts as signals. Once the in silico dish could replicate the lab's in vitro dish results, it created a different kind of distributed problem-solving system wherein all the researchers now directed their efforts toward trying to understand bursts as signals that could be controlled and made use of in supervised learning. The computational visualization served as a generator of many types of lab activity and coalesced the various research thrusts (open and closed loop), which when taken together enabled them to get the grip on the in vitro dish that had eluded them in interacting with it alone. With the new concepts, they could think about how to undertake a range of new investigations, such as how to control the embodied dish model-systems (hybrot and animat) and, to a lesser extent, how to control epilepsy in patients.

Next, I consider the role of analogical inference. As we saw in chapter 2, scientific innovation based on analogy is often more complex—being distributed across time, artifacts, and people—than studied by analogy researchers in philosophy and cognitive science. In the case at hand, we see once again the power of building the analogical source as a creative problem-solving practice in frontier science where there are no ready-to-hand analogies to map and transfer (Nersessian 2008), and we see also the power of a visual analogy. Once the built analogical model is understood, it can be used to conceptualize and get a grip on puzzling target phenomena, as we saw in this case.

Customarily, analogical processes of mapping and transfer are understood to proceed from the source to the target, but in cases of building the source, there is often an iterative interaction between target and source, wherein constraints from the target are also built into the source. The source, in this case, is a hybrid analogy, incorporating features from several domains, including the target, which thus provides a novel synthesis. D11's stated goal in building the in silico dish model was to build an analogy: *"I thought that [computational] modeling could give us some information about the problem [bursting and control] we could not solve at the time [using the in vitro dish]."* In the process of building the in silico dish model, for instance, D11 drew experimental data from single-neuron studies, brain slice studies, and other dish experiments, which he justified by assuming that, in the respects that matter, the behaviors exhibited by these neurons should also be exhibited by the neurons in the lab D in vitro dishes. The in silico model also incorporates the engineering constraints of the structure of the in vitro dish (grid and electrode placement), as well as constraints from the CSIM neural network modeling platform. The finished model incorporates experimental data from the lab D dish. In general, how a model is built determines the nature of the analogical comparison it supports. The warrant for believing that reasoning with a built analogical source can provide understanding of the target stems from its having been constructed in ways germane to the target problem and from a determination of how well it exemplifies features cognitively relevant to the problem. In the case of a computational model, the relevant features are largely behavioral: the model needs to enact the behaviors of the target system in order to warrant transfer. We can see this in the following way.

Through iterative building processes, a computational model merges information pertinent to the behavior of the target system. For one thing, the model provides a synthesis of data, in this case from the experimental neuroscience and neuroengineering literatures, relevant to the goals of the problem situation. Each replication of the various experiments reconfigures a model's underlying parameters. Each of these replications infuses data into the model and adds to its complexity in a cumulative fashion, with each replication of the model building on the previous replication, until it comes to exemplify the observed behavior of the target system. In the case at hand, the final in silico dish model was built on a synthesis of all the data drawn from the literature, including from the lab. In an important sense, this model (and all such simulation models) creates a global structure that provides a *running literature review*, which synthesizes and makes manifest behavior, explicit and implicit, in the data on the related target systems. Through that synthesis, novel combinations of structures and behaviors can emerge. The model is fine-tuned by the replications until it gains sufficient complexity to enact not only the target behavior, but also counterfactual cases the modeler creates by changing parameters to explore the global behavior of the model. The model cannot create new data, but it can uncover potential consequences of the synthesized data—novel implications hidden in the data.

Analogical transfer is warranted because the model is built to exemplify the behaviors of systems of that kind. In this case, the in silico model was built to exemplify the behaviors of a target class of systems that comprise the in vitro models. Further, on the supposition that the brain is also a system of that kind, the researchers had sufficient warrant to transfer, provisionally, their new understanding of bursts to investigate the burst-like phenomena seen in some in vivo neurological disorders, as D4 did in her epilepsy research. In sum, the analogical model is a generic representation, and representations developed from it, such as CAT, are generic and pertain to members of a class of phenomena. Part of the research is to figure out the range of phenomena that constitute the members of the class. The lab D researchers developed concepts initially for the computational simulation and then, by analogy, with suitable mapping modification, for the target in vitro dish. In the final analysis, however, the researchers determined the CAT to be a generic representation that captures the center of activity of *any* patterns with spatial extent.

3.3 Summary: Concept Formation "in the Wild"

A substantial literature in the philosophy and history of science draws from historical cases to examine conceptual innovation and the roles of concepts in investigative practices, to which I have contributed. The case of conceptual innovation I have detailed here provides an opportunity to reflect on what more we learn about concepts by studying research-in-action. For one thing, our cognitive ethnographic investigations contribute to a general conclusion that model-based reasoning is productive of conceptual innovation and change across a wide range of sciences and historical periods and on into present-day science. Of course, the specific kinds of modeling possibilities have enlarged over the history of science, bringing with them new epistemic affordances, for instance, those of dynamical simulation and visualization of the sort afforded by computational modeling.

I did not go into this ethnographic research with the intent to apply in these new studies the analyses of concept formation and change that I developed in my earlier cognitive-historical research. However, as we collected and analyzed the data pertinent to physical and computational modeling processes, features emerged that paralleled my earlier analyses of model-based reasoning in conceptual innovation. Many members of our research group were not familiar with my previous research. Together, we discovered, in particular, that the process of incremental and iterative analogue model-building, in this case physical and computational models, promotes conceptual innovation, just as I had found to be the case with conceptual models. To use a notion drawn from ethnographic analysis, such processes *transfer robustly* across different sciences and time periods, as well as across several sources of data and methods of analysis. So, the ethnographic studies lend support to the interpretations developed from the less rich historical records. Further, as has been established in historical cases, the ethnographic cases underscore that model-based reasoning, across the range of modeling platforms, is closely connected with visualization, analogy, and simulation. And, as in those cases, exemplification is an important criterion by which researchers provide warrant for transfer from the source model to the target phenomena.

Most importantly, the ability to collect field observations and interviews surrounding problem-solving practices during the research process provides a wealth of insight into creative scientific practices, and the integrative nature of cognitive-cultural dimensions of these. Indeed, cognitive ethnography captures aspects of such practices that would never make it into the historical records. The reasoning of the researchers and the considerations in play at the time they are working on the problem, the evolving dynamics of the interactions among the members of a research group, and between them and the modeling artifacts, and the evolution of those artifacts are the most prominent among these aspects. Even for the most detailed concurrent records (which are rare in contemporary, if not all, science) there are numerous relevant data points about such processes that are unlikely to be archived. The computational visualization that enabled them (and me) literally to see the burst patterns as they were occurring provides an example: a sentence in a publication remarking that "burst patterns were noted" conveys neither its cognitive impact nor the change it sparked in group dynamics that led to integration across the three research projects. My research group was, of course, not able to make all the observations and collect all the records that are pertinent to these conceptual innovations, since ethnographic data collection is complex and time consuming and, of necessity, selective. However, once it became apparent that significant scientific developments were starting to come from lab D's research (nearly two years into our research), we did have sufficient data to mine and could collect additional data to document and enrich the most salient aspects of the innovation processes, which I have been analyzing in this chapter.

The case here examined underscores *that for philosophers and cognitive scientists to understand the nature of conceptual innovation in science requires an analysis of the interacting components within an evolving distributed system of researchers, artifact models, and practices*. This analysis need not be ethnographic, but can also be based on historical records, as I have demonstrated (Nersessian 2008). My main point is that conceptual innovation stems from the interplay among specific problems; the conceptual, material, and analytical resources provided by the problem situation; and reasoning and representational practices. Concepts and models have a dual existence as mental and artifactual representations, and inferences derive from a coupling of these. As we saw, building the simulation model enabled D11 to think—and then the group (including the director)—of the dish at the level of individual neurons in networks of neurons and how these interact dynamically to produce system-level behavior. This simplified in silico model enabled the researchers to *"see into the dish,"* which was opaque in its in vitro complexity. The network visualization reinforced thinking of the neuron culture as a network and provided a dynamical simulation (captured in movies that could be examined more carefully and repeatedly) of the real-time propagation of the activity across the network. This kind of visualization enabled D11 to see that there were similar-looking burst patterns—and to infer that they were limited in number. Further, he could show these to the others who could also see these phenomena. It enabled them, as they said, *"to look inside the dish,"* which had thus far been a black box.

As noted, many affordances of a computational model are not available with either an in vitro or a mental model, such as being able to run numerous simulations of the network under various conditions (for instance, resolution and parameter settings) or to stop and examine the state of the simulation whenever desired. The specific affordances of computational models for dynamic visualization, however, create a different kind of distributed model-based reasoning system that was paramount for this research. The in silico model with its visualization altered the problem-solving environment, helped define the problem(s), and furthered the incorporation of the researchers and artifacts into a distributed problem-solving system. The manifest nature of the in silico dish network, through its visualization, enabled the group to exploit its affordances and make judgments about its limitations communally. In particular, it facilitated the group in making inferences about potential mappings to the in vitro dish, in rejecting false leads, in developing extensions, and in coming to consensus—all of which led to conceptual innovation.

4 Interlude: Building "the Lab"

In the preceding chapters on the BME labs, and in the ones on the ISB labs to follow, I examine each lab's practices around modeling and how models are built toward accomplishing the lab's epistemic aims. We have seen, in the BME case, how labs in different fields use the practice of in vitro simulation modeling to build the cognitive-cultural resources, primarily material and conceptual, they need to investigate specific aspects of complex biological systems. In this chapter I have a different objective. In the course of our BME investigations we came to realize that the devices themselves drive much of the direction in which a lab develops, especially through posing problems or opening new avenues of research, which in turn leads to new technologies and practices, all of which shape the student researchers as scientists. Getting a grasp on the evolving, historical dimension of this kind of a distributed cognitive-cultural system requires a different kind of analysis—of how the system, in effect, builds itself.

"The lab" is often associated with those physical spaces that house the research-specific technologies, instruments, artifacts, and workbenches. "The lab" is also used to designate a research agenda: the problems and groups of people associated with it. In the latter case, it is often referred to as "the X lab" where "X" is the name of the principal investigator/director. And, as in the case of lab A, the tissue engineering lab, it can also be referred to, internally and externally, by a salient research object, in this case, "the flow-loop lab." This designation directs attention to the kind of epistemic practices through which the lab carries out its research, while also signaling that these practices are sufficiently known in a broader community to be a meaningful designation. For lab A, this is the epistemic practice of in vitro simulation modeling of mechanical forces in blood vessels by means of in vitro devices. In chapter 2, I dubbed such objects "signature artifacts" and examined in detail how they and the warrant for their use as in vitro simulation models are built in both labs. In this chapter, I examine the function of these devices as providing what William Wimsatt (2013a,b) has called "structuring constraints" for future development within an ecosystem. In particular, I examine the role the signature artifacts in lab A played in *building the lab* into a distributed cognitive-cultural system comprising researchers, artifacts, problems, and practices, all with intersecting developmental trajectories.

As discussed in chapter 1, our analysis of research labs cannot simply apply the framework of D-cog as developed initially through studies of highly structured problem-solving environments. In such environments, participants carry out largely routinized tasks that use existing technologies and bring to bear knowledge that is relatively stable, even as used in novel situations. In contrast, the BME research lab is an innovation community where researchers do not have established methods, technologies, and well-defined problems in advance of beginning the research. Although there are loci of stability, there are equally important features of these labs that are continually undergoing development and change. These features include the ongoing development of the technologies, methods, and problems; the formation of social practices and systems; and the development of the researchers as they learn to be bioengineering scientists in the processes of carrying out a research agenda of a lab director at a stage of his or her research program. At each slice in time, "the lab" comprises the current state of these, its features.

As I discussed earlier, D-cog's customary use of the adjective "distributed" is past tense, which signifies a process of distribution already completed. To study scientific practice, however, requires we attend to how cognition is actively distributed as a system is built, that is, attend to what Rogers Hall has called the dynamic processes of *distributing cognition*. As he explains, drawing on his research group's examinations of the research practices of scientists and mathematicians, "the word *distributing* is a verb, operating in an ongoing present, and shifts our attention to studies of how cognition . . . is produced historically out of human activity" (Hall et al. 2010, 226; emphasis original). In this chapter I examine the research lab as a dynamic environment that *builds itself* as a D-cog system "historically out of human activity" with specific affordances and limitations for problem-solving as it furthers its epistemic aims. I focus, in particular, on the role of the signature devices in these processes.

In the following sections I examine how the signature artifacts of lab A not only provide a platform for current problem-solving, but also create structural constraints and affordances for research potentialities not yet envisioned, and how these build "the lab." I then discuss the wider ecosystem that has been designed to turn the student researchers into the hybrid bio-medical-engineers envisioned by the senior researchers to populate and build a novel twenty-first-century version of the field of BME.

4.1 Creating Epistemic Infrastructure: The Laboratory for Tissue Engineering

The director of lab A, as with many of the pioneers in biological engineering, had an unusual career trajectory. In an interview I conducted with him as he was closing down the lab (approximately ten years after we concluded our research), he characterized that trajectory as *"from astronauts to stem cells"*—a trajectory inconceivable to him at the outset. Starting in the late 1950s, the future director of lab A trained as a mechanical engineer and then worked in an aeronautical engineering lab for the space program. Since he received funding from NASA for his research, they drafted him to help study how the effects of vibration along the axis of the Saturn launch vehicle and during reentry in the Apollo capsule ("pogo stick vibration") affect the cardiovascular system of astronauts. They tapped him because of his knowledge of the physics of launch and reentry forces. He reported that he did not know *"anything about biology and medicine,"* but that he felt an obligation to try to help them, and the problem was interesting. He discovered that no one had examined the effects of even the natural physical forces of blood flow through the cardiovascular system. He came to suspect that the mechanical forces, in the first instance, shear, would most likely impact the endothelium—the innermost layer of cells in a blood vessel—and decided to shift his own research program to this focus. First, though, he needed to learn some cell and vascular biology.

He spent a year as a visitor in a vascular biology lab that conducted research with an interdisciplinary team of medical and engineering researchers, and then left research in aeronautical engineering entirely for biomedical engineering. By then he was a tenured full professor, and he leveraged his engineering faculty position to begin research into how natural and aberrant blood flow through the arteries could affect the blood vessels in animals and to learn as much about the biology of endothelial cells as needed to conduct research on the effects of the flow on the cells. Since no biologists would collaborate with him, he conducted this research in a veterinary lab at his institution (research discussed in chapter 2). He then moved to a new university, where he had a largely administrative role, while continuing the animal research long distance. The laboratory for tissue engineering (lab A) dates from 1987, when the director moved to another university to take advantage of the opportunity to begin research in the emerging area of tissue engineering and to create a new department of biomedical engineering. Importantly, the change provided the opportunity to move out of animal studies and *"take the research in vitro."*

When we met the director in 1999, he was widely recognized as a senior pioneer in biomedical engineering. As we have seen in chapter 2, the epistemic practices of lab A center around creating in vitro devices, assembling them in various model-system configurations, and performing in vitro simulations under various controlled experimental conditions. The lab began with the flow loop, at that time a large device, cobbled together but precisely engineered in its flow components, which simulated blood shear forces on cells on slides. Flow experiments were still a significant portion of the lab's activity when we arrived, but by then the initial device had been replaced by a redesigned compact version that could be assembled under the sterile hood to limit contamination and placed in an incubator to keep the cells alive. A new tissue-engineered device, the construct, had been introduced, and it was playing a central role in building the lab at that time, and throughout our investigation.

4.1.1 Ontology of Artifacts

Shortly after we entered the lab, we conducted a sorting task in which we asked the researchers to put the names of the material artifacts they used to conduct their research on index cards and sort them into categories of their own devising. Their agreed-upon classification in terms of "devices," "instruments," and "equipment" is shown in figure 4.1.

Based on additional ethnographic interviews and observations, we formulated working definitions of the categories, with which the researchers concurred. *Devices* are hybrid bioengineered facsimiles that serve as in vitro models and sites of simulation. *Instruments*, some shared with other labs, extract and process information and generate measured output in

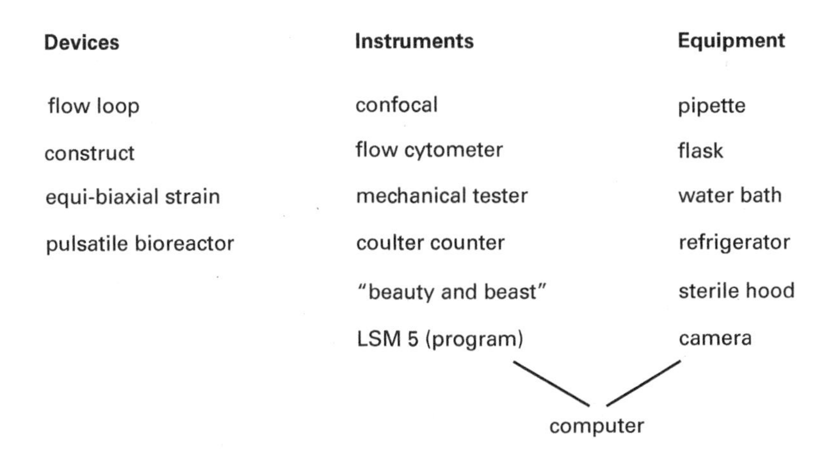

Figure 4.1
Sorting task of lab A artifacts

graphical, quantitative, or pictorial form. *Equipment* assists with manual or mental labor. I focus on the devices here, since they drive the lab-building process, but the equipment and instruments also play important roles in the cognitive-cultural system. In fact, when we asked the researchers to tell us what is the most important artifact, the novice researchers settled on the pipette, to the astonishment of the lab director. He, and the more experienced student researchers, said it was a toss-up between the flow loop and construct. When I discuss cell culturing practices, it will become apparent why the novices made this choice. This ontology also underscores the fact that all of the cognitive artifacts are part of the material culture of an epistemic community, though the reverse does not hold. Here, only the devices and instruments perform cognitive functions.

The flow loop and construct were the signature devices of the lab. These devices, which were built in-house, provided structuring constraints for the evolution of a range of cognitive-cultural practices in the complex system that is "the lab." These artifacts are what Wimsatt called "generatively entrenched" in the evolution of a complex ecosystem (Schank and Wimsatt 1986; Wimsatt 2007, 2013a,b; Love and Wimsatt 2019). Briefly, an artifact or entity is generatively entrenched if it acts as a constraint on the future direction of development of a complex dynamical system, which, in Wimsatt's account, includes cultural systems as well as the biological systems for which the notion was developed originally.[1] As we will see, the requirements of, and opportunities afforded by, these two devices led to the development of virtually all of the lab A cognitive-cultural structure as we encountered it and during the course of our investigation.

Most off-the-shelf purchases made by the lab fall into the equipment category. All of the equipment, except for the computer and camera, was used for cell culturing, which is critical to the research. The researchers noted later that they had forgotten the incubator, which is essential to keep the cells and constructs alive. The water bath and incubator are designed with biological knowledge of the requirements to keep the cells and tissues healthy—or as the researchers would say, "*happy*." Unhappy cells become contaminated or dead, which can spell disaster for the research project under way. Researchers set the incubator's temperature and atmospheric content to what is optimal for growing cells. The water bath, which is the medium that surrounds the constructs ("*water for cells*"), includes nutrients that are optimal for cell life. Importantly, culturing cells is a prelude to building the construct models needed for most research projects. Because all the researchers need to build their own constructs, learning to culture cells had supplanted learning how to operate the flow loop as the entry point into the lab research and culture when we began our investigation. Cells on slides could be prepared by a lab manager, but constructs need to be built by a tissue engineer. So, as a senior researcher told us, learning to culture cells was "*baseline to everything*." That is, it provided entry into the problem space and cognitive-cultural practices of the lab and, so, incorporated the new member into the community.

Learning to set up and manipulate the flow loop is a relatively easy task for an engineer; learning to culture cells and build constructs is not. For new researchers, this is often the first contact they have had with biological materials, concepts, and procedures. How much care and maintenance are needed to support the viability of a cell culture amazed and constantly frustrated them. The consequences of failure are high: when cell cultures die, experiments are ruined. As a result, mentoring within the lab usually began around learning to culture cells, which started with harvesting them from the animal arteries donated to the lab from a nearby veterinary school. It was quite common for us to observe the new member and the mentor huddled close together scraping cells off arteries or learning embodied techniques of culturing under the sterile hood. Although there are written protocols for the steps, culturing is also a performance art that needs considerable practice to acquire. It is a highly embodied skill in which an incorrect angle of the hand with the pipette can lead to disaster.

The discourse of the lab frequently centered on keeping the cells *"happy,"* calling them *"pets,"* being told to *"think of them as children"* and bemoaning long weekends lost to *"babysitting"* them. As one way to address the tragedies of failure with cells, more senior researchers shared war stories about the recalcitrance of cells to respond in the ways they desired, and how they had emerged victorious, eventually. In such episodes, we witnessed how the researchers began to build resilience in the face of obstacles or failures, which are a constant in their pioneering research, and in all the labs we studied. The general ethos of the lab reflected the attitude that failure or impasses provide opportunities to learn. This attitude was reinforced by a broader community that purposefully promoted opportunities for structured and unstructured interaction among students and faculty from different labs where research impasses as well as successes could be shared and discussed. What is especially interesting is that we did not encounter any situations in either lab where an in vitro model, once broadly envisioned and in the process of being built, or once selected from the existing ones as the means to pursue a research problem, was abandoned in the research. Impasses or failures usually led researchers to make modifications to the model or model-system or to the scope of what might be investigated using it. Significantly, everyone in the labs when we were there conducted sufficient research to graduate, and they went on to academic or industry positions. Although lab directors encouraged practices for cultivating resilience in many ways, the most unusual was lab A director's policy of handing out a compilation of what he called "The Rules of Life: The Planet Earth School," which listed aphorisms about how to thrive in research and in life, to each new member—and to me, as I started this risky research on his lab.

During the process of learning to culture cells and make constructs, new members explored the research of the lab and the roles of various instruments and devices in informal conversations with more senior researchers. Through these conversations they came to understand how the research largely involves working with the constructs as *"modeling tools."* This learning experience, too, began to build an interdisciplinary epistemic identity, shifting it from engineer to bioengineer (Osbeck and Nersessian 2017). As one researcher put it, when we witnessed the scene of her joyously dancing around the lab with her first success in getting cells to do what she wanted after nearly a year of repeated failures in every approach she had tried, *"I'm a bio-bioengineer."*

In the instrument category, the confocal microscope, flow cytometer, and Coulter counter are large, expensive instruments that have been purchased by the department for all the labs to use. Everyone in lab A uses these instruments. LSM 5 (laser scanning microscope) is the program used with the confocal microscope and enables user-directed image manipulation and analysis of how the cells behave and change as a result of the in vitro simulations, which is likely why the researchers singled it out. "Beauty and the beast" was a nickname they gave to the large computer (beast) and camera (beauty) setup that had been designed for analysis (including the software) by a researcher who had just wrapped up her research when we entered. Her project had been to develop a better substrate for proliferating endothelial cells on slides, which she thought might also be used to help the cells migrate in the constructs. It provides an example of technology still residing in the lab but no longer used. It remained, taking up lab space, throughout our research. The other researchers could explain what it does. Old technologies tended to hang around since they have the potential to be repurposed in new lines of research.

The mechanical tester is the most interesting instrument for understanding the central place of the construct model in building the lab, and provides an example of such repurposing. The design of the construct was continually under revision toward being both a better model and a viable implant. The properties of every new design needed to be examined and evaluated. The mechanical tester was used to examine the mechanical strength of various iterations of constructs. The original mechanical tester was an unused instrument in another lab that conducted tests on the strength of native tissues. An enterprising lab A researcher saw its potential, with suitable redesign, to be used to evaluate the mechanical strength of their engineered tissue. When we entered, it was a clunky, cobbled-together instrument that had been modified incrementally ever since the construct had been introduced. The testing process, at the time, was as follows. After constructs are "stimulated" in tubular form by the forces produced by various simulation devices, they are cut into rings and beads are glued at various intervals to examine local distension. The rings are placed in the liquid chamber of the tester, and each side of the ring is attached to the tester's hooks. At this point the ring is pulled apart until it breaks, while the process is videotaped (a quite recent modification). The tester can measure stress, strain, and ultimate tensile strength.

Interestingly, while we were there, they decided to buy a machine made by Instron because of its increased range of forces and sensitivity, but despite having spent a considerable sum on it, the researchers never used it for mechanical testing. Although it could do much more than their tester, to them Instron tester was a black box that did not, in particular, fit into their practice of placing the rings into the liquid chamber—a feature missing from the Instron—to keep the rings from sticking together before attaching them to the hooks. The machine had been built by Instron to test a range of materials, but it would need to be modified to work with the lab's practices with constructs. The researchers noted they had been *"avoiding this thing [Instron], because no one wants to design something that'll work."* To make it work they would either have to redesign the Instron to hold open the rings or redesign their practices for preparing the constructs to be tested. With many mechanical engineers in the lab, the former is something they could likely have done quite readily. However, despite being a jumble of parts and difficult to use, their mechanical tester had evolved alongside the lab's practices and had become entrenched in ways the Instron proved unable to dislodge. Later, researchers new to the lab would appropriate the Instron and modify it for the completely different purpose of developing a device to simulate the effects of compression on stem cells (Harmon and Nersessian 2008).

All the in vitro simulation models fall into the category of devices. The flow loop preceded the construct; together they constituted the primary model-system of the lab. As we saw in chapter 2, the researchers considered it unnecessary to undertake a considerable redesign of the flow loop to accommodate the construct's tubular design. Instead, the constructs were cut open and flowed flat when subjected to shear stress forces of the liquid. To accommodate the thickness of constructs, as compared with the cells on slides for which the flow chamber had been designed, a spacer was added to the flow chamber. Although "just" a spacer, it did require some redesign of the chamber to comply with the physics of the behavior of the flowing liquid. The baboon model-system experiment, discussed in chapter 2, was the first time it became necessary to maintain the tubular form of the construct. The researchers anticipated they would undertake a significant redesign of the flow loop, but as we saw, A7 was able to attach a shunt to the flow chamber and connect the construct to the shunt tubing, just as she had done with the animal—an ingenious method that saved them considerable time and expense.

The lab's evolving understanding, goals, and problems in relation to the construct opened new lines of research and led to building new in vitro models through which to manipulate and examine construct properties under various conditions. The researchers built the other devices listed in the ontology to explore mechanical properties of the tubular constructs other than shear (stress, strain, pressure). The research into these properties was directed especially toward strengthening the construct to meet the requirements of the application goal (vascular implant) opened by its introduction into the lab. This research led to new conceptual resources related to arterial stress, strain, and pressure.

To be either a functional model or an implant requires (among other things) that the cells that are embedded in the scaffolding material replicate the capabilities and behaviors of in vivo cells so that higher-level tissue functions can be achieved, such as expressing the right proteins and genetic markers. Further, a vascular implant needs to be strong enough to be able to withstand the in vivo blood forces of a pumping heart, and so understanding what creates mechanical strength and integrity in native tissue became prime concerns. All of the experiments with the tubular construct required a silicon sleeve because it could not withstand the forces itself. The sleeve could be made to varying criteria that included thickness, elasticity, "stickiness" in holding onto the construct, and with or without a collagen coating. The researchers hoped through their investigations into mechanical strength to find a way not to use the sleeve, both because it *"added a level of doubt,"* to their simulation results and, of course, because *"a surgeon would actually want to suture the construct [sans sleeve] into the patient."* To address the issues of mechanical strength and integrity, researchers created two devices to simulate mechanical forces of pressure in the tubes (the pulsatile bioreactor) and strain on the cells (the equi-biaxial strain device, or EBSAD).

Finally, a vascular implant requires a high-yield source of endothelial cells, most desirably derived from manipulation of the patient's progenitor cells or marrow stem cells to prevent immune rejection. We saw in chapter 2 how A7 built a model-system with the construct and an animal model to examine a hypothesis about how progenitor cells function (more on that in the next section). The researchers further speculated (they explicitly denied it was a hypothesis) that compression forces might be the mechanism through which endothelial progenitor cells differentiated into endothelial cells in vivo. During the last part of our research (and so not on the ontology list), two newly arrived researchers were working on two compression bioreactor devices using confined and unconfined compression to examine the effects of these forces on progenitor and bone marrow stem cells. This provisional research aimed to determine whether a speculation could be turned into a hypothesis worthy of pursuit and, ultimately, to see whether the needed endothelial cells could be created by either of these methods. For the confined compression bioreactor, the researchers modified the rejected Instron mechanical tester. The unconfined bioreactor was in the planning stage when we concluded our research. This emerging research project provided us with a glimpse of how a new line of research and the requisite infrastructure derive from the then-current cognitive-cultural system of the lab.[2]

4.1.2 Configuring the Problem Space

The representation drawn by the director when we asked him to draw a picture of the lab research partway though our study (figure 4.2) depicts, in our terms, the configuration of lab A as a distributed problem space. We gave him no instructions for how to do this. He declined to draw it while we were present, but said he would think about how to aproach it and would work on it during a flight he was about to take. When he gave it to us, he said he had wanted to depict how his research "*barriers*" (listed at top), researchers (middle section), and technologies (listed at bottom) are interconnected. The diagram on paper is a static representation, but the word "*being*" marks his intention to capture the configuration of the lab's ongoing research. In line with it as a dynamic depiction, I interpret it as a schematic of "the lab as an evolving distributed cognitive-cultural system with epistemic aims"—a dynamic constellation of interrelated problems, researchers, simulation models, methods, instruments, and other technologies. The picture depicts the hybrid nature of the interdisciplinary problem space of BME, as we will see in unpacking it. Each of the barriers references the interlocking of biology and engineering.

The diagram references the five graduate student rearchers and one postdoctoral researcher (A8) in the lab at the time. A notable feature of the lab,

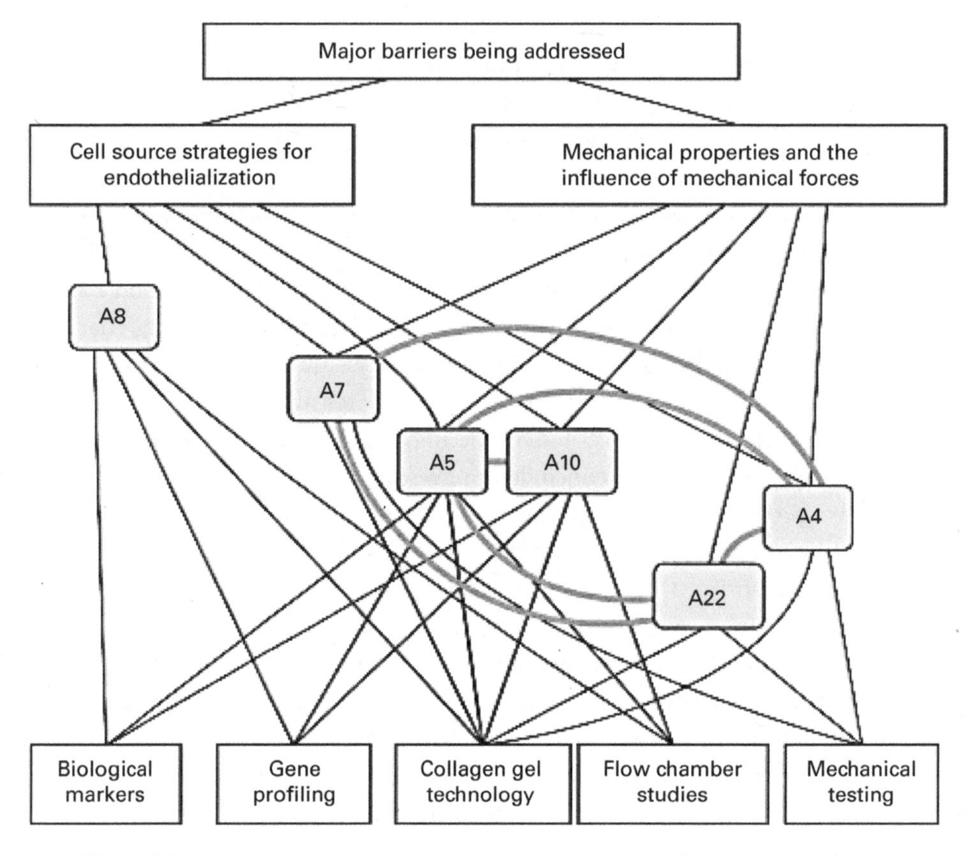

Figure 4.2
Lab A director's representation of how he envisioned "the lab." The original figure was drawn by hand, and we have redrawn it without alteration to make it easier to see.

not depicted, is that five of the researchers were women (A10 the lone man). I found the composition of the lab striking, coming myself from physics and philosophy, where women are underrepresented, and I asked the director about it. His response was that he chooses only "*the most qualified applicants*." We were later to discover that although the proportion in lab A was unusual, BME has a high number of women in the field.[3] Although the director did not include himself on the diagram, he is, of course, an integral part of the system even though his visits to the physical space of the lab were rare. He spent a significant amount of time on the road to promote the research and to obtain the financial resources to conduct it, in addition to administering an interdisciplinary center, and building and promoting an interdisciplinary educational program. He held lab meetings at varying intervals whenever he was in town, as well as individual meetings with each researcher when needed.

In this section I use the diagram to examine some of the research lines and their associated epistemic infrastructure that participated in evolving the lab as the researchers addressed the barriers.

At the top of the diagram the director categorized the "major barriers" with which the research was dealing. In lab A, the researchers addressed "barriers" by formulating research problems that interconnected the basic biological research of the lab and its medical application aims, and pursued these problems through developing technologies for in vitro investigation. To address the barrier of "mechanical properties and the influence of mechanical forces" required researchers to formulate problems directed toward understanding basic biological processes of arterial shear, stress, and strain and their role in normal and disease processes. It also required all the researchers to address the problem of developing a construct with desired in vivo biological and mechanical properties. The lab did not aspire to create an implant, but rather to solve the problems that would further that application goal. In particular, this meant bringing the research closer to understanding the requirements to create an implant with the requisite functional properties. As part of that goal, the lab undertook research directed at the barrier of "cell source strategies," by which the director meant research to find an appropriate source of endothelial cells, such that a tissue-engineered implant would not be rejected by the recipient's body. This research took the novel direction of investigations into the possible role of mechanical forces on cell differentiation and maturation. It opened lines of basic biological research for the researchers, such as on the role of forces in adult stem cell differentiation (A8) and in the maturation of progenitor cells (A7), which in the latter case led to the lab's first animal model-system.

The lab-built devices are designated by "collagen gel technology" (construct), "flow chamber studies" (flow loop), and "mechanical testing" (pulsatile bioreactor and equi-biaxial strain device). "Mechanical testing" also indicates the lab-built instrument for testing mechanical strength of a construct. The kinds of investigations along the bottom of the diagram implicate both the devices and the technologies through which researchers could examine simulation outcomes from experiments conducted with them. For instance, after a flow chamber simulation in which the construct is subjected to various controlled shear stresses, the researchers examine the effects on the endothelial cells for various biological markers with instruments or through gene profiling. These kinds of studies implicate a range of technologies, many external to the physical space of the lab, such as the confocal microscope used to study morphology and migration or the DNA microarray technology used to study gene expression.

The director used the thick lines to denote interconnections among the individual research projects, especially with respect to the researcher (A7) designated to build the animal model-system that would integrate findings from all these projects. He represented a postdoctoral researcher (A8) as unconnected to the students because she had just started the new line of lab research into the possibility that stem cells might be made to differentiate into endothelial cells by means of mechanical forces, and thus provide a cell source. Her project did become more central in the research after she was successful, and later led to the new researchers' project to examine compression effects mentioned in the previous section. At the time of the diagram, she did interact with other lab members about her and their research through conversations in the course of the lab activities and at lab research meetings. Although the research projects were carried out by individual lab members (sometimes assisted by an undergraduate or MS student), we witnessed frequent joint problem-solving episodes within the lab and at all the lab meetings. Each individual research project and its associated problem-solving processes formed a D-cog system. Each of these subsystems contributed to and was constrained by the lab's dual basic and applied research problems and goals. The diagram depicts the interconnected subsystems that constituted the configuration of the lab's problem space at that time and the direction of the lab's evolution. Together they constituted "the lab" as an evolving distributed cognitive-cultural system with epistemic aims.

When the researchers noted in figure 4.2 entered the lab, the flow-loop model was a well-established technology of research, but several of them formulated research problems that would require some redesign of it. The construct model was a recent addition, and all the researchers played significant roles in furthering its design in directions related to their specific projects. A22's research focused on developing the collagen gel technology toward improving the mechanical strength of constructs, although at the time she was still in the process of figuring out how she would approach the problem. She was the newest member in the lab, who started as an MS student, and she decided to transition to a PhD student only shortly before we concluded.[4] A4's research was to examine specific biological markers in relation to controlled mechanical stimulation of constructs by stretching with the pulsatile bioreactor, as compared with their behavior in native tissue, thought to be stimulated by pressure forces in vivo. A5's research was to correlate the development of arteriosclerosis with the genetic behavior of the endothelial cells and progenitor endothelial cells that circulate in the bloodstream by simulating normal and abnormal flow conditions with the flow loop. She, along with A10, introduced new biological methods and tests related to gene profiling.

Its interesting for our purposes to have a glimpse of A10's project because he introduced a new kind of construct, which in turn required him to build a new simulation device to pair it with. A10's project was to investigate the effects of shear stress on the function of the aortic valve. Stenosis in vivo is a frequent problem, and there has been a long history of largely unsuccessful attempts to replace the valve. A10's research aimed both to understand normal and diseased valve functioning and to contribute to the goal of a tissued-engineered replacement. He built a novel aortic construct and used valvular endothelial cells harvested from animal valves. Valvular endotheliel cells experience forces different from those that line the arteries; in particular, they undergo significant stretching due to their proximity to the pumping heart. Understanding what creates mechanical integrity and strength in native valves was a primary concern for him. He hypothesized that the effects of stretching on the cells might be what strengthens the extracellular matrix. To follow out this hypothesis he decided to build a new device. The only device in the lab that simulated repeated stretching of the construct was the pulsatile bioreactor, which was an inadequate design because it simulated stretching along only one axis of the tube, so different parts of the construct, and thus the cells, experienced different stretches. A10 wanted to look at cellular behavior where *"it's critical to make sure you are doing the same things to every single cell."* He saw a design for a biaxial strain that another mechanical engineer at a university in distant state had built for different purpose. A10 initiated a collaboration with him, and they redesigned and built the EBSAD for use on valvular constructs. This device could simulate the strain (deformation from stress) experienced by a vessel as blood flows through it.

Interwoven with the engineering task, A10 worked to develop the biological knowledge and expertise to determine whether the cells, when exposed to the simulated in vivo stretching by the EBSAD, would produce biological markers that indicated strengthening. He struggled for quite a while to figure out what to analyze as markers, deciding on the proteins that make up the extracellular maxtix, which binds the cells in the tissue. He reasoned that *"the cells secrete protein. . . . I surmise that the valvular cells, because they are in a highly dynamic flexing environment . . . have to constantly remodel the matrix they're in to kind of repair it."* His research, thus, again led to new biological methods being introduced into the lab practices: the gene microarry studies to compare protein generation in stretched and nonstretched cells. He chose that method because *"there may be characteristics from the gene profile that suggest that they [proteins] will interact with the matrix in a certain way that may strengthen it."* Both the time and financial investment of the lab into building the EBSAD and the costs of establishing a collaboration with a nearby university to conduct the complex gene array studies represented a gamble. His hypotheses about strengthening through stretching and the gene profile characteristics *"very well may not hold."* As with virtually every research project the lab undertook, it represented a significant risk to invest in building out in a specific direction.

As indicated by the thick lines on the diagram, all of the system's components are connected to A7. All of the research projects undergird the construct-baboon model-system designed by A7, which she called interchangeably an ex vivo (meaning outside the animal's body) or in vivo experiment, that we discussed in chapter 2. In an early interview A7 noted that she had been designated as *"the person who would take the construct in vivo."* This meant that she would need to create a model-system in which a construct would be connected to the vascular system of a living animal. To be successful, as she said, the project would need to *"obviously integrate the results of colleagues here in the lab."* At the start, she was quite unclear about just what she would study with the model. Once she decided on a specific animal, she devoted condiserable time to designing a means of connecting the fragile construct to the animal without it rupturing (and in an animal-friendly way). As her research project evolved, it became clear that it would connect the two "barriers" by investigating whether shear stress conditioning of endothelial progenitor cells with the flow loop would make them function as mature endothelial cells in the production of thrombomodulin (a protein that prevents platelet formation) when attached to an animal circulatory system.

For her investigation, she designed a model-system (figure 2.6) that could connect the teflon-scaffolded construct to the bloodstream of a baboon by means of an exterior shunt between the femoral artery and vein of the animal. The ex vivo simulation was designed to be run in real time through a gamma camera to provide functional imaging. Conducting a simulation with this model-system under the requisite experimental controls was the most complex problem the lab had undertaken. As A7 noted, "*In the lab we can control . . . exactly what the flow is like. . . . But when we move to an animal model, it's more physiologic—the challenge then is that it is a much more complex system.*" Importantly, she was able to determine that precondtioning the progenitor cells with flow-loop shear at the normal human in vivo blood flow rate enhances the ability of progenitor cells to express anticoagulant proteins within the model-system, but not at lower rates. This finding made a significant contribution both to the research community's understanding of the effects of arterial shear, along with further articulation of that concept, and to the problem of finding endothelial cell sources for a vascular graft. With respect to the latter, it demonstrated that mature endothelial cells can be created by mechanical forces from progenitor cells, which gave a boost to the lab's research in that area. A7's research was completed just at the end of our follow-up investigation, so it took approximately five years of concentrated work, but it was predicated on nearly thirty years of building the lab.

Although not represented explicitly on the diagram, the barriers, technologies, and researchers implicate both lab history and research potentialities. One potentiality is seen in the, then, less-integrated line of A8's research. Her novel investigation into the effects on mechanical forces on adult stem cells as a strategy for producing endothelial cells—if successful—could lead to more research along those lines. In fact it was successful, and it did open other lines, including the line we mentioned above in connection with the Instron mechanical tester. The two new researchers would be connected to A8 in carrying on her project on cell differentiation by means of forces but, in their case, compression on bone marrow stem cells and progenitor cells. They modified the Instron to carry out confined compression studies and were building a new device to carry out unconfined compression studies. Their research also required they make modifications to the collagen gel technology.

Our analysis shows that lab history is implicated in current problems, technologies, methods, and researchers. A7's account of how her understanding of how it was possible she could now use the construct in an animal experiment provided in an interview in her third year in lab illustrates the *hands-on role of history* in the research: *"One of the main limitations of the collagen gel construct is its mechanical strength. And like over the course of research in our lab, A1 had looked at things like mechanical conditioning to increase the strength, and of course A12's work has focused on how he could integrate elastin. Well, with his integration of the elastin sleeve we've now actually made enough progress in the area of mechanical strength that we have a strong enough construct to put in an animal."*

This account characterizes constructs as products of communal activity around a problem, the lack of mechanical strength. Her sketch of its historical trajectory thus far in the lab gives the artifact meaning through its relationship to two prior members and their roles in this ongoing problem-solving effort. One person looked at mechanical conditioning as a possible source of strength, while the other added a new component to the construct. A7 identified the current problem situation and her future work and lab role as yet another chapter in the building of the construct and of the lab. The historicity of the construct served to create a thread that binds the activities of lab members within its developing cognitive-cultural fabric. Such accounts by members of the lab-built technologies were commonplace in our interviews and informal discussions. In chapter 2, I provided their account of history of the flow loop, as recounted by several members. These accounts led us to understand the importance of the historical dimension of building the lab is a resource for current and future research (see, e.g., Kurz-Milcke et al. 2004). The agenda of design and redesign makes history a resource that is intellectually hands-on; that is, history is meaningfully related to present work with lab technologies, devices in particular. Devices, inherited and new, need to be (re)designed for the current problem situation. To avoid past pitfalls requires, among other things, knowing why and how a certain problem situation has led to the realization of certain design options and what about those options worked or did not. The historicity of the artifacts is a resource for novel design options in the present. In practice it is not an easily accessible resource, but becomes more available as a researcher's participation in the community develops.

Finally, a central component of the epistemic and sociocultural infrastructure of the lab is not explicit in the diagram. It is the educational infrastructure at the institutional level that was under development at the same time and was directed specifically at creating a new kind of interdisciplinary researcher in BME—a program designed to move the research field beyond the problematic collaborations of researchers in different disciplines by designing hybrid BME researchers. A central dimension of our research was to use our findings about their epistemic and learning practices in the context of research to aid in the development of their curriculum to enhance that kind of research. We dubbed our approach a "translational strategy." I had long worked informally with K–12 science education researchers, which reinforced my strong belief that philosophers of science could—and should—make a contribution to the improvement of science education. This research provided an exciting opportunity to contribute to building a practice-informed educational program from the ground up. But I also saw that the funding we received from the educational directorate at the US National Science Foundation for developing a novel, practice-informed education in BME would also provide the means to collect the data of the sort needed to address the problem of cognitive-cultural integration in the epistemic practices of science.

4.1.3 Designing Educational Infrastructure for Hybrid Researchers

In the BME labs, graduate students are simultaneously learning to be scientists and pioneering researchers. Thus, the development of student learners into BME researchers is a significant component of building the lab. These BME communities see themselves as conducting cutting-edge research on the frontiers of science, engineering, and medicine. The lab ethos is infused with an open-ended sense of possibility, as well as a tinge of anxiety about how little is known in their area and whether PhD research projects will work out. The researchers place a high value on innovation in methods, materials, technologies, and applications. Obstacles and impasses are omnipresent, as are lab-devised support structures for dealing with them. These structures help student researchers to see failures along the way are viewed as opportunites for learning. During our investigation, we saw several instances where "big gambles" led to high payoffs, which sustained this attitude, despite the fact that most of the researchers engaged in high-risk research are doing it for their dissertation projects. The sociocultural fabric each lab built, along with the supports developed in their local community, has been successful in helping students to graduate.

Our labs resided in a BME community that decided to place high value on what it calls *"interdisciplinary integration"* at the level of the individual researcher. For them this meant to move beyond problematic collaborations, which stem from the numerous differences between the practices and epistemic norms and values of engineers and of bioscientists, to the extent possible, and cultivate the individual researcher as a hybrid bio-medical-engineer from the outset. The nature of the research requires lab members, who arrive predominantly with engineering backgrounds, to develop equal facility with wet-lab techniques and in vitro engineering design, as well as to develop a selective deep knowledge of the biology of their research targets. Although it was clearly possible to, as the lab A director expressed, *"learn the biology as they go along,"* the lab directors knew from their own experience that this was often an arduous and haphazard process, and so sought to develop an educational program to facilitate systematic hybrid learning.

The lab A director and other senior colleagues felt the lab context and interaction with wider research communities were not sufficient to provide the infrastructure for students to develop fully as researchers. They saw it as their challange to design and build a new educational environment in which to develop their students into a new breed of reseacher, better-equipped to meet the demands of an emerging field and become leaders in it. This new breed would move beyond the faculty's own experiences of being educated as engineers who later moved into biomedical research—they would be educated as hybrid biomedical engineers from the outset. In conducting research, they would be able to integrate engineering and biological concepts, methods, and materials to address, mainly, medical problems.

As a consequence, the faculty determined they would build a pioneering educational program that would firmly establish BME as an "interdiscipline" that integrated all three components in its research and education.[5] So, in this framing, to address biomedical problems within an engineering framing did not require the BME researcher to establish collaborations with biologists or integrate them into the research, although they could do so. The graduates of this program would be able to collaborate fluently with other hybrids or with disciplinary colleagues in each area, thus mitigating much of the "interactional complexity" of interdisciplinarity (Wimsatt 1974). They would be equally able to move into academia, medicine, public health, industry, or government.

This was the vision. They translated it into an explicit decision for how to build that vision with three main components: (1) two new buildings with architecture designed to promote interdisciplinarity among bioengineering, biosciences, and medicine, with one building dedicated entirely to the envisioned BME department; (2) a new joint department of biomedical engineering across two universities, with one university providing largely engineering and bioengineering expertise and the other medical expertise, with the biosciences drawn from each university; and (3) a new educational program (starting at the graduate level, but quickly adding an undergraduate degree) that would integrate the three components of the field throughout its curriculum and cultivate student identities as bio-medical-engineers. Together these components would serve to articulate and institutionalize the kind of interdisciplinarity they broadly envisioned. This pioneering educational program has since attained national and international recognition, as well as garnered major awards.[6]

When we became involved, the first two components were well under way and provided the institutional, material, and financial structures from which to develop an educational program. They were raising funds and consulting with architectural experts in building spaces for labs and offices and for developing community activities that would promote interdiscipinary interaction and community-building. They had few ideas, however, about how to build an educational program to achieve their vision, and there were no established curricula or textbooks that could be adapted to that vision in their estimation. Through a serendipitous circumstance they became interested in understanding what cognitive science might have to offer as a resource. At that time the US National Science Foundation had a requirement that any grant that included an educational program also needed to include a cognitive science dimension. The NSF, as with other funding agencies, often includes such requirements to further their own objectives, in this case to improve the quality of science and engineering education through incorporating cognitive science research on learning. The leaders of the BME initiative were applying for an engineering research center (ERC) that would include graduate training. I was director of the

Program in Cognitive Science, so they contacted me and asked if I could explain to them why the NSF would have such a requirement, which I interpreted to mean to explain what cognitive science has to offer education. My response to their request created a partnership between them and me and my colleague Wendy Newstetter, whom they would hire into the new department and who became the co-PI on our NSF-funded research.

Our NSF funding, in turn, led to our creating a research group to conduct the investigations into the cognitive and learning practices emerging in frontier bioengineering sciences research labs. CLIC: the Cognition and Learning in Interdisciplinary Cultures research group continued, with varying composition, for fifteen years. We proposed a "translational approach": to study their cognitive and learning practices in authentic settings of research and translate our findings about the requirements to carry out BME research into classroom and instructional lab educational experiences. Our proposal to create what they called "a cognitively informed educational program" was a novel conception consonant with their novel objectives. If successful, it would put them on the map as leaders in education as well as research. Indeed, eighteen years from the time we began to develop it, the program received the highest award in educational innovation from the United States National Academies of Engineering, as well as other awards along the way.[7] The project, as envisioned, would also contribute pioneering research to cognitive and learning sciences, as well as to philosophy of science, since it provided the opportunity to examine cognitive-cultural integration as it occurs "in the wild" of science, as well as to investigate novel model-based reasoning practices as they emerge in interdisciplinary practice.

Much cognitive and learning science research has established that making students active participants in their learning is more effective than simply lecturing to them, and in the sciences especially, if they are engaged in attempting to solve authentic problems. In the K–12 area, there was by then a long history of educational initiatives based on "problem-based learning" activities. Given this and what we were finding about problem-solving in the labs, we were predisposed to find a way to make problem-based learning (PBL) central to the developing curriculum. Our choice was reinforced further by the fact that the method is widely used in medical education as a means of preparing students for the clinic, and thus familiar to the medical faculty. With medical PBL, small groups are presented with problems—rich and complex real-world medical cases—that enable them to engage in the authentic practices of the field, with "scaffolding" created by the teachers (who act as "facilitators" to student problem-solving) to support the development of expertise in diagnostic practices. In the course of working to solve authentic medical diagnostic problems, students develop a deep understanding of the human body, diagnostic capabilities, and an identity as medical problem-solvers.

PBL, as used in medical schools, however, was designed to scaffold the kind of hypothetical-deductive and inductive reasoning needed for diagnosing ailments. Our research determined that problem-solving in BME is model-based. To scaffold biomedical engineering model-based reasoning (Nersessian 1992a, 2002, 2008, 2009) we needed to develop a different kind of scaffolding in collaboration with the faculty who would run the courses. The faculty, at first, did not understand what we meant by model-based reasoning, but given a few examples, they agreed our characterization of their practices is apt. To distinguish our problem-solving objectives from the medical field, we called the new PBL-informed method for BME education "problem-driven learning" (PDL). Over time, through several iterations, this method has become woven into the BME curriculum. It is still a dynamic curriculum, which has continued to evolve since our research grants ended. At that time, the graduate level had two core PDL classes, and at the undergraduate level there were three core PDL courses, two classes and one instructional lab, we helped to create in collaboration with the faculty (Newstetter 2006; Newstetter et al. 2010; Osbeck and Nersessian 2019). Notably, as the undergraduate level developed, the education provided began to create an outstanding pool of undergraduate researchers for the labs. Much of the rest of the curriculum at both levels, which we did not ourselves design and develop with them, contains significant PDL elements incorporated by individual faculty members who have been inspired by what they experienced as facilitators of the introductory PDL course (all faculty facilitate). They did continue to consult with Wendy Newstetter, our project co-PI, who had become a member of their department and who was also a facilitator in the introductory undergraduate course, in developing these courses. Thus PDL, as a method, has become generatively entrenched, in that it provides structuring constraints for course design.

The introductory course is taken by all incoming students, who work in groups of eight on the problem outside of class, and with one faculty or postdoc facilitator during the class periods.[8] The problems they work on are carefully designed by the faculty, with the assistance of Wendy, to present complex, ill-structured health-care problems drawn from the real world, which encourage students to develop, integrate, and anchor their bioscience and engineering knowledge in the context of medical applications. For example, in a problem about cancer screening, student teams need to formulate and address questions concerning the biology of cancer, current screening technologies (e.g., CT scans or MRI), as well as envision future screening strategies (e.g., at the nanoscale), and to develop statistical models, among other topics of investigation. There is now a substantial repository of problems that faculty can draw from and modify to keep updated, as well as add new problems to.

It is important to underscore that the curriculum development is not a linear process. Hutchins has characterized learning as "adaptive reorganization in a complex system" (Hutchins 1995a, 289). The development of the BME educational program, too, fits the notion of *building* we have been using: designing, constructing, experimenting, evaluating, and redesigning incrementally through numerous iterations. This kind of iterative course development is called "design-based research" in the cognitive and learning sciences areas, and was poineered in K–12 education (Brown 1992; Collins 1992). Our research group and the BME faculty were also learners, and much "adaptive reorganization" took place in the early years of this curriculum development. We were pioneers in attempting a translational approach to curriculum development. Further, there had been little cognitive science, educational, or philosophical research on the emerging research practices of biomedical engineering (or any field of engineering) when we began. Although university research laboratories are the main training grounds for future researchers, they have rarely served as sites in which to study situated learning. Our program of translational research focused on turning our findings about the nature of the epistemic practices and of the effective strategies that support problem-solving and learning in the setting of the research lab into educational experiences in the instructional settings.

In both our philosophical and cognitive science research we sought to understand the ways in which the social, cultural, material, and cognitive aspects of practice and learning mutually inform, and are informed by, the research setting. We analyzed the ecological features of the research labs—the cognitive, investigational, and interactive practices—that invite and support complex learning and used them to guide design principles for instructional settings. Our findings led us to characterize the research labs as *agentive learning environments*, where student reseachers are made agents of their own learning, unlike traditional passive instruction via lecture and the canned, recipe-driven instructional lab (Newstetter et al. 2004; Newstetter 2005). These findings reinforced our initial choice of problem-based learning as a pedagogical method through which to implement the design principles through numerous iterations. Presently, learning scientists[9] and experienced faculty work with incoming faculty, which, together with the repository of PDL problems, constitute what we call a "faculty incubator." The environment of the incubator provides the cognitive-cultural saffolding for new faculty to rapidly participate in what, for them, is usually a novel pedagogical approach and learning-centered BME ecosystem. Finally, through the engineering education outreach efforts of Wendy Newstetter, the BME faculty, and the PhD students of the program who have gone on to university appointments, significant elements of our novel PDL approach have become generatively entrenched in other BME programs in the United States and internationally.

4.2 Summary: Lab A as "an Evolving Distributed Cognitive-Cultural System with Epistemic Aims"

The brief glimpse of lab A practices sketched in this chapter and chapter 2 provide an illustration of our findings about how the devices researchers build in the course of specific problem-solving efforts in a lab largely drive the building of the lab as a distributed cognitive-cultural system. These artifacts possess possibilities that researchers can exploit to evolve the system further. At the outset, the lab A director did not envision his lab engaging in tissue engineering to make vascular construct models or conducting stem cell research and gene profiling. His initial epistemic goal was to understand the effects of the force of arterial shear on endothelial cells, which in turn might help to inform understanding about disease processes of the vascular system, such as arteriosclerosis. At the end of his career he expressed wonderment at the fact that his research program had spanned "*astronauts to stem cells.*"

The director began his research program with the problem of the effect of vibratory forces on the cariovascular systems of astronauts by using his engineering knowledge to create mathematical models. Later, as he transformed into a hybrid biomedical engineer, he developed those models with experiments on animals. The animal in vivo research provided insight into important dimensions of the effects of shear on the vessels, but lack of control and other limitations led him to build the first in vitro model-system, which comprised the flow loop and endothelial cell cultures on slides. The flow-loop device afforded more control and opened the possibility to examine selected features of arterial shear in relation to endothelial cells, which were isolated from other features of the in vivo system. Specifically, this model-system configuration both enabled and constrained the research to focus on structural properties and proliferation behavior of cells under shear. Significant problems in conducting the flow-loop simulations, especially with respect to contamination, led lab members to redesign it into a compact artifact that could be assembled and run in a sterile environment.

The researchers realized all along that the cell cultures provided a limited model of the vascular wall in relation to the blood mechanical forces, as well as that the flow loop offered the possibility to examine the relationships among different kinds of cells in the blood vessel wall, if they could engineer a living three-dimensional tissue model. With the advent of new technology for tissue engineering, the lab undertook to design the construct family of models, which provides a range of models that instantiate more of the physiological functionality of the blood vessel to use in flow-loop simulations. The construct device opened the application potential to create a vacsular graft to repair diseased arteries and led the researchers to investigate the requirements of such a graft, which, in turn, opened the new problems and avenues of research into mechanical strength and integrity and began the quest to find out whether it is possible to use mechanical forces to develop a high-yield endothelial cell source. Importantly, the tubular shape of the construct supported the researchers in formulating new epistemic goals with respect to understanding the functional properties of blood vessels in relation to a range of mechanical forces. These goals required the researchers to build several new devices, for instance to examine pressure and strain, and an instrument to test mechanical strength, as well as to introduce new methods and technologies to examine experimental outcomes. Eventually, all of this led to the lab's ability to create a completely different kind of animal model than that of the director's initial research.

In sum, in vitro simulation devices provide structuring constraints for articulating the cognitive-cultural system that constitutes lab A as it develops over time. This system comprises researchers, goals, problems, models, methods, concepts, and epistemic norms and values, together with technologies for experimentation, visualization, and analysis and with sociocultural practices. The material infrastructure, in particular, both drives the direction of and becomes *incorporated* into the D-cog system and subsystems of a research lab, and is essential infrastructure for its epistemic goals and accomplishments. In the BME labs we studied, signature devices, in particular, contain the potential for development of future cycles of building, which often proceeds in novel and unanticipated ways.

In an important sense, then, a core activity of the lab is building itself as a distributed cognitive-cultural system directed toward achieving the overaching epistemic and application goals of the research. The initial and persistent goal of lab A had been to understand the role of physical forces on biological processes in the vascular system. The flow loop was particularly generatively entrenched in that it served as a structuring constraint on nearly all of the research of the lab for all the years of its existence. It made possible taking the research in vitro because, with it, normal and pathological in vivo forces on cells could be replicated to a first-order approximation. It also had the potential to simulate higher-order effects, if these proved important. It was generatively entrenched on two levels. On a physical level, as a device, it has formed a component of most experimental model-systems. On a metalevel, it entrenched the practice of importing engineering concepts and methods of analysis pertaining to mechanical forces into the study of biological phenomena. Most importantly, its affordances and constraints served to direct the researchers in forming new problems and building novel technologies. What kinds of experimentation the researchers envisioned could be done with the flow loop led, for instance, to the novel construct family of models. The construct needed to be designed to interlock with the flow loop in experimental situations, which in some instances required modifications to the design of both. The construct device provided the lab with a more physiologically realistic model and opened an application possibility and, with it, a line of stem cell research. These features generatively entrenched the construct in the remainder of the lab's existence as it opened and drove new directions of research.

Although I have looked only at the tissue engineering lab in this chapter, the features of processes of "building the lab" I have discussed transfer robustly across lab D, and likely those other BME labs that use similar practices of in vitro simulation modeling. The signature artifacts of a lab provide the structuring constraints that afford ways of evolving the research program without rigidly specifying in advance what moves can be made. Further, frontier research areas, such as those in twenty-first-century bioengineering sciences, often require researchers located in universities to build educational infrastructure. The BME educational program, built to facilitate a specific kind of integrative interdisciplinary research, provides a demonstration of "the manner in which epistemic integration interacts with organizations and institutions" in interdisciplinary research (Gerson 2013, 515; see also Caporael 2014). Existing institutions adopted the idea that innovative BME research requires a more directed and richer epistemic integration of biology, engineering, and medicine than collaboration alone could produce. Following out this idea, in turn, required the creation of new institutions, new kinds of architecture, and new modes of organization. Most notably, it led to a novel educational program, generatively entrenched in a new kind of cross-university department aimed at creating hybrid researchers, themselves poised to work at the forefront of biomedical engineering and to extend the frontiers for the next generation.

5 Managing Complexity: Modeling Biological Systems Computationally

In the previous chapters we have examined modeling practices that isolate and selectively focus on specific entities and processes, separate from much of their contexts in biological systems, in order to develop understanding and control of specific behaviors. Research in the burgeoning field of computational—or integrated—systems biology (ISB) aims to get a grip on how the higher-level functionality of complex biological systems emerges from a multitude of interactions among the elements of a system. The modeling practices in this field attempt to use as much information about the biological system as the modeler can find, while keeping the computational model computationally and cognitively tractable. Although ISB is a diverse field, the modeling practices in labs we have been studying are representative of a major area that draws on the resources of engineering fields that model human-made complex dynamical systems, such as electrical engineering, control engineering, systems engineering, and telecommunications engineering, as well as on mathematical and algorithmic resources from the computational sciences to model complex biological systems.[1]

ISB researchers investigate systems that comprise a range of biological phenomena that extends from intracellular interactions to those within organs or ecosystems. There are many objectives of the field, but in general, and especially in the bioengineering stream, the overarching objectives are (1) to build large-scale models that draw out the dynamics of biological networks and enable prediction and control with respect to phenomena of interest and (2) to use models to investigate what they call "the design principles"—or organizational principles that characterize the subcomponents of the biological systems. Understanding these principles, it is hoped, will provide the basis for a general mathematical theory of biological systems, as well as aid efforts by researchers and clinicians to control and intervene on systems. Much of the research in the field is directed toward interventions in health and the environment, such as to design new classes of antibiotics, create personalized cancer therapies, produce biofuels, or develop protective strategies for ecosystems.

ISB researchers position themselves in contrast to traditional biological fields, especially molecular biology. Although biological experimentation can reveal local causal interactions among molecular elements, biochemical functions are coordinated and controlled through large-scale networks, which are networks that have wide boundaries and that involve many interacting elements. These networks tend to function by means of nonlinear interactions (for instance, feedback loops) such that the causal properties of an element of the network depend on interactions happening upstream and downstream in the network. Further, these complex networks generate robustness and redundancy, and have nonlinear sensitivity to changes in their parameters, which give rise to variability across individual cells and organisms. All of these features make biological systems difficult to understand and control, and explain, in part, why systemic diseases such as cancer or cystic fibrosis have proven so difficult to treat (see, e.g., Hood et al. 2004). Only quantitative simulation models of such networks have the potential to capture network intricacies at the scale and size required to identify variables and predict network behavior in response to perturbations with accuracy sufficient to determine how to intervene on them effectively. As the lab G director stated, systems-level modeling *"allows us to merge diverse data and contextual pieces of information into quantitative conceptual structures; analyze these structures with the rigor of mathematics; yield novel insight into biological systems; and suggest new means of manipulation and optimization."*

Although the desideratum and philosophy of a systems-level understanding in biology has a long history (see, e.g., O'Malley and Dupré 2005; Trewavas 2006), many researchers, including the directors of the labs we investigated, look to the Human Genome Project as the origin of the contemporary field. As the lab G director stated in our initial interview, *"So if you were to put a point there, it was the Human Genome Project . . . and at the same time you had the microarray and all that stuff started to come out. They [bioscientists] said 'Wow! You can do then thousand data points in one pop. Who wants to look at all that data with the naked eye? That's not possible to do, so we need computers'—whatever that meant."* The confluence, around the turn of the twenty-first century, of engineering developments for biological experimentation, especially high-throughput technologies that produce reams of data from one experiment; the widespread availability of powerful computing (including, but not just high-performance computing); the development of sophisticated mathematical and algorithmic methods for solving equations computationally; and the development of Internet browsers and search engines that enable rapid searching of scientific literature and databases all have contributed to making computational modeling and simulation of complex biological systems possible (see, e.g., Kitano 2002; Krohs and Callebaut 2007; O'Malley and Dupré 2005).

The labs we studied prefer to use the descriptor "integrated," rather than "computational," to emphasize the integrative effect of putting all the pieces together in a computational structure—"*like an integrated circuit.*" At the conceptual level, "integration," as one researcher noted, means "*the tasks on this new frontier require thinking beyond linear chains of causes and effects—[rather]thinking in terms of integrated functional entities, thinking in systems, networks, and models.*" This kind of thinking is about the dynamic behavior of complex biological systems and requires computational modeling and simulation to carry it out. Based on our research, we would also add that ISB is integrative in another sense: it incorporates and adapts engineering concepts and methods, for instance from systems theory and control theory, computational algorithms and methods from computer science and applied mathematics, and experimental techniques, concepts, and data from experimental biology. As we have discovered, "integration" in this sense is not smooth, since the concepts and methods drawn from different fields carry with them, among other issues, conflicting epistemic values and norms, as I discuss further in chapter 7.

ISB is a heterogeneous field that brings together researchers in biosciences, computational sciences (including applied mathematics), and engineering sciences in various configurations. Although some researchers have developed into hybrids over the course of their careers, the field of ISB does not aim at the kind of hybridization through education that I discussed with respect to BME. Instead, solutions to the problems the field poses create an essential *epistemic interdependence* among the participating fields. Although there are some ongoing attempts to develop hybrid modeler-experimentalists, the nature of the problems ISB addresses, arguably, requires both specialization and collaboration. The norm in the current state of the field (and some would say, in principle) is for modelers to be trained in engineering or applied mathematics and for experimentalists to be trained primarily in molecular biology or biochemistry. However, to function most efficiently demands a symbiotic relationship. But, with little knowledge of one another's methods, concepts, technologies, and epistemic norms and values, at the present time symbiosis is more a desideratum than a reality. Our research has focused on the modelers, who by and large are driving the field, although we did conduct interviews with their experimental bioscience collaborators, when that was feasible (most were located at distant universities or in industry), and those provided important insights into collaboration issues from their perspective.

Researchers in ISB, as well as philosophers analyzing the field, have identified two broad strands of modeling in systems biology, namely *top-down* and *bottom-up* (see, e.g., Bruggeman and Westerhoff 2007; Krohs and Callebaut 2007). The top-down strand relies on high-throughput technology that generates large quantities of time-series data (dynamic data, as opposed to steady-state) for many elements of a system, such as chemical concentrations within cells. Computational methods, especially machine learning algorithms, are then used to attempt to "reverse-engineer" the system structure through making correlations among those elements. The bottom-up strand, on the other hand, aims to "reproduce" ("simulate") the behaviors of systems with dynamic computational models built using PCs. To build a model is an intensive process that draws on what can be pieced together of the network structure of the system and such features as kinetic and physicochemical properties of its components. The initial data for building the model usually come from collaborators, especially in molecular biology. These initial data often lack much of the information modelers need, such as on the concentrations of metabolites, and are often not a time series, which requires the modeler to interpolate data she can extract from the wider literature (including in databases). Building a model in these circumstances usually also requires modelers to use, and sometimes develop, sophisticated algorithmic techniques to estimate parameters (numbers) that provide the best "fit" of the model to the real-world data.

The labs we have investigated are both situated closer to the bottom-up strand, although both build models that they consider mid-scale or "mesoscopic." Simply put, these models contain modest details of system composition and organization, in that they simplify both the target mechanisms used to build the model and the underlying system functions they seek to represent (see, e.g., Voit et al. 2012). Such models can be informative in themselves, but they also provide the basis for incrementally and iteratively building out the system representation, in ways that can enrich both the lower (mechanistic) level and the higher (systems) level. Both labs work in the area of biochemical systems biology. Research in this area is directed toward representing, understanding, and controlling intracellular metabolic and signaling pathways. Both labs aim to build models of these kinds of pathways individually, as well as those that integrate these pathways. Lab G gets its modeling problems from collaborators, and so works on building models of pathways of a wide variety of phenomena, including, during our investigation, dopamine regulation in Parkinson's disease, biofuel production from plants, yeast response to heat shock, and arteriosclerosis. Such modeling problems also provide material for the lab's own agenda of developing novel algorithms for parameter estimation. Lab C's modeling focuses solely on pathways in complementary processes of reduction and oxidation (redox), which are thought to produce inflammation, including immunosenescence, cancers, and arteriosclerosis. These labs have quite distinct methodological practices, but they both share the feature that the researchers come predominantly from engineering backgrounds. It is an important statement about the nature of the field that, while claiming to do systems biology, the researchers did not refer to themselves as systems biologists, but rather identified themselves and their biological collaborators functionally, as "modelers" and "experimentalists" (alternatively, "experimenters"). This contrasts with BME, where we found, despite differences in subfields (tissue engineering, neural engineering), researchers identified as biomedical engineers.

We cast the differences in methodological practices between the two labs as different accommodations to numerous constraints we have identified these researchers to be operating under. These constraints are so challenging, that the reader might wonder how in the world researchers in this domain can accomplish anything. As we will see, modelers in this field develop such effective strategies to manage the complexity of building models of biological systems that they routinely produce novel and valuable insights into the behaviors of these systems and into how to manipulate, control, or modify them productively, and—as the specific case of G10 (section 5.2) demonstrates—sometimes make quite spectacular biological discoveries. "Managing complexity" is a major theme we associated with the codes we developed with respect to the methodological practices in each lab. As we will see, epistemic aims and cognitive needs intersect to shape problem-solving practices around managing the complexity not only of the biological systems, but also of the model-building process. I begin this chapter by laying out the constraints (section 5.1), then focus on how lab G modeling practices accommodate these constraints, in general and in a specific case (section 5.2), and then examine the epistemic and cognitive affordances of the methods (section 5.3) as they enable researchers to gain epistemic access and achieve their aims of getting a grip on complex biological systems.

5.1 Adaptive Problem-Solving in ISB

A major feature of problem-solving in the labs we investigated is that the research lacks the reasonably well-structured task environments that characterize established sciences such as molecular biology and bioinformatics. Nearly every step in the processes of model-building requires the judgement of the researcher to determine how to proceed, including how to (re)structure the problem, what modeling method to use, how and what portions of the biological pathway network to construct, what literature to rely on, what programming software to use, how to determine reliable parameters, and so forth. There is little available in the way of routines or protocols. Ultimately, what is produced in the form of a computational simulation model is a *strategic adaptation* to the constraints that model-building in ISB, in general and in the specific case, operates under in its present form. We have determined many of these constraints from our lab G and lab C investigations, but our claim that these are in effect across the wider field stems from widespread discussions about similar issues in the systems biology literature, including on education; by responses to our analyses by ISB researchers in audiences we have addressed and to our publications; and from findings in other ethnographic research I have conducted in ISB beyond this study.

5.1.1 Overarching Constraints on Model-Building

Modelers in ISB rarely can simply apply a formalism or preestablished principles to build a model that accounts accurately for a biological phenomenon. In effect, they face a multidimensional problem-solving task. Any model is the result of numerous choices about what to model, and how, with whatever resources are available. Some of the constraints on model-building we have observed operating in the labs are as follows:

1. The biological problem: Biological systems possess features that produce nonlinear behaviors, with many elements playing multiple roles. For instance, cells contain networks of genes, proteins, and metabolites that interact in feed-forward and feed-backward loops and create myriad biochemical interactions. A modeler must restrict the considerable complexity of the biological system so as to formulate a tractable problem to model, while at the same time representing it in sufficient detail for the model to simulate the target behaviors and yield predictions.
2. Knowledge constraints: Modelers usually have no familiarity with the biological system prior to starting on the problem. Today they might have to model lignin production in plants, and next, drug resistance in a cancer. They know little about biological entities and experimental methods in general, which limits their understanding of what is biologically plausible and what reliable extrapolations can be made from the available data sets. By and large, there is no reservoir of theoretical models and laws of the biological phenomena to provide the structure and dynamics from which to articulate a model, such as there is in physics-based modeling.
3. Infrastructure constraints: Comprehensive databases of experimental information for most biological systems, while growing in number, are still limited. There is little in the way of standardized modeling software, or of generally accepted routines and formalisms to apply in building a model. There are few textbooks and little in the way of educational infrastructure directed toward computational systems biology, although several initiatives are under way.
4. Data constraints: The kind of experimental data (time series) needed for building dynamic models and parameter fitting is often not available or difficult to obtain, and the available data are usually noisy. Model-building is data-intensive and routinely relies on data beyond what are collected by bioscience collaborators in small-scale experiments, leaving modelers to forage for pertinent data in the literature and databases on their own.
5. Cost constraints: New experimental data are quite costly to obtain. Experimentalists often do not see the cost-benefit of producing the specific data modelers need. On the computational side, it can be costly in time and money to update old software.

6. Computational constraints: Most biosystems modeling is carried out on PCs, not with high-performance computing resources. Although significant improvements in their speed and efficiency have facilitated the rise of simulation modeling, computational constraints still figure into the level of complexity a model can have to keep such processes as simulation and parameter fixing manageable.
7. Time-scale constraints: Processes of generating experimental data and of model construction, simulation, and testing operate on vastly different time scales. Modelers can wait for months for data to build or test a model.
8. Collaboration constraints: The significant differences in epistemic practices and educational backgrounds of experimentalists and modelers limit their ability to communicate effectively and to understand and fulfill one another's epistemic needs. Thus, it is difficult for modelers to obtain the kind of data or expert advice they need from their collaborators.
9. Cognitive constraints: Modelers need to be able to track many relations at the same time and, especially, monitor indirect influences in the system. The need to keep multiple constraints and other factors in mind as one builds a model is a multidimensional problem. In general, human cognitive constraints, such as on memory and mental modeling and simulation capacities, limit the ability of modelers to manipulate and reason about the models, and therefore limit the scale of the models they can manage.

The labs we have studied have adopted different methodological approaches to deal with these constraints. Many of the constraints on this list indicate a problem situation in which there are limited data for building a model. We began our research with lab G, and it was immediately notable how often modelers started off discussing their work with complaints about how hard it is to find sufficient data of the right kind to build their models. These complaints were frequently expressed, with considerable emotion, as the model "*needing*" data, which led us to code such expressions as a concern with "feeding the model." We frequently heard the expressions "*parameter estimation*" and "*parameter fitting*," with modelers expressing considerable worries about finding parameters (due to insufficient, inadequate, noisy data). Parameters, roughly, are the constants in the equations and are needed to control the behavior of the model, such as the rate constants of an enzyme reaction.

Model-building in ISB is not guided by theory the way the physics-based modeling that philosophers have usually studied is guided. "Theory" is, of course, a multifarious and contested notion. In positioning ISB model-building with respect to accounts of physics-based modeling that have become standard in philosophy, we take "theory" to mean a reservoir of laws, canonical theoretical models, principles of representation (such as boundary conditions), and ontological posits about the composition of the phenomena under investigation that guide, constrain, and resource building models in diverse disciplines across a wide spectrum of physical systems. Model-building in ISB starts without such a reservoir. As the lab G director noted, in the absence of the kind of theory available in physics, a "*big problem is where do we get functions from*?" Instead, modelers have to make what they call "*educated guesses*" as to the functions, guided by mathematical notions, such as growth functions, and by principles developed in molecular biology such as Michaelis-Menten enzyme kinetics (a model of the rate at which enzymes catalyze in a specific reaction), often referred to by biologists as "partial theory." There are no correlates, for example, to Navier-Stokes equations, which describe the movements of gasses and liquids, used by climate modelers. Such equations also help modelers determine significant parameters, in this case, temperature and wind speed. In physics-based modeling, theory is a resource that can inform the modeler how to go from a data set to a good representation. The models of lab G often have large numbers of unspecified or "open" parameters. Without experimental data of sufficient or good-enough quality, researchers have to rely on mathematical and computational ways to determine parameters so as to fit a model, such that it simulates the system behavior with sufficient reliability to make predictions. For this reason, a major methodological enterprise in lab G is to develop new algorithms to advance what researchers call "*the art*" of parameter estimation.

Lab C's methodological approach is to have modelers also conduct wet-lab experiments to supply data for their models—what we have called the *bimodal strategy*. This is the director's adaptation to data limitations. We rarely heard modelers in this lab talk about parameter estimation problems, since the models they built were smaller in scale, and they would conduct biological experiments to determine many parameters as they were building the model. Thus, lab C models tended to be closer to the data, and open parameters in need of estimation were few, though the larger-scale models they built had the fitting problems encountered in lab G. Lab C modelers did experience challenges around the need to master and coordinate model-building and wet-lab experimentation on their system in the course of developing a model, as we will see in chapter 6. The lab G director's methodological choice to collaborate with experimentalists rather than produce their own data is the predominant choice in ISB at present. The differences in methodological approach with lab C mark what the lab G director calls "*a philosophical divide*" in the field, which I discuss in chapter 6.

Both labs practice forms of what we called "adaptive problem-solving." All problem-solving is adaptive to some extent, but what is remarkable about the practices we witnessed in these ISB labs is the extent to which routine problem-solving depends on the researchers' ability to think innovatively while managing a range of constraints that create a significant cognitive load. Researchers in both labs specialize in building ordinary differential equation (ODE) models of gene regulatory, cellular metabolic, and cell signaling networks. Their efforts to integrate metabolic and signaling networks are novel (at least when we began our investigations). The variables in the ODE equations represent concentrations of individual metabolites in the network in a cell. Systems of equations are used to build dynamic models that can be run to simulate the changes to the concentrations of metabolites in a cellular network over time, where each metabolite pool interacts with specific other metabolites, represented as its neighbors in the network. Running the computational model under various conditions ("simulation experiment") shows how dynamic patterns emerge through the interaction of the pathway components over time. In general, the modelers aim to produce models that, when run, make reliable predictions of the dynamic relationships among specific variables in the model and perform robustly with respect to variations in parameter and initial conditions.

All aspects of the process of building a model are open to decision or modification, including the scope of the problem, how to represent the biochemical reactions, what data sets to use, what pathway elements to include, and how to estimate and fit parameters. As I noted previously, every model is a strategic adaptation to the constraints the modeler is working under and the resources she has at hand. The main, interrelated kinds of adaptations made by the modelers in the labs we studied have to do with the scale of the models they chose to build and with how to adapt problems to make them tractable, both of which are situated in the context of determining what kinds of conceptual and methodological adaptations to make to apply engineering and mathematical resources to biological problems.

5.1.2 Mesoscopic Modeling

As I noted at the beginning of this chapter, the overarching aspiration of the field of ISB is to build large-scale high-fidelity models of biological systems that should, in principle, facilitate understanding of the design or organizing principles of systems or predict the consequences of manipulating the systems towards desired outcomes, such as to produce biofuels efficiently or to design personalized medical treatments. The current state of the field, though, as Eberhard Voit et al. observed, is that "the vast majority [of ISB models] are neither small enough to permit elegant mathematical analyses of organizing principles not large enough to approach the reality of cells and disease processes with high fidelity. Instead, most models contain between a handful and a few dozen variables, which firmly positions them in a grey zone far outside both declared goals of systems biology" (Voit et al. 2012, 23). They call such models "mesoscopic." We agree that to attain specific goals, a mesoscopic model might be the most informative, and therefore desirable, choice in itself (see, e.g., Batterman and Green 2020; Bertolaso 2011; Bertolaso et al. 2014). However, we consider the prevalence of this kind of modeling, which falls short of the epistemic aims of the field, to be a largely pragmatic and rational response to the constraints of managing the complexity of modeling these systems.

A mesoscopic model provides a "coarse structure that allows us to investigate high-level functioning of the system at one hand—and to test to what degree we understand, at least in broad strokes, how key components of a biological system interact to generate responses" (Voit et al. 2012, 23). Such broad understanding can enable modelers to make substantive predictions—sometimes of major significance—but also, importantly, provide insight into how to expand the model in both directions ("middle-out strategy") to provide a more comprehensive representation (Noble 2006; Voit et al. 2012). The initial model creates an affordance in the problem-solving environment that modelers can use to guide and structure their investigation in stepwise fashion. Understood in this way, mesoscopic modeling is a strategy for gaining epistemic access to complex biological systems by building out the system representation so as to be able to enrich both the lower (mechanistic) and higher (system) levels. This expansion can be carried out by the builder(s) of the initial mesoscopic model(s) or by others in the field leveraging their insights.

Interestingly, Voit et al. advance a cognitive argument for the mesoscopic strategy, which they liken to hierarchical learning in human development: "This strategy of locally increasing granularity has its (ultimately unknown) roots in semantic networks of learning and the way humans acquire complex knowledge. . . . Hierarchical learning is very effective, because we are able to start simple and add information as we are capable of grasping it" (Voit et al. 2012, 23). We have advanced an additional cognitive argument that the mesoscopic strategy is a bounded rational response to handling the complexity of the constraints under which a modeler works (Macleod and Nersessian 2020). Herbert Simon (1957) argued that when faced with complex decision-making problems, people do not seek optimally rational problem solutions, but rather settle on solutions that are good enough to make progress ("satisfice"). The mesoscopic strategy enables the modeler to make progress, while holding out the promise of producing larger-scale models as modelers gradually gain understanding and control. A further cognitive argument, developed in section 5.3.2, is that as part of a coupled inferential system, the complexity of these midsize models remains at a level at which the modeler can still develop insight and intuition about the model's behavior and therefore make inferences about how to proceed in the model-building process.

As part of the mesoscopic strategy, modelers usually find ways to adapt the problems they tackle to simplify or get better traction on the specific problem. One way to adapt the problem is to keep the network representations relatively small through careful selection of what networks to include in the model at the start. Rather than attempt to model an entire complex network, modelers tend to focus on what their experimental collaborators indicate as potentially significant subsets when selecting the experimental literature to consider. An experimental collaborator of lab G relayed an example of this kind of adaptation when he told us of the reaction of the lab director after he had come to him with a large network to model: "*I think he's been in the real world long enough doing this systems stuff—long enough that he knows to start small. . . . So, when I came to him, I had these proteomics systems. We've seen about 10% changes in all the systems of the CF [cystic fibrosis] cell vs non-CF cell. Now when you think about the number of systems that are in the cells, 10% changes in all of those systems . . . is a lot of information. So, he's like 'you are deluding yourself.' So, then we decided to start with glycolysis and the pentose phosphate pathway of the Krebs cycle . . . to narrow it down to energetic pathways that are very well modeled."* In this example, instead of trying to build a model of a large, intractable network, the director—a highly experienced modeler—moved the model-building process toward the strategy of adapting the problem in the direction of using small models, already established, and building outward from those.

5.1.3 Engineering Transfer

One important kind of problem adaptation is to use strategies and heuristics from engineering to alter the dimensions of a problem.[2] For instance, modelers might situate a network under study within a broader network, on the basis that the broader network can reveal connections between parameters in a subnetwork in ways that have a significant effect on the behavior of the subnetwork. This strategy helps to elucidate confusing dynamics. Another quite common strategy is to use an engineering method called sensitivity analysis to isolate the elements of a network that play the most significant role in the dynamics of the network. Sensitivity analysis basically targets the uncertainty in a model by examining the change in output produced by the change of specific parameters. This method can be used to simplify the network representation or to identify parameters that do not have too great an effect on the dynamics. These parameters, then, can just be assigned arbitrary values to reduce the parameter-fixing problem. Additionally, modelers often black-box component systems or component interactions to reduce the complexity of the network and, conversely, de-black box them if it appears that a subsystem is having a nonlinear effect on the network. Further, if the modeler cannot see a means of directly building a good model for a specific network, she will work on an alternative network for related phenomena that is simpler or for which better data are available. The modelers we studied often switched systems in this way or switched cell types for the sake of better data, with the hope they would be able to modify that model in the direction of the original problem. In general, to adapt problems, modelers employ strategies that incorporate and integrate engineering methods into systems biology. These methods, themselves, have to be adapted for the new subject matter and research environment, along with the engineering epistemic values that favor precision.

Throughout the course of model-building, modelers import concepts and methods that have been used in engineering for building models of human-made systems to transform biological problems into a form appropriate for mathematical and computational analysis. Such transformation strategies range from adapting the individual problem to designing methods for classes of problems. ISB modelers, in general, draw concepts and methods primarily from engineering fields, and especially control engineering, which has developed techniques for measuring and deciphering electronic signaling networks. The lab G director claimed that modelers can tackle a range of biological problems about which they have no prior knowledge because their training in engineering methods and concepts gives them "*the right mind set*"; that is, "*the flexibility to recognize shared features of control/regulation across disparate domains.*" Many of the methods used in the labs are borrowed from these fields, including, but not limited to, simulated annealing methods of parameter-fixing (approximating a global optimum of a function), and nonlinear network analysis techniques. They also used standard computational modeling tools that are used more widely than in engineering, such as Monte Carlo methods of parameter estimation (an approximation technique using random samples of numbers).

In borrowing from engineering, ISB modelers are following a practice that pre-dates modern computational systems biology. Biologists have a long history of borrowing concepts such as circuit, system control, modularity, redundancy, noise, and sensitivity to conceptualize system-level phenomena (see, e.g., Wimsatt 2007). For instance, metabolic control analysis, which began in the 1960s, is based on engineering analysis of network control, which derives from sensitivity analysis in engineering (Westerhoff et al. 2009a; Westerhoff et al. 2009b). Modelers in the labs we investigated continue to extend these practices by experimenting with their own adaptations from their different engineering backgrounds. For example, one lab G researcher we followed, who had a background in telecommunications engineering, was trying to figure out whether, and if so, how she could adapt wave-smoothing techniques from signal processing to smooth noisy biological data. We found the degree to which both lab directors allowed graduate student and postdoctoral modelers the flexibility to choose how to go about trying to solve their problems—what background methods and concepts to rely on—to be quite remarkable. A successful adaptation can require considerable ingenuity. Failed attempts along the way are par for the course, but these are seen to provide invaluable insights into the problems, as well as information on what to try next. Importantly, as we will see, modelers rely on the model-building process, with its ongoing simulations, to develop an understanding of their systems and figure out how to adapt them to their specific epistemic goals.

It needs to be noted, though, that although engineering methods and techniques can facilitate the model-building process, some biological understanding is required to help discriminate good moves from bad ones. Modelers talked all the time about the need to get a sense of *"what is reasonable and what is not reasonable"* biologically. This is a problem for which collaboration constraints are strongly felt. As the lab G director noted, *"really good biologists have a feel for things. . . . They know what to look into, how difficult it's going to be. . . . This intuition . . . is very hard to mimic or acquire."* In our investigation, we found that graduate student and postdoctoral modelers relied heavily on the lab directors, both of whom had considerable breadth and depth of biological understanding, to help them determine whether their moves were reasonable. This was so even for modelers in lab C, who conducted wet-lab experiments on the biological systems they were modeling. Sometimes researchers could ask experimental collaborators, but lab G modelers, especially, usually found it hard to get the attention of their collaborators. They frequently expressed to us a desire, along the lines as one modeler put it, to have *"a biologist in my desk drawer,"* to back up their judgment on the moves they were making. She made that comment as I sat beside her to watch how she determined the steps she took to forage for data online. On the whole, the process of model-building is mainly the responsibility of an individual modeler rather than a well-coordinated process between modeler and collaborators. This is largely because the computational model is a "black box" to most experimentalists, and modelers, especially those who have done no experimental work, do not know how to convey model details or what they require for the building process to their collaborators.[3] We made the collaboration problem the focus of the educational experiences we promoted for ISB, as discussed in chapter 7.

The lab G researchers were fortunate to have, in addition to the director, a long-time experimental collaborator who often visited the lab as he transitioned to being a modeler, with whom they frequently consulted. He found it amusing that just because he was an expert in yeasts, the modelers thought he could answer their questions about any area of biology. He also found it problematic that often they *"don't know the right questions to ask."* However, as we will see, the model-building process, itself, although not a substitute for a biologist's intuition, also helps the modeler build some biological intuition. Modelers claimed to develop this intuition through the extensive searching in and reading of the biological literature required to build out the pathway network and from their examination of the biological system's behaviors through simulations under various conditions (including counterfactual). Such intuition is particularly important when it comes to fitting the model (a process described below). As one modeler claimed, over time *"you get a feel for what might work and what probably doesn't"* with the system under study, and eventually more broadly. The graduate student modelers usually work on specific systems for four or five years—the director much longer.

As we will see, accounting for problem-solving and discovery processes in these labs requires analysis of a D-cog system comprising modeler, experimental collaborators, lab mates, and various artifacts, including computational models, pathway representations, diagrams, graphs, pen and paper representations, and data sources (publications, databases, search engines). I turn now to the model-building practices of lab G, first providing a general overview and then examining some of the details of one of the long-term modeling projects we followed.

5.2 "Where Numbers Come to Life": Getting a Grip on Systems Computationally

As mentioned earlier, lab G's practice is to obtain their modeling problems from experimental collaborators and, hopefully, obtain experimental data for building and testing the model from them. The director is a senior pioneer in the field of ISB and he portrayed the situation when we arrived as *"Biologists and clinicians come to my office and say, 'we have some data, so you want to work with us?'"* He also pointed out that this was a drastic change from when he started out: *"Twenty years ago that would have been utterly, totally impossible."* When he was a student in the 1970s, he found it a *"nightmare"* to figure out how to combine his interests in biology and math, because the combination *"was not only not supported, but outright considered ridiculous by biologists, and even more so by mathematicians."* He managed to get a PhD in developmental biology (*"it was really theoretical biology, but there was no degree in that"*) by developing some rudimentary computational simulations of predator-prey relations and scar patterns on budding yeast cells. The work on yeast led to a postdoctoral position with an electrical engineer working on developing methods for how to model biological systems, who, himself, had managed to get a faculty position in a microbiology department. Together, they built mathematical models of yeast and developed tools of mathematical and computational analysis for the emerging field. The future lab G director's first faculty position was in an interdisciplinary unit of epidemiology in a medical school, where *"the chair . . . was a visionary guy. . . . He hired engineers, he hired some people doing signal processing and AI."* In his lab there, among other projects, he continued to model yeasts with local and international collaborators. He also continued to develop methods to analyze biochemical systems and for parameter-fitting for nearly twenty years, before he moved to his current position to set up lab G approximately six years before we entered.

Lab G, as we encountered it, had modelers who were working on a range of biological systems, including metabolism in yeast, atherosclerosis, neurodegenerative diseases, and sustainable biofuels production, at the request of experimentalists external to the lab (and most, external to the university). They also worked on novel algorithms for parameter estimation and structure identification in biochemical systems modeling in general. The general practice of the lab is that everyone works on an individual problem in collaboration with the lab director. The lab director has numerous one-on-one meetings with every researcher and contributes actively throughout the model-building processes. Because the researchers work on problems from quite diverse areas of systems biology, the lab director said he felt that lab meetings would not be useful. At his own initiative, though, he did arrange several group meetings to introduce us to the research of the lab. Interestingly, the researchers all expressed a desire to continue to have lab meetings, since they found it useful to see in detail what modeling issues the others were struggling with, but the director did not continue. We did witness—and were told—that it was standard for researchers to discuss specific problems with one another as they arose, which usually proved fruitful despite project differences. As one member noted, *"When I have to discuss, I grab hold of somebody and we start working on the board. It's as simple as that."* The lab space consists of open cubicles with desktop PCs and is often empty of people, since most work from home and come in when they have a course, a meeting, or a need to find someone to discuss modeling problems with.

The cognitive-cultural artifacts of lab G comprise computational and mathematical resources that are essential to achieving its epistemic aims. In our initial interview, the director framed their epistemic aims in terms of the overarching aims of ISB: "*We want to put pathways together and we want to predict what they do and then see if they do what is predicted. To do this rationally correct, you need to understand the types of design principles . . . regulation, adaptation, whatever. . . . So, ultimately, we need to understand these types of design principles and operating principles. We want to understand them because a) we are academicians and b) because we want to muck around with these things and change them.*" In this statement, the director expressed both objectives of getting a grip on complex biological systems: to understand how and why they exist a specific way in nature (mechanisms and design principles) and to determine what possibilities there are to manipulate them in a desired direction. Achieving these goals will both further the development of biological theory and enable bioscience and medical collaborators to manipulate the systems, for instance to produce biofuels from plants, tailor drug treatments to cancer patients, or manage bacterial populations in lakes. In the current situation, though, the kinds of in silico simulation models (mesoscopic) it is possible to build are, usually, of a scale and complexity that can provide only limited understanding of the mechanisms underlying the behaviors of complex biological systems. However, the insights they do provide are often sufficient to make novel predictions and can lead to successful experimental manipulations by collaborators. Further, as the lab G director noted, even such limited computational models can provide insights into "*why you have this one design in nature*" by comparison with a model as a "*hypothetical alternative,*" which allows the modeler to examine counterfactual designs that, in principle, could exist in nature.

In practice, we found that modelers in both labs had much more limited goals. They tended to focus on modeling a system (1) to discover robust mathematical relationships among specific input and output variables in order to manipulate them in the in vivo system and (2) to infer the potential role of a specific molecular or component process in a network and its interactions, and to use this information to predict the effects of manipulations of these on system dynamics. In lab G, these goals were usually connected with requests from experimental collaborators to generate hypotheses about what might be missing or wrong in their data, or to discover new relationships in the data. Such information could help them better direct their experiments or manipulations.

In both ISB labs, developing skills at building biosystems simulation models makes one a part of the lab cognitively and socioculturally. Many features of the basic model-building process are similar across the labs, but, as we will see, because lab C works with a smaller number of reactions, modelers are often able to use off-the-shelf modeling tools, and, significantly, the modelers conduct wet-lab experiments to collect additional data to build and test their models, which reduces the parameter-fitting problem significantly.

5.2.1 "I Always Start from Zero": Overview of the Model-Building Process

To grasp the complexity of the problem of modeling complex biological systems, a picture is indeed worth a thousand words. Figure 5.1 is the picture the lab G director gave us to illustrate the biochemical systems modeler's challenge. The left figure is the metabolic pathway (network of elements and interactions) of sphingolipid yeast, a budding yeast such as used in brewing and baking, that he has worked on for years. The pathway diagram represents, spatially, sequences of molecular interactions in the cell. It depicts a chain of reactions that result in the performance of some biological function. The right figure illustrates the limited portion one could model, and the abstraction of the elements and interactions a tractable model could handle.

The first step in model-building is to develop a representation of the biological network that shows the main reactions among the targeted elements in a system, called the "pathway diagram," which provides the basis for building the model. Most experimentalists work with only a specific subsystem within a network—often a tiny fraction of the overall pathway. It falls to the modeler to build out the pathway relevant to the biological system in the detail required to model it. Figure 5.2 provides an example of a pathway a lab G modeler was working on. In general terms, the pathway diagram is a conceptual model that represents, spatially, sequences of molecular interactions (metabolic and signaling, for our modelers) in living cells. In essence, it maps out a chain of reactions that result in some biological function being performed. The diagram also captures positive and negative regulation effects, which specify the influence of metabolites on different reactions. For the modeler, the configuration of the pathway elements specifies

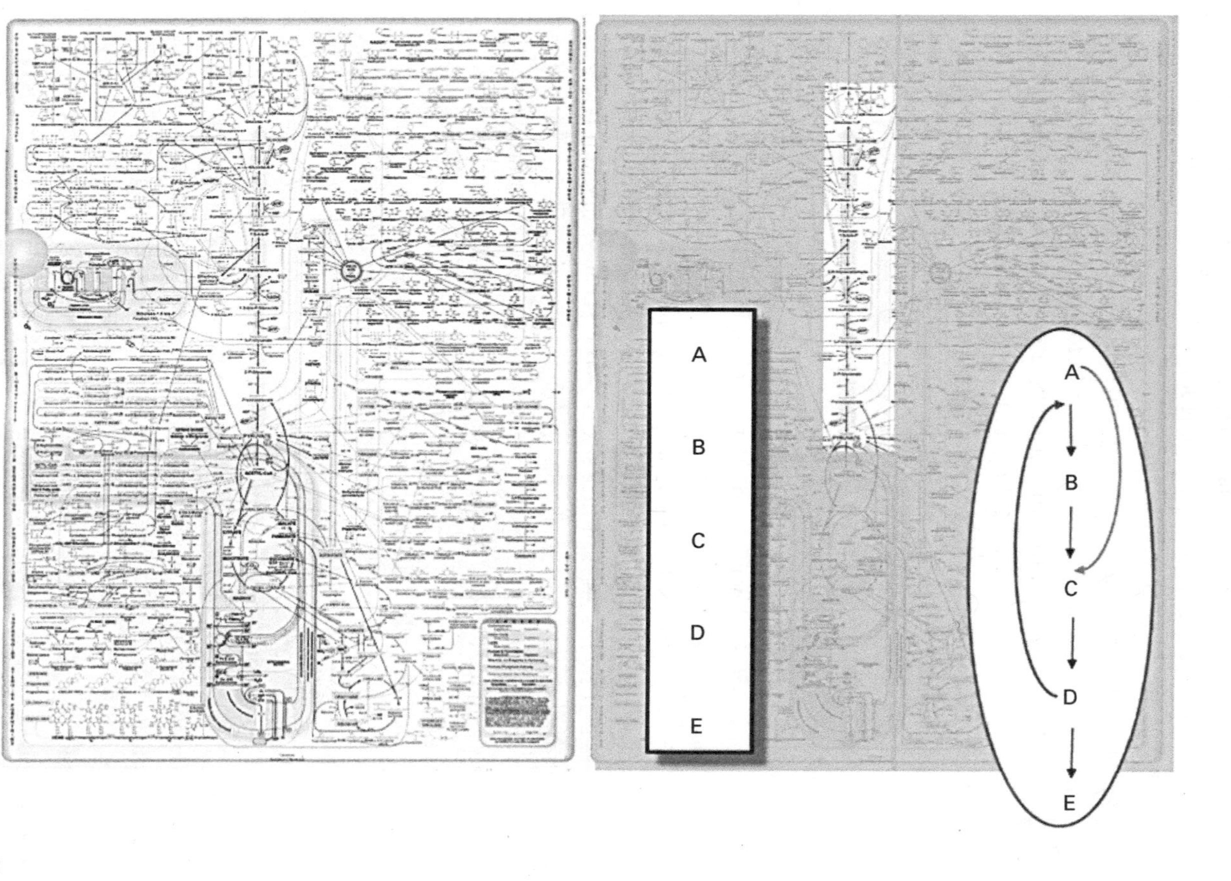

Figure 5.1
Pathway diagram of the sphingolipid yeast metabolic network. The left diagram is the pathway as currently understood and the right diagram illustrates the limited portion and relations that can be managed in a tractable model.

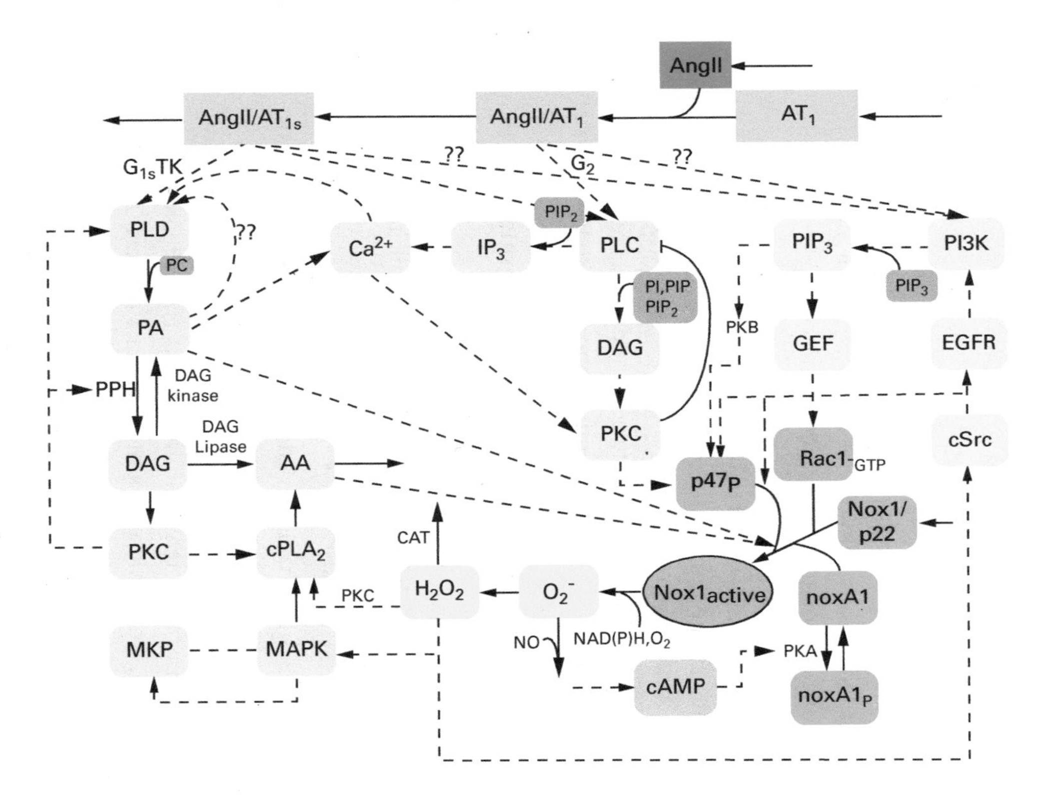

Figure 5.2
G12's preliminary pathway diagram for the angiotensin (AngII) hormone, which causes vasoconstriction, at an intermediate stage of development. Metabolite names are in the boxes. The dark lines indicate connections where material or information moves across nodes; the dotted lines indicate regulatory connections. The arrows indicate the direction of the processes. Note the question marks over some of the connections that are *"guesses"* by the modeler, which she will check with her collaborator to determine if they are *"reasonable"* biologically. The color coding indicates modules she modeled in different configurations.

the logical structure of the model. Among the affordances of the diagram format are the possibility to color code segments of the pathway, which the modeler might use to block out various combinations to examine in model-building, and the possibility to annotate it in various ways, such as to indicate uncertainties or to provide additional information on reactions.

The pathway diagram is often likened by systems biologists to "a roadmap where we wish to understand traffic patterns and their dynamics" (see, e.g., Kitano 2002 2). Lauren Ross (2018) explores this analogy in an illuminating analysis of the pathway concept from the perspective of biologists. Since developing biological pathways are a critical part of the modeling process, it is useful to elaborate on it here, before moving to lab G practices. As Ross notes, "when biologists use the pathway concept they often imply that some system can be understood in terms of causal routes or roadways. These causal routes capture interconnected paths that track the movement of some entity or informational signal through a system" (Ross 2018, 9). Her objective in that analysis is to explicate how the pathway represents the causal relational structure of biological entities and processes rather than the underlying mechanisms producing the biological phenomena. On her analysis, the pathway develops a fixed order of causal relationships that "capture the 'flow' of some entity or signal through the system. . . . Cell signaling pathways track the flow of a signal through molecular and cellular systems, metabolic pathways trace the flow of chemical substances through stepwise changes" (11). The language biologists use in discussing pathway representations—"flow," "flux," "connection," "blockage," and so forth—indicates "something that is carried over from one causal step to the next . . . something that travels along causal connections" (Ross 2018, 11) She also notes, importantly that, as with a road map, a significant amount of causal detail (for instance, temperature and Ph) is not represented in the pathway diagram, which makes it a more abstract representation than would be needed to represent causal mechanisms. This does not mean that the pathway diagram is devoid of mechanistic information, for instance, regulatory processes.

Ross does not discuss how pathway representations are used in computationally modeling biological systems, but her analysis accords with the features of those representations we have seen modelers in both labs emphasize when they build pathways to model the behavioral dynamics of a system. Modelers use the same descriptive language, but in addition, from the modelers' perspective, they say *"the mathematics is in the arrows"* that represent the causal flow. The map itself is a static representation, but the modelers see the dynamics of the system in the arrows, which provide, as one modeler stated, the *"functional dependencies from which [they] derive . . . differential equations."* The model is built to capture the flow of interactions among the pathway components over time so that the model produced from it will act out the dynamics of the system-level behaviors of the target system that produced the data. In effect the pathway representation provides an analogue model that exemplifies the causal relational structure of the processes that produce system behavior, without specifying the underlying causal mechanisms that produce the behavior. From this representation, modelers can use the representational affordances of mathematics (such as power laws that represent relational changes) to create models that enact the causal dynamics. Understanding the dynamical behavior does not require knowing the underlying mechanism.[4]

What is remarkable about the process of developing this network map is that, usually, modelers receive only a small piece of the diagram from their collaborators, if anything at all, and it is their responsibility to build this network of reactions through foraging in the literature and available databases. The modelers we followed were engineers with little biological knowledge, and no knowledge of the specific biological systems when they started to model them. As one modeler noted, *"I always start from zero."* After sitting with this modeler to watch how she collected data and other information through Internet searches to build her model, I nicknamed this process "google biology." In publications modelers often insert numbers in parentheses at locations where specific references from the literature have been used to build out the pathway. Importantly, the primary means by which modelers begin to learn about the biological systems they are responsible for modeling are through literature searches, from which they not only gather data, but also develop conceptual understanding, and through building and simulating partial models in the course of building out the pathway network.

Our interviews with modelers and experimental collaborators show the pathway diagram to function as a boundary object (Star and Griesemer 1989) in that it is a representation used with sufficient flexibility in interpretation so as to provide a means of communication among the different communities. Importantly, the pathway diagram provides a tool to identify and track, visually, causal movements within the network, for both communities, as Ross's and our analyses show. These movements provide a basis for reasoning about the network both in molecular biology (see, e.g., Sheredos et al. 2013) and in mathematical modeling. When we began our investigation, we had thought the computational model would perform the function of a boundary object, but soon realized bioscientists have little to no understanding of it. To them it is largely a black box. From what we witnessed, collaboration is rarely smooth, but one important interaction modelers have with experimentalists is to check whether a modification they have made to the pathway is "*reasonable.*" This important check is possible because, as we saw above, for experimentalists the pathway diagram represents a set of qualitative causal relationships among molecular elements, and they usually can infer whether the proposed modeler modifications are plausible within the causal structure. Sometimes they are even aware of additional experimental literature that will aid the modeler in confirming their modifications.

For modelers, the pathway diagram represents a mathematical structure. As one lab G researcher expressed it, modelers "*in some sense translate it [the pathway] into a map we can deduce math from.*" For the modeler, the nodes are variables and they put "*[mathematical] meaning into the arrows,*" by giving them precise quantitative values for the rates of a reaction. This process necessarily involves much simplification and abstraction of what modelers often refer to as "*messy,*" "*dirty,*" and "*noisy*" biological systems so that they can be modeled quantitatively. The model is built on a generic pathway, such as the lignin pathway in plants, and is a generic dynamic representation in that it simulates the behavior of systems of that kind—of a generalized target.

It requires considerable effort and judgement on the part of the modeler to find the data in the literature and databases and evaluate which ones are relevant to and important for their problem. As part of their judgements, modelers need to determine what data sources are "*trustworthy.*" That is, the modelers need to exercise judgment about the source, quality, and relevance of the data. The modelers we studied pointed out that they try to select data from labs that they consider to "*produce reliable data,*" based on their lab's experience with them, especially those of the director. Similar judgements are made about databases, which are developed and curated in significant sociocultural negotiations (Leonelli 2016). There are many missing pieces (e.g., the pink portion of the lower right quadrant of figure 5.2 was built out entirely by the modeler) and many open questions (note the question marks), which require the modelers to guess potential reactions. When possible, they present their guesses to experimental collaborators in the form of hypotheses to determine, as G12 explained, if the addition is "*reasonable*."

Notably, our investigations have shown that *building the pathway is an iterative and incremental process in which model simulation is itself a critical resource*. That is, the pathway structure is assembled in an exploration that involves preliminary simulations. Often small pieces of the pathway are simulated by the modeler, for instance by running through specific values for variables, using pen and paper (or marker and whiteboard) and their imaginations, before running segments in a computational simulation (see figure 5.3).

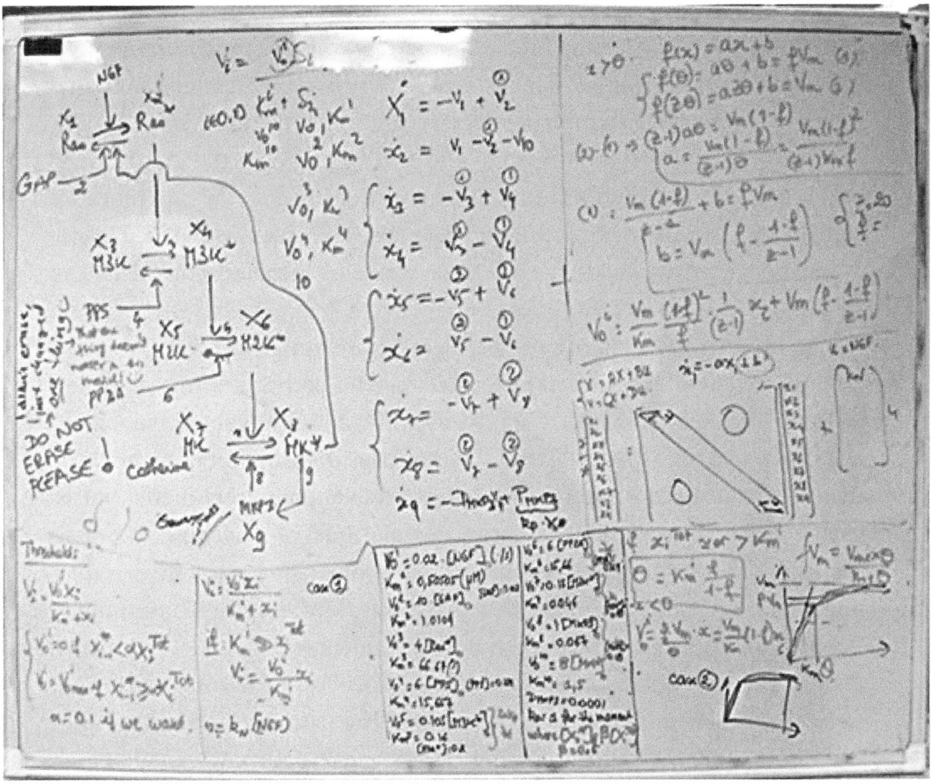

Figure 5.3
A lab G whiteboard on which a modeler is working out pieces of the pathway under investigation "by hand" and imagination.

Once the pathway is developed in sufficient detail, preliminary models usually are built in a modular fashion, with and without pieces of the pathway (such as the different color-coded sections of figure 5.2). Simulations of these help to determine, for instance, what can be trimmed or where there are possibly missing pieces. Through these simulations the modeler builds up an understanding of the dynamics and relevance of specific pathway elements. This enables her to make judgements about adding elements that might be playing a role but are not discussed in the literature on the system, or about feedback relationships that are not documented, or about what can be safely left out of the pathway. Such determinations are often based on elements in the literature that are thought to be related to the system, such as from different species and different cell lines. Modelers also use simulations to determine values of parameters often missing from the literature, such as the speed of the reaction (rate constant) and the sequence of reactions to the product (kinetic order), which experimentalists usually do not measure. Determining parameters from the literature is itself a complex process in which modelers have to reverse engineer the graphs they encounter into the numbers they need for their models. As one modeler explained, *"There might be graphs that have trends or what not, and then I have to quantify the graphs and then either figure out slopes or things of that nature to get at a particular number."*

The processes of building the pathway create a unique composite network of metabolites and parameter values. The pathway brings together pieces of information that are spread over a wide set of papers, databases, and unreported experimental data. The pathway diagram not only provides the basis from which the modeler builds the computational model but is itself a visual representation of a conceptual model of a network of causal interactions. The computational model built from it creates a synthesis that is, in effect, a *running literature review* that exists nowhere else. Thus, simulation is not used only to "sound out the consequences of a model" (Lenhard 2007, 181), but, notably, also *to learn and assemble the relevant ontological features of a system.* The process of adapting the pathway network continues throughout the model-building process until pathway, experimental data, and parameter fit coalesce into a model (or small set of models) that simulates the behavior of the target system (model output matches experimental data), at which point the model can be diagnosed and tested until it is considered validated. If the model fails testing, diagnosis is, as one modeler explained, "*a big problem . . . because you don't know if the pathway structure may be wrong. Second, maybe the parameter is wrong. So, maybe the algorithm is wrong. So, I have to check every part of it to make sure of everything if something goes wrong. . . . It's actually a cycle, an iterative process—so we go back and forth.*"

For the modelers in our labs, model-building is a labor-intensive process that usually takes several years. Building the model requires the modeler to make numerous choices along the way. Modelers can choose a variety of formalisms to build the model. Choices include, for instance, whether to use phenomenological models, such as agent-based models, or mechanistic models, discrete or continuous models, spatial (partial differential equations, PDEs) or nonspatial (ordinary differential equations, ODEs) models, stochastic or deterministic models, and multiscale or uniscale models. Lab G most often chooses to represent the interactions in sets of coupled ODEs that capture how the concentration levels of different metabolites in the pathway change over time. The number of reactions investigated by the lab G modelers during our study ranged between fourteen and thirty-four—a number that the director characterized as "*just a handful,*" when compared with those in the actual system (figure 5.1). The number of equations needed to capture these reactions varies with the specific questions the modeler is exploring, the nature and availability of data, and the computational resources. The advantage of ODE models is that they are both relatively simple conceptually and have the potential to be highly informative. In addition, there exists a wide range of computational and mathematical resources for analyzing system dynamics and for estimating parameters for ODE models.

Selecting an ODE framework opens another range of choices about whether, for instance, to model the system as steady state (static) or away from equilibrium, whether to use a mass-action stoichiometric model (based on rate of chemical reactions), or to use a canonical mathematical template such as biochemical systems theory (BST) that averages over the details of the interactions, or a mechanistic model that sticks closer to the molecular details in the form of rate laws of individual enzymatic reactions.[5] The choice depends on the nature of the problem, the goals of the modeler, and the nature of the available data. In the culture of lab G, BST plays a major role in ODE model-building. Even so, there are no set choices, and much depends also on the preferences of the modeler. As one modeler told us in discussing the model-building process, for many choices, *"It's a pragmatic choice. That's why modeling is still an art—it's a choice people make. I make one choice and another one would make a different choice."*

The modelers usually split the experimental data into two sets, one used to develop and fit the model (training data) and the other, to validate or test the fitted model (test data). The complexity of the tasks of fitting and testing a model is highly dependent on the nature and quality of the experimental data available. Rarely do lab G modelers have access to rich, dynamic data (time series). Most often, they have steady-state data that show how an experimental manipulation led to a change in metabolite level from a baseline. These data are reported by experimentalists usually as a single data point going up or down or holding steady ("steady-state" data), which provides the experimentalists with all the information they need, but not the modeler. This difference in needs again points to how differences in epistemic aims create problems in collaboration. A common lament we heard about experimental collaborators was expressed by one modeler as: *"They just care up/down. . . . They don't care time series . . . how this dynamically changed. They just care what is the result."*[6] I used the word "lament" because this complaint was always expressed emotionally, with considerable frustration. In the absence of good dynamic data, the modeler faces considerable uncertainty because a range of parameter values can generate model results that fit sparse data, so the fit is not unique. The modeler can use algorithmic techniques and various computational tricks to figure out how the parameters might be changed and at least narrow down the range of acceptable fits. But it is often unclear whether the lack of a unique solution is because the parameter estimation is poor or whether some elements are missing in the pathway.

If the data generated by the model do not fit the test data, the modeler tweaks the parameters (*"tunes the model"*) until the results provide a satisfactory fit. The modelers we studied do not use real-time dynamic visualizations of model behavior (as in lab D). Rather, they generate graphs that plot the concentration value of a molecule in the pathway across time for the model and for the experimental data, and compare a stack of graphs for different parameter values to judge how good the fit is. All of the modelers we interviewed pointed to parameter estimation as the most difficult part of the model-building process. In lab G, modelers often use optimization algorithms to estimate a significant number of open parameters. But just as often they need to use novel reasoning about the problem to develop fitting options, such as to determine what reactions might be set to zero or what kinetic orders the free parameters might be. One technique we saw was to use data available on the same metabolic elements from other cell lines, such as using neural cell data to get parameters for a metabolite in smooth muscle cells. Modelers justify this move on the basis of their judgement that the systems in the diverse cells are reasonably homologous. Other common techniques modelers use include sensitivity analysis, which enables them to set parameters that do not affect network dynamics (insensitive) to a default value, or to explore the dynamics of different parameter values and ranges by running through random numbers with Monte Carlo simulations. In lab G, modelers sometimes create new algorithms for parameter estimation as part of the fitting process for the specific case, which, if useful, they will try to extend to other cases. All of these processes for parameter estimation and fit involve running numerous simulations (on the order of hundreds of thousands). Thus, simulation is not simply the end phase of problem-solving. *Simulation is a resource for iteratively building the simulation model itself.*

Once a satisfactory fit is achieved, the model is run through a series of diagnostic tests, including for stability (does not crash for a range of values), sensitivity (input is proportional to output), and consistency (reactant material is not lost or added). If these diagnostic tests fail, the modeler can tune the parameters again or modify the pathway. In general, modelers employ strategies that adapt and integrate engineering modeling methods into systems biology. These labor-intensive processes, as well as others I have not mentioned, continue until the model fits the available experimental data, as established by the data output of its simulation runs. The simulation model created by these means is generic in that it makes manifest the dynamics of all the available data on that type of system, including natural systems, in vitro systems, and engineered or modified systems. In an important sense, the "system" modeled is an abstract general system, and the dynamical behavior the model exemplifies is that of a system of that kind. As such, it enables the modeler to examine a range of behaviors, including counterfactual cases, which can provide insight into, and predictions about, how the pathway might be reengineered for specific purposes.

That the model produces a satisfactory "fit" does not mean it provides a point-by-point replication of the data, but, rather, that the behavior of the model replicates trends (metabolite production going up or down) for most of the variables. The three main elements of the model—data fit, parameter values, and pathway structure—are mutually constraining, since they are tuned together in an incremental and iterative process until a model is considered validated. The pathway representation, for instance, is both tailored to fit the capacities of mathematical frameworks and shaped by parameter fitting in terms of available parameters and of the estimation tools used. All of these elements are kept in dialogue throughout the model-building process. Every version is *"just a version of your knowledge at the time—of what you think is going on. And it will keep changing as you learn more and more about the system."* In the end, this modeler noted, *"the best your model can do—is a verifiable hypothesis about what you think is going on."* That is, the objective of ISB modeling is to build a computational model of the target biological system that exemplifies its behavior under selected conditions, which in this case means that it replicates the existing experimental data and predicts new data that experimentalists can verify. At the completion of the model-building process, the goal is to have a robust model, stable for a wide range of parameter values, from which to derive novel behavioral predictions that have sufficient warrant to transfer as hypotheses to the target system, and, hopefully, will be tested experimentally by collaborators. As all of the modelers pointed out, making predictions—not just fitting the available data—is the only way to get past the underdetermination of a model.

In sum, modelers assemble the structure and local dynamics of the system being modeled largely from scratch by gathering empirical information from a variety of sources and piecing it together into an effective representation using a variety of assumptions, abstractions (modelers noted especially simplifications and approximations), and mathematical and computational techniques. Each modeler chooses the methods and strategies he or she thinks best to solve the problem without any formal procedure governing the selection process. Similar to the way a bird will gather whatever is available to build a stable nest, a modeler pulls together bits of biological data and understanding, principles developed in molecular biology, mathematical and computational theory, and engineering principles from a range of sources in order to create stable robust simulations of the behavior of a biological network ("bird-nesting process"). Modelers rely on the building process, especially their ongoing simulations, to come to understand their systems and adapt their representations of them to their specific epistemic goals. Thus, modelers rely on the dynamical behavior of the model, itself, to make inferences about how to proceed in building both the pathway network and the model. This important role of simulation for the modeling-building process has not received sufficient attention in the literature on the epistemology of simulation.[7] A major benefit of ethnographic investigation is that it can uncover the hidden creative work modelers carry out with the choices they make in model-building, as well as the ongoing processes of developing epistemic warrant for the model and the model-building practices, which are unlikely to be included with the formal analysis presented in a publication.

Building computational simulation models in lab G requires a sophisticated grasp of mathematics, computational methods, and systems engineering analysis methods. It is notable that students begin their research with little to no prior experience in biosystems modeling. We followed three of the graduate students intensively, two (electrical engineering background) from near the start to the finish of their dissertation research (~four years) and one (telecommunications engineering) during the course of her first year, in which the lab director gave her projects to help out on, which was his usual training procedure. We also conducted numerous interviews with the other graduate students and the postdoctoral researchers, including about the algorithm development work, and with two experimental collaborators. We were able to grasp enough of their model-building practices to inform our research questions. Section 5.2.2 briefly outlines the model-building processes of one graduate student we were able to track from start to finish, to provide an exemplar of how researchers in this field achieve their epistemic aims. We did not anticipate that he would make a significant biological discovery. G10's model-building process is typical of the nature of the problems lab G modelers address and the strategies they use in handling problems. As noted earlier, there are a wide variety of modeling practices in ISB, but, in general, lab G practices are representative of practices in the field that use ODE models. It is a remarkable feature of their modeling practices that engineers with little knowledge of biology, and none of the system under study, are able to construct models that not only replicate the available data but also produce highly specific verifiable predictions about complex biological systems. The exemplar demonstrates the need to examine the *processes* of model-building, which to a large extent cannot be gleaned from published papers, in order to develop an account of the epistemic and cognitive affordances of computational simulation (section 5.3).

5.2.2 A "Model-Based Signal Postulate": Finding a Remedy for Lignin "Recalcitrance"

G10 has an undergraduate degree in electrical engineering and a masters in bioengineering. For his MS degree he had worked on a bioinformatics modeling project for which he took a couple of biology courses (without labs), but he was not familiar with systems biology modeling when he arrived at lab G. He had read a biosystems modeling text by the lab director before deciding to apply. G10 had started on his dissertation project shortly before we entered the lab and finished in four years. In our initial interview, G10 stated that his "*engineering background contributes a lot to my way of thinking to solve a problem.*" He contrasted his engineering perspective, derived, in particular, from control theory, with that of a biochemist as evidenced in "*the biological journal literature*": "*[Engineers] look at things more at the systems level than the individual level, . . . Biochemistry look at the single protein or the single pathway—they don't really look at the whole system and how each pathway will interact with each other.*" He considered the systems perspective essential "*if you really want to understand how the human works or how the plant works.*"[8] G10's project started with a request from biofuels industry researchers for the lab to help them figure out how to tweak the lignin pathway in alfalfa to develop transgenic plants with lower lignin, so they could more easily extract sugars for the production of biofuels. Lignin is a natural polymer that hardens plant cell walls and enables the plant to grow upright. It is difficult to break down (it exhibits "*recalcitrance*") when biomass is processed into fermentable sugars using enzymes or microbes. The experimentalists had been developing genetically engineered plants with lower lignin content but were finding it difficult to determine a balance that would keep the plant structurally sound. Further, their transgenic species decreased only one of the three lignin monomer building blocks (called monolignols H, G, S). Although they had not collaborated with modelers before, they felt that modeling might be able to help them understand something about the mechanisms underlying lignin production, which would enable them to develop transgenic species with low lignin content and good growth. They also hoped, at the very least, that modeling would provide information that would enable them to develop plants with different ratios of lignin monomers, especially a lower

S/G ratio, which would improve the extraction of sugar from plant cellulose. This was a new modeling area for systems biology. Other modeling efforts in the biofuels domain were in the area of bioinformatics, or were models of organisms that are used to break up the plant mass. G10 expressed the hope that *"model-based insights will become the foundation for the rational design of metabolic engineering strategies"* for biofuel production.

G10 described his own "bird-nesting process," generally, as follows: *"We just search the literature and find the necessary data from it. . . . But most of the time you don't have much data. . . . I need to, you know, add other components from other theories, for example, the flux balance analysis . . . and I combine that with biochemical systems theory to build a model."* One unusual feature of this case is that G10 had a fairly well-established lignin pathway in the literature to start from and, in the end, lots of data for the fitting process, though he still had a considerable number of open parameters. To carry out parameter fitting for the specific lignin system he was working on, he needed to develop several novel modeling strategies for his analysis, one of which he also published separately as a potential community resource for handling systems of this kind.

In the beginning, G10's collaborators gave him few data, and what they did give him was of poor quality for modeling. As he noted, *"They don't measure the concentration, for example. And they have few kinetic data. . . . Most of the data they have is just output, the final output."* This created a significant problem because there was little literature on alfalfa, the plant they were working with. Complicating things further, they were unresponsive: *"Sometimes you want to ask question, and he would get back to you in a month—or even two months—or even don't reply. . . . That's a problem because we are not expert in the field. . . . They have more information than we know from the literature."* This is not an unusual "collaboration" situation for ISB modelers. Even when the bioscientists request the modeling, it often is low priority for them. Importantly, these bioscientists were unwilling to part with unpublished data, which constituted the bulk of their data. They seemed not to understand that the modelers would use it only for building the model and would not publish the data: *"Right now they just give us the data they have published. . . . They told me they need to publish it first—and then they can give me the data later."*

The collaborators projected it would be about six months before they would give G10 the additional data, so he decided to build a model of lignin biosynthesis in poplar—a related species for which there were ample data in the literature, because it is the preferred biofuel species in Europe. His idea was to build the poplar model as a *"proof of concept"* for biosystems modeling in that domain, which would also help him understand the lignin pathway better. He assumed some of what he did would transfer to the alfalfa case. It turned out, unexpectedly, that to build the poplar model he needed to develop what he called *"a new two-step modeling approach"* to deal with the mathematical complexity and parameter estimation for the lignin pathway. He thought this novel method might then provide a template for modeling in the lignin domain. His approach was to integrate dynamics models with fluxes (the rate at which a metabolite is processed) derived from constraint-based models. The two steps were first to build a static, constraint-based model, which assumes the metabolic system is in steady state, and then use flux information derived from that to build the dynamic, kinetics-based model. The static model used the flux balance analysis (FBA) method, which assumes that the metabolic system is in a steady state in which, for each metabolite, the sum of fluxes coming into the pool equals the sum of fluxes coming out of the pool. The dynamic model made use of the BST modeling framework, where each differential equation in a model represents the time-dependent change in one metabolite as the sum of production fluxes minus the sum of degradation fluxes. For this model, G10 used the BST framework's generalized mass action (GMA) representations, which model the flux as a sum of the inputs minus the outputs.

The attractiveness of the BST framework in data-poor modeling is that it can account for a variety of dynamics by modeling the flux of dependent variables as a product of power law functions (relative change in one quantity gives rise to a proportional change in another), with each individual flux represented separately with one power law function. This means that even if the nature of the interactions among elements is not well known for the system, there is a good chance the model will capture the underlying causal regularities in the system, and so account for the system dynamics within the range of the realistic parameters. The parameters used are the rate constant, which determines the turnover rate of the process, and the kinetic order, which characterizes the influence of one variable on a given process.

The two-step process still left G10 with twenty-seven open parameters and required using optimization strategies to fit. To reduce the parameter space, G10 set all but the parameters considered significant (small change leads to large change in S/G) to what he considered *"biologically reasonable"* values (determined to be so from reading the literature and discussion with the director). He then optimized the significant parameters by using various computational techniques. The results of this approach generated an ensemble of models (there was no unique model) with minimal error (SSE: sum of squared error) between model results and experimental data. Our modelers use "ensemble" to refer to a small group of models with different parameter settings that they settle on to cover uncertainties in the parameter values.[9] These models enabled G10 to identify key reactions that influence the S/G ratio in poplar, and he was able to make some predictions about how the pathway might be tweaked by knocking down specific enzymes to lower the S/G ratio.

The process of building the lignin model for poplar prepared him to deal with the more complex modeling problem presented by the alfalfa lignin system. The alfalfa model would contain twenty-four ODEs. In addition, the collaborator data included points in the growth of the plant over time (eight different internodes), where the lignin levels were different for each of these points. G10 used a slightly modified two-step procedure to analyze several internodes simultaneously, while interactively building the model and modifying the pathway in an incremental and iterative process.

Once G10's collaborators had published the relevant alfalfa research, they gave him the Excel files for all their data—which meant that, unlike the typical case, he had "*many data . . . for seven transgenic experiments and each experiment generate about seven sets of data. . . . They have more data than we need to know.*" The collaborators did not give him any pathway structure, but again he was fortunate: "*The [generic lignin] pathway structure is from the literature—everybody is using it.*" But, as he discovered, species-dependent data would be important in building out—and significantly altering—that lignin pathway, which had been established for twenty years. At the outset, his own literature search led him to add new elements to the pathway network, noted in red in figure 5.4.

The first model he built was for a wild-type system at steady-state, using the modified pathway (left diagram, figure 5.4). G10 discovered that this model could not produce accurate data when inputs were perturbed out of equilibrium, which suggested to him that some regulatory mechanisms controlling excess flux needed to be figured into the pathway. He tried out several pathway variations from studying the model structure with simulations, and then selected those that were the most "*biologically reasonable*"

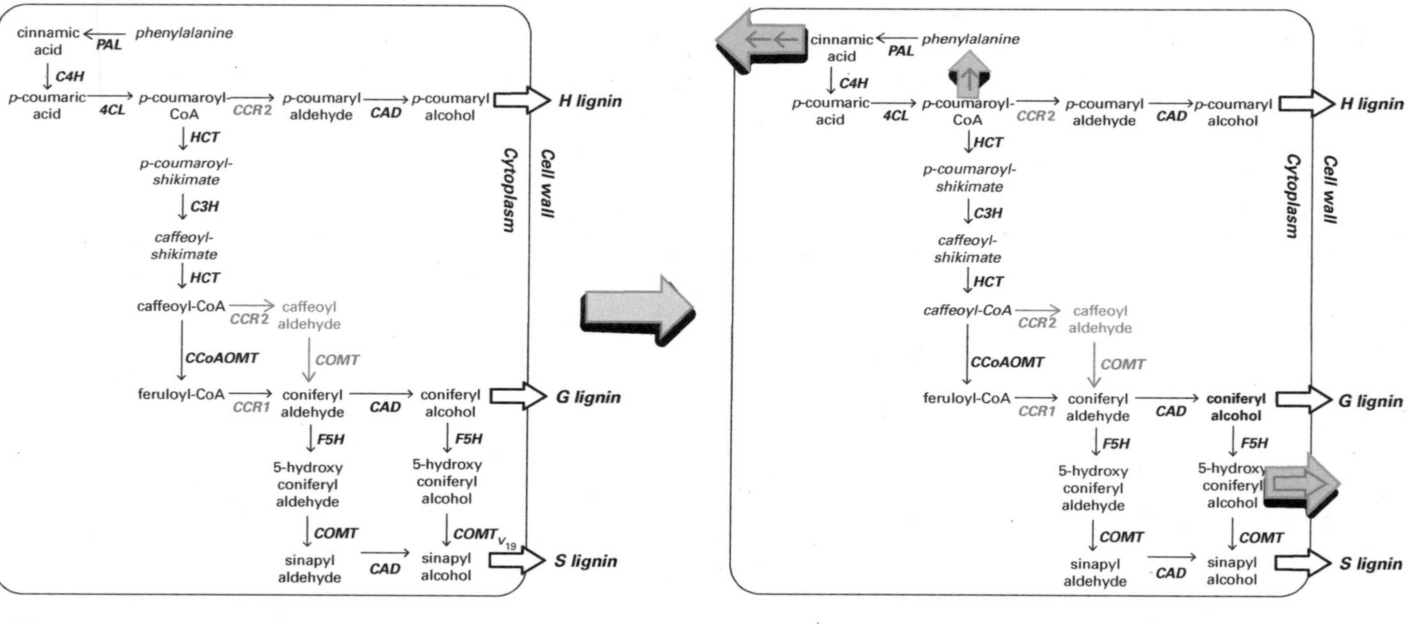

Figure 5.4

G10's initial modifications to lignin pathway. The left diagram shows additional reactions (in red) that lead to the H, G, and S monomers. In the right diagram, the highlighted blue arrows connect the extra flux to the environment at the points G10 hypothesized it leaves the system.

ways of removing flux (*"overflow fluxes"*: highlighted blue arrows in the right diagram, figure 5.4). He translated these into precise mathematical modifications that would relieve the system. As he explained his process, *"We have data from our collaborators and we analyze it with very simple linear models, and based on our analysis results, we suggest there—this original pathway needs to be modified so that this data can be explained. . . . This is an important piece of knowledge that comes from the model,"* that is, through the understanding of its system dynamics provided by the model. With the new pathway structure, G10 was able to build a dynamical model for each internode in each wild-type or transgenic plant and make hypotheses about the metabolic control of this pathway.

He used the data for each of the seven transgenic plants to build models on the biological assumption that the genetically modified strains would function as close to the wild-type as possible, within the limits imposed by the modification. Fitting the models was again a complex process, with numerous open parameters for each model, which he handled in a manner analogous to that of the poplar model. In the end, for the final modified alfalfa pathway, G10 arrived at a consistent convergence of five optimized models that tested well, and each gave similar predictions. He argued that the fact that this ensemble of models converged on similar mathematical relations for the target variables *"provides validation"* for the model. These models provided specific new causal information about which enzymes could potentially be targeted to decrease the S/G ratio, but did not provide an overall mechanistic explanation for the system behaviors. Altogether, G10 arrived at what he called seven *"model-based postulates,"* which are mapped out on his final representation of the pathway (figure 5.5).

Two important postulates are, first, the reversibility of some reactions (straight arrows pointing upward in figure 5.5A and B) in the path where he had earlier removed excess flux. Second, he hypothesized the possibility of independent pathways ("channels") for synthesis of G (blue) and S (red) monolignols. Channeling can make a metabolic pathway more rapid and efficient, and the potential role of these channels in the lignin pathway was an important new hypothesis. He offered this second postulate as a solution to what he called a *"puzzle"*: given the data he had on up-regulation and down-regulation of specific variables, the S to G ratio was considerably higher in transgenic plants than in the wild-type. But now he claimed the model enabled him to *"see what happens inside the pathway,"*

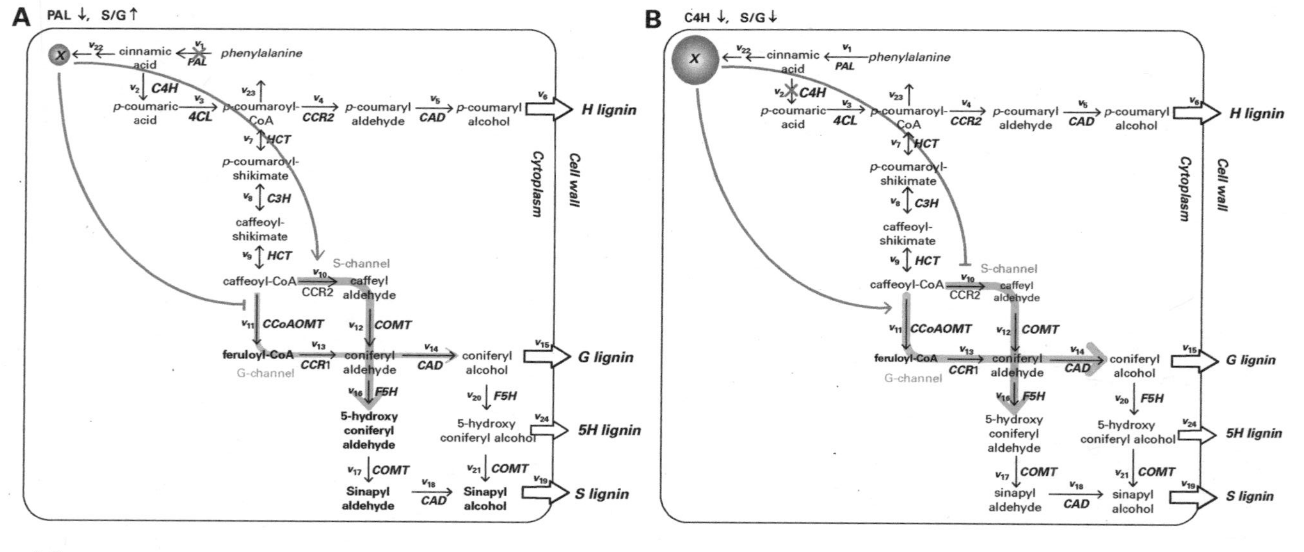

Figure 5.5

The final lignin pathways for the alfalfa model, which includes G10's most significant "model-based postulates." The upward-directed straight arrows on each indicate the reactions are reversible. The red and blue thick lines indicate channels that lead to the S and G monomers. The thickness of the lines indicates the size of the channels, which determines the rate of the reaction. The purple lines indicate the influences of the hypothesized entity X. The curved purple lines ending in arrows indicate a positive influence; those ending in a horizontal bar indicate a negative influence. The size of the circle around X indicates the concentration of X. The red cross indicates the reaction is blocked.

and this postulate made biological sense in addition to making the model work. However, it led to a significant new problem: based on hundreds of thousands of simulations of possible variations of the channelization, these channels appeared unlikely to be regulated by enzyme kinetics. G10 carried out this examination of all possible experimentally supported pathway designs with another novel method he developed of computational enumeration *"to permit an expedient and exhaustive assessment of alternative regulatory schemes."*

Based on the dynamical behaviors of his model, G10 inferred that the easiest way to resolve the problem in the model was make a spectacular biological hypothesis: the established lignin pathway of twenty years is importantly incomplete in that there appears to be an element *outside the current pathway* that has a significant regulatory effect on its behavior. This element would selectively regulate (figure 5.5 curved purple arrows) the pathways (channels) responsible for generating S (figure 5.5A) and G (figure 5.5B) lignin. Because of his limited knowledge of biology, he called the element "X" and had no way to hazard a guess as to what it might be. The postulate is warranted on the basis of the model: if excess cinnamic acid produced a substance X that both up-regulated the G channel flux and down-regulated the S channel flux, then the model produced highly accurate dynamical behavior. His postulation of a heretofore unknown metabolite in the lignin pathway derived from the understanding that the model-building process provided of the quantitative dynamics within the network and of how to control the parameters effectively. As he stated, *"So this is actually the biggest finding from our model. So, by adding this reaction you can see that we hypothesize there is another compound that can give a regulation . . . to other parts of the pathway. And this finding will not be possible if we haven't done any modeling—because, well, if you just look at the data, the data only tells you the composition of these three lignin."*

The model-building process gave G10 a comprehensive view of how the existing data on lignin fit the pathway structure and led him to question that structure as a last resort, because the model dynamics appeared to require it. This prediction finally got the attention of his collaborators, and they conducted experiments that confirmed the hypothesis, identifying "X" to be the signaling molecule salicylic acid. They determined that the molecule acts as an inhibitor of monolignol biosynthesis, which was hailed as a significant biological discovery. G10 also predicted, *"I guess this finding will give them more confidence in what we are doing so maybe in the future they could be more willing to give us—to share more data."* This prediction was also borne out, in that they went on to collaborate further with G10 in postdoctoral research.

The G10 case leads us to a more general question about computational model-based reasoning in ISB: *How is it possible for an engineer with a few months of biosystems modeling experience and little knowledge of biology to make fundamental discoveries in biology?* To answer this question, we need to fathom how "discovery" is the outcome of processes that create cognitive-cultural systems that extend the capabilities of scientists beyond their basic human limitations to more effectively probe the natural world. Our investigations into discoveries made by in silico simulation modeling, as with in vitro simulation modeling, show these are the epistemic achievements of complex evolving distributed cognitive-cultural systems. For lab G, the distributed problem-solving systems comprise the modeler, model, lab director, other lab members "*grabbed*" for discussion, model-building resources specific to the culture of the lab (here, PCs, ODEs, BST and so forth), conceptual and methodological model-building resources from engineering and computational sciences, epistemic norms and values, experimental collaborators (even if interaction is limited), Internet resources (search engines, data bases, literature), diagrams (pathway, graph), "pen-and-paper" representations, and presentation and publication venues, which provide community feedback. As with all D-cog systems, these systems have properties that are different from those of the individual.

In what follows, I focus on epistemic affordances of specific components of the distributed model-based reasoning system, namely, the coupled system of interaction between two kinds of models, researcher mental models and computational simulation models. I consider ways in which the processes of building the artifact model enhance the inferential powers of the researcher. The back-and-forth interactions between these components of the coupled system create changes that are particularly important to account for the ability of the D-cog system to improve its investigations of a given biological system. I then address the nature of the warrant for believing the outcomes of sufficiently credible models are worthy of pursuit as hypotheses about target systems, which all the modelers noted is an important epistemic aim of these D-cog systems.

5.3 Computational Model-Based Reasoning: Building "a Feeling for the Model"

Interest in the methodology of computational modeling and simulation in science has been growing in the philosophy of science. There is now a substantial body of philosophical research that focuses primarily on physics-based modeling, such as conducted in quantum mechanics, nanoscience, and climate science (see, e.g., Galison 1997; Humphreys 2004; Lenhard 2020; Parker 2009, 2010a,b; Winsberg 2010). These analyses have produced important insights, some of which do pertain to what we have been learning about computational simulation across the board. However, as we argue, there are important differences, as I indicated earlier.

In general, the characterization of computational simulation models Eric Winsberg (Winsberg 2001) has formulated as downward, motley, and autonomous is widely accepted. "Downward" signals that established scientific theories provide the starting point from which to develop a computational simulation model, and that they contribute to the credibility of the model and to the warrant for the belief that modeling outcomes can be transferred, provisionally, to real-world phenomena. "Motley" indicates that the model-building process introduces arbitrary elements that work against any claim that the model is fully derived from theory. To build a stable, robust model requires using a range of such elements, which include abstractions, parameterizations, ad hoc assumptions, mathematical tricks, numerical methods, and much trial and error. In view of their motley nature, in particular, Paul Humphreys (2004, 148) has dubbed complex physics-based simulations as "epistemically opaque" (see also, Lenhard 2007) This means that although they begin from theory and depend on it, the ingredients needed to make a simulation work obscure the operations of the theory's laws and make analytic solutions to equations impossible. Thus, a model can be theory-driven, but in an important sense it is autonomous from theory (see also, Morgan and Morrison 1999). "Autonomous" (or, better, "semi-autonomous") in Winsberg's characterization, also underscores that simulations, customarily, are used in situations where data are sparse because real-world experiment and observation are quite difficult or not possible, and thus simulation provides a source of predictions and understanding that often cannot be checked against—or warranted fully by comparison with—data from real-world sources.

As we have seen in the previous sections, our research on computational modeling and simulation in ISB agrees with the motley characterization (which we likened to a bird building a nest) and the autonomous nature of computational simulation. However, as we also have seen, model-building is not a "downward" process; rather, lacking a theory of the system phenomena, models are built "from the ground up" (MacLeod and Nersessian 2013).[10] There are differences to be discerned from practices that lack a theoretical basis from which to draw resources to build models that are important for understanding how modelers achieve their epistemic aims.[11] One such difference derived from our analyses of ISB practices is to bring out additional, different roles for simulation than have been discussed in the physics-based literature. An important insight from our investigation is that simulation in this domain contributes to building the pathway representation and, so, to the model-building process itself. These and other findings I discussed in the section on general lab G modeling practices underscore the benefit of collecting ethnographic data on the model-building process as it is going on, rather than just relying on published scientific literature, augmented possibly with archival records, retrospective accounts, and anecdotes. There is much that is important for understanding how computational simulation affords epistemic access that is omitted from final reports or not recalled retrospectively.

Much of the recent philosophical literature on computational simulation focuses on issues about whether a new epistemology of science is needed to accommodate computational simulation as an investigative practice or on whether simulation experiments are the same as or different from wet-lab experiments (see, e.g., Beisbart 2018; Frigg and Reiss 2009; Winsberg 2009). These are interesting and important issues, but rather than address them, I consider a largely neglected issue that is important to the discovery question I raised at the end of the previous section. This important issue has only been hinted at in the philosophical literature: the need to bring considerations about human cognition into the epistemology of simulation. Humphreys, for instance, has cast the situation in which science is conducted at least partially by computers as a "hybrid scenario," by which he means "one cannot completely abstract from human cognitive abilities when dealing with representational and computational issues" (Humphreys 2009, 616).[12] Witness, also, the main title of his book, *Extending Ourselves*. Humphreys argues that computational methods and simulation belong to a long line of "technological enhancements" that scientists have developed as tools to extend human capabilities. Some of these have targeted a specific modality, such as microscopes and telescopes, which have enhanced our native abilities to see. Computational simulation was developed to deal with the problem of processing vast amounts of data, which human cognition cannot. Computational technologies, as Humphreys claimed, provide "enhancements our native cognitive abilities required to process this information" (2009, 8). I agree, and would add, to make inferences from it. However, his claim is not backed up with any account of the nature of the native cognitive abilities that are enhanced by computational technologies—and how they are enhanced. "Extending" is left only as a metaphor.

As I indicated in chapter 3 and as we will see even more so here, there is a significant difference between Humphreys' tool view of extension by means of computational simulation and the coupled system view we have been advancing. In the tool view, over the course of science, scientists have been extending their sensory and cognitive abilities by creating instruments (e.g., telescopes) and analytical tools (e.g., models) that allow them to *use* the artifact as a tool to perform new operations, such as the fast numerical solutions to complex equations performed by computational simulation models. This view suggests the cognitive-cultural divide, from the cognitive side: the individual is able to perform different cognitive tasks—or do them better—using the new artifacts. When our perspective shifts from using a simulation model to building this artifact, we come to understand how the back-and-forth interaction with the human agent incorporates the computational model, along with other elements of culture, such as conceptual and methodological resources and epistemic norms and values, into a *hybrid, coupled human-artifact model-system that accomplishes simulative model-based reasoning*. This construal is compatible with the notion of models as "epistemic tools" (Knuuttila 2005) but focuses attention on the processes of building and incorporating the tool rather than using the final product. "Extending ourselves," in our account, is an iterative and incremental process that incorporates humans and the epistemic tool into a cognitive-cultural system with properties different from those of the individual. This process provides an example in the domain of science of what Hutchins meant more generally by "humans create their cognitive powers by creating the environments in which they exercise those powers." These system-level properties of the "hybrid scenario" facilitate epistemic access to otherwise inaccessible processes in complex biological systems. We interpret, then, Humphreys' claim that "in extending ourselves, scientific epistemology is no longer human epistemology" (2009, 8) as meaning that *scientific epistemology is the epistemology of a D-cog system* (not only in the case of computational modeling, as we have seen in previous chapters). In the following sections I elaborate on the epistemic and cognitive affordances of building in silico simulation models that I outlined in chapter 3. Specifically, I consider how the modeler's inferential capabilities ("cognitive powers") are extended in model-building processes and the warrant for their claims that these processes can provide epistemic access to the behavior of the biological systems.

5.3.1 Extending the Capacity for Simulative Model-Based Reasoning

As I have discussed in previous chapters, analyses within the D-cog framework customarily cast the human component of a system as "off-loading" cognitive functions to specific artifacts and "coordinating" among system components to accomplish a task. These metaphors, even when explicated in terms of specific tasks, are insufficient to understand how the scientific D-cog system *improves* its ability to investigate target phenomena. Such improvement is driven by learning on the part of the human component, which in turn leads to the further development of the artifact model. We have argued, based on cognitive science research and our own data, that this kind of learning involves building more accurate mental models. We have, thus, cast model-based reasoning with in silico models as a system of interaction—a coupling—between two kinds of models (mental and artifact), which creates changes in the D-cog system that improve its ability to investigate, in the case at hand, complex biological systems.

The reasoning by the modelers captured in our interviews and observational studies, as well as self-reports of their reasoning processes, provide evidence that many of the inferences they make in the course of building a computational model, especially with respect to how and where to modify it, rely on simulative mental modeling. The modelers we have studied across both ISB labs articulate their reasoning in terms of causal interactions in the biological networks, which, we claim, allow them to simulate and perturb limited aspects of the network dynamics mentally. They have walked us through how these simulations—often performed in conjunction with pen and paper or whiteboard representations (see, e.g., figure 5.3)—enable them to perform various kinds of reasoning tasks, such as to identify possible errors, explore hypotheses about network structure or parameters in a limited fashion, and identify dominant variables. These simulations enable the modeler, in particular, to screen plausible candidates for fixing errors in the structure of the model before implementing them. This ability is important, because errors in the structure of the system model can have numerous causes and be in numerous locations, so many different manipulations of the computational model might resolve them. The modeler's ability to screen candidates by limited mental simulations cuts down on the work of parameter fitting, which, as we have seen, is a highly labor- and time-intensive process. Our findings are in line with cognitive science findings about how scientists and engineers use mental simulation as they try to solve problems in their research (Christensen and Schunn 2008; Trafton et al. 2005; Trickett and Trafton 2007). Of particular note, that research shows that the use of such mental simulations increases in cases of inferential uncertainty when scientists are trying to develop a general grasp of the phenomena under investigation.

There are three aspects of the character of the simulative mental models our study participants build that we have analyzed as especially significant. First, from the way they reason out loud with their models, we infer that their mental models are qualitative. Modelers, for instance, track qualitative effects of specific variables on other variables using terms like "*increasing*" and "*decreasing*" to describe these relations, such as "an increase in variable A produces a decrease in variable B." Modelers often do sketch out on paper or whiteboard some quantitative details of what they are thinking, but they do not compute precise numbers and values in these activities. Our characterization of their mental models as qualitative is consistent with a range of cognitive science research, especially studies of causal-mechanical reasoning by physicists and engineers (see, e.g., Roschelle and Greeno 1987; DeKleer and Brown 1983).

Second, also in accord with the cognitive science literature, modelers reason about the pathway networks and models in piecemeal fashion in interaction with pen and paper representations (Roschelle and Greeno 1987; Hegarty 1992, 2004; Schwartz and Black 1996). For instance, they track only a limited number of interactions in the network mentally or make inferences about the consequences of manipulating the values of a limited set of variables to explore what might be the effects of specific modifications to the computational model. As one modeler recounted, the modeler *"has to visualize the pathway in his head and divide it up into parts and write codes for each part,"* which is why the modeler *"draws so much and uses so much paper."*

Third, and likewise in accord with cognitive science research, modelers appear to reason by carrying out simulations with these piecemeal models. Research on nonexpert reasoning about simple mechanical pulley systems (Hegarty 2004; Schwartz and Black 1996), for instance, establishes that participants reasoned by carrying out simulations of intermediate pulleys in the system, which facilitated their ability to reason over a larger scale. This strategy is consistent with constraints on working memory that limit how much information can be processed at a time. For modelers, these constraints mean that they should be able to track and manipulate only a limited number of variables at any one time, which accords with our research. We have seen modelers use selective and piecemeal representations of the system, for example, to identify and bracket nonlinear relations into separate behaviors and simulate each separately to make inferences. In their mental simulations, modelers usually focus on elements of the pathway network that interact directly, but are not necessarily contiguous. These qualitative simulations of pieces of the network help them to understand the qualitative effects of the quantitative mathematical relations represented in computational model as they build the model. In the cognitive literature, such qualitative simulations have been called "envisioning" (DeKleer and Brown 1983). As one modeler described her envisioning process in building an intuition about her model, *"So the thing is—when you want to solve a mathematical problem . . . sometimes you use numbers and try numbers, something to give you a feel of—like intuitively how this, for example, equation works and all. So, I'm trying out numbers and then trying to make the steps kind of discrete—like sort of a state machine, kind of thinking like we're in this state. And then, now this much is going to this other metabolite pool and then, at the same time, we have less of that. So, I'm trying to see what the constraints are by actually like doing a step-by-step sort of thing."* While she was describing this to us, she was also using her finger to point to and trace out her sketches of these *"steps"* sketched in her notebook.

As I discussed in chapter 1, some cognitive scientists have proposed the way to understand how mental models and external representations work together during reasoning is as *coupled inferential processing* (see, e.g., Greeno

1989a,b; Zhang and Norman 1995; Gorman 1997; Hegarty 2004; Nersessian 2008). However, unlike the case of coupling between mental and static artifact representations considered in this literature (mainly diagrams), in the case of computational representations, both kinds of models have their own simulation capabilities. Our extension of the coupling proposal to include in silico models proposes that the incremental and iterative processes of building and simulating the computational model create a key change in the D-cog system. Namely, the process builds a close dynamic coupling between the modeler's mental model and the artifact model that incorporates modeler and model into a powerful *simulative model-based reasoning system* that significantly enhances the limited human capability to reason about the behavior of complex biological systems (Chandrasekharan and Nersessian 2015; MacLeod and Nersessian 2018). Notably, the coupling enhances the human cognitive powers used in mental modeling, such as memory, information synthesis, visualization, simulation, abstraction, imagination, and intuition. In the way we propose to understand the model-building activity, *cognitive functions are not off-loaded to the computational model, but are enriched and extended into a coupled system by virtue of it.*

As we saw, the computational model can integrate a vast amount of information from disparate sources. Further, computers have the capacity to process complex systems of quantitative representations, such as the twenty-four equations needed to build G10's model. Their speed and manipulability enable the modeler to implement changes quickly and efficiently so that he can run through pathway options or hypotheses in quick succession. The computational model can generate many kinds of visual representations, such as graphs to track only specific relations or three-dimensional visualizations to track dynamic system behaviors. The choice depends on what the modeler thinks most useful for the problem. The computational model can, also, be put through thousands of simulations of many configurations in a matter of seconds. Configurations that use, for instance, different time points or parameter values produce different network behaviors that the modeler can partition into families of mental models, which can be used to build her intuition about the behavior of the model and develop insight into how to proceed with the building process. In addition, numerous and diverse simulations enable the modeler to develop a holistic, global perspective on the system dynamics. Model simulations, in addition, enhance the modeler's ability to think about possible worlds and make counterfactual inferences in ways that outstrip her capacity for thought experimenting alone.[13] The model's representation in variables, in particular, promotes such counterfactual explorations. Thinking in variables, too, helps to build what the lab G director called *"the flexibility to recognize shared features of control/regulation across disparate domains."* This kind of cognitive flexibility allows the modeler to move with relative ease from modeling yeast to modeling cancer, and so forth.

The overall effect of the back-and-forth exchange between these components of the D-cog system is to extend human inferential powers such that the modeler can make reasonable inferences about how to build out the pathway or improve the parameter fit of the model. The model-building process is not always successful. However, a stable and robust model (or model ensemble, such as G10 developed) can lead to hypotheses about how to understand the system-level behavior in the target or how to manipulate it, such as the significant and novel *"model-based postulates"* made by G10. G10's major biological discovery did not involve gaining expertise in biology (thus the designation "X" for the unknown biological entity) but did involve developing confidence in his judgement in the warrant for the inference that computational model did enact the behavior (exemplify) of the in vivo system (hypothesis transfer)—a confidence buoyed from numerous iterations of model-building and simulation. In the end, he could postulate with confidence that some heretofore not considered element is part of the regulation of the lignin pathway, because the addition of it produces stable dynamic behavior in the model. The warrant for making such a bold move derives from the processes in which G10 created and examined numerous model variations and found that every plausible biologically reasonable change other than this one fails to provide a good fit.

To verify, hopefully, this hypothesis and determine what "X" is required action by the collaborator component of the D-cog system, and, as we saw, they were scarcely involved beyond supplying data. Here we see that another epistemic affordance of simulation modeling in ISB is to enhance collaboration, when biologically plausible hypotheses intrigue experimentalists sufficiently to pursue them. Every experimental collaborator we interviewed expressed a degree of skepticism about computational modeling, even when they had sought out the collaboration. The collaborators often complained that modelers seemed to want to build models *"for their own sake,"* and were content with just replicating data—sometimes very old data (*"who cares about that?"*)—which the experimentalist characterized as a *"tautology."* As an experimentalist stated in discussing another lab G modeler's work with us, *"I think it's absolutely essential for anybody who is going to model to build in a step of their modeling where they test its predictive power. . . . If we get some answer [experimentally], as she did, I'm going to have a lot more confidence in your model."* Having confidence in the possibilities of modeling on the part of the experimentalist is prerequisite to effective collaboration. As we saw with G10, once his collaborators had that confidence, they actively pursued further, more engaged, collaboration.

In all our investigations, we encountered the claim by computational modelers that it is of great importance to develop *"a feeling for the model."* Our analyses interpret this "feeling" as having several dimensions. One aspect refers to the intuition and confidence modelers develop about the behavior of the model through the coupling process, as well as about their ability to correct deficiencies in the desired direction of a stable and predictively robust model. The central role modelers ascribe to developing a feeling for the model in order to make progress underscores, for our analysis, that although model-based reasoning is carried out by a coupled inferential system, specific attention needs to be paid to the human component. Ultimately, it is the human agent who has to draw the inferences about how to proceed to improve the model or to flesh out the potential implications of the model's behavior, as well as possess confidence in the direction they choose. As such, the level of complexity a modeler can handle likely constrains the size of the models that are productive for the modeler to attempt to build. This consideration provides an important additional rationale for the mesoscopic modeling strategy that Voit et al. 2012 have observed to be prevalent in ISB.[14]

We think this phrase, too, is an important indicator of how the processes of building the simulation model provide insight and understanding about the target biological system; the "feeling for the model" in turn provides the modeler, by analogy, with a "feeling for the biological system." The model-building process gradually builds intuition about the dynamics of the target behavior through a large number of iterations of simulations wherein a range of factors such as sensitivity, stability, consistency, computational complexity, and so forth are explored. In the process the modeler, interactively with simulation, builds out the structure of the pathway network, which delineates a sequence of causal interactions among the elements of the biological pathway. The simulations of the model's behavioral dynamics build an intuitive understanding of how the pathway generates the existing experimental data and what interventions might be made in the target while its stability is maintained. Simulations can also be used to explore why other sets of values are not (or have not been) seen in real-world systems, which can provide the modeler some insight into "design principles" that underlie the values seen in in vivo experimentation. These repeated interactions with the model, pathway, and biological literature develops the modeler's capacity to judge the biological "reasonableness" of the hypotheses or predictions they make about the system.

In order for the experimentalist to intervene on a target system, the model does need to provide specific causal information about the system. As we saw in the case of G10, he made causal predictions, based on the behavior of the model, about what enzymes might be knocked down to lower the S/G ratio but maintain structural integrity of the plant and predicted by what percentage these knockdowns would decrease the natural ratio. How to tweak the lignin pathway was the objective of the modeling, but the model also enabled an unanticipated, even more significant, causal prediction (figure 5.5): if cinnamic acid (postulated flux leaving the system) produced a compound ("X") that both up-regulated the G-channel (pathway) flows and down-regulated S-channel flows, then the model would produce highly accurate dynamic behavior in accord with the existing experimental data. In our discussion with the lab director about this case, he pointed out that such mesoscopic models can provide *"a certain level of explanation . . . something causal you didn't know before,"* pending, of course, experimental verification.

In general, as seen in the G10 case, the director claimed, *"if you can trace out a causal pathway, then it's an explanatory model even though you may not know every single detail."* However, this kind of explanation is not mechanistic because *"with every [such] explanatory model, you have some regression in there or some association. . . . It's not pure."* That is, the top-down abstraction strategies—such as those associated with canonical mathematical frameworks, shrinking and fixing the parameter space, and global fitting algorithms—used to build the model wash out or obscure details of the mechanisms underlying the causal connections. Nevertheless, the causal information provided by the model about the pathway structure (*"trace out a causal pathway"*) does provide a *"certain level of explanation"* about the dynamical behavior of the in vivo biological system, and, in particular, how, possibly, to manipulate it.

In most instances, the kind of understanding mesoscopic modeling provides is largely pragmatic—understanding about the target system sufficient to propose ways to manipulate or control it to attain desired outcomes, but not sufficient to explain its behavior fully. Lenhard (2006) has argued, with respect to a case he investigated in nanoscience, that there are instances in this kind of understanding in physics-based computational modeling too. In such instances, models begin from theory but the equations produced for the complex phenomena are impossible to solve analytically. Instead, "simulations squeeze out the consequences in an often unintelligible and opaque way" (612), because of abstracting and averaging techniques that fit the equations to the data, as well as the numerical methods that render equations computable. He argues that, even though such simulation models do not provide the kind of explanatory understanding one derives from laws, these models do provide understanding about how, possibly, to intervene, control, or manipulate the phenomena, and thus, pragmatic understanding.

Systems biologists often write that the goal of the field is to attain "systems-level understanding" of complex biological phenomena, which they cast in terms of theories that capture general mathematical features and properties of biological systems, from which models of individual systems can be derived (to the extent possible in physics) (see, e.g., Kitano 2002; Westerhoff and Kell 2007). We interpret this aspiration to mean systems-level understanding, eventually, should be not just pragmatic, as a capacity for manipulation and control alone, but a genuine theoretical or mathematical form of understanding from which the ability to manipulate and control would follow. This is the ideal scenario. In practice in the current state of the field—at least from what we have witnessed—the complexity of the systems and the constraints on model-building are such that modelers pursue more limited goals with respect to what they can learn about their systems. They make the pragmatic decision to pursue less detailed and robust models that, in principle, can be predictively accurate for only certain elements of the systems. The understanding such models provide is pragmatic also in content, in that they provide neither a higher-level mathematical/theoretical understanding nor a mechanistic explanation (MacLeod and Nersessian 2015). Thus, the situation Lenhard describes in some physics-based modeling is the current state of computational modeling in at least the area of ISB we have investigated, if not more widely.

Another aspect of the modeler's claim to have a "feeling for" the model is that it indicates a belief in the credibility of a validated model's predictions. In our interviews with modelers and in the presentations of their research that we witnessed, computational modelers (in lab G, lab C, and lab D) exhibit a high degree of confidence that their models, when fitted and rigorously run through diagnostic and cross-validation testing, produce simulations that do exemplify the dynamic behavior of the target systems. What warrants that confidence? In some cases, it will not be possible to conduct experiments on the system to back up this belief, and where it is possible, it often requires a considerable investment of time and resources on the part of experimental collaborators, so the modeler's confidence that predictive inferences are credible and worthy of pursuit needs to be quite high. Of course, this belief is fallible since models can be wrong even if they fit the available biological evidence, but modelers do express confidence that a validated model is correct "*in respects that matter.*"

5.3.2 Building Epistemic Warrant

The most detailed philosophical account of the epistemology of computational simulation modeling is that of Eric Winsberg (2010), based on an analysis of physics-based modeling largely as recorded in the published literature. As he points out, models are built in data-scarce situations to serve as alternatives to real-world experimentation, which in many instances cannot be carried out (think of colliding galaxies or climate change). He proposes that the credibility or epistemic warrant for a model rests on two pillars, which are related to his characterization of simulation models as downward, motley, and autonomous. The first pillar is the credibility of the theory of the phenomena, such as fluid mechanics, that informs the building process (downward). But the methods required to build a stable and robust model always introduce extraneous and arbitrary elements into the process (motley), which give it autonomy from theory. So, the epistemic warrant also, importantly, derives from the second pillar, the credibility of the methods used in building the model. As we have seen, the case of ISB modeling is importantly different with respect to these sources of credibility. There are no guiding theories of the biological phenomena under investigation. Instead, modelers assemble the network of reactions and regulatory relations among elements (in our case, metabolites and signaling molecules) of the system in conjunction with literature searches and preliminary simulations of the model as they are building it. To build a computational representation, they use bits of what they sometimes refer to as "theory," such as enzyme kinetics, and canonical frameworks, such as BST, that provide a possible structure by which to glue together the lower-level information. With respect to the various methods used to build the model, these have, largely, been developed to model human-made systems. These methods have considerable credibility for modeling those kinds of systems, but are, here, being used on living systems—natural, modified (e.g., genetically), or engineered (e.g., synthetic).

In the physics-based modeling fields Winsberg considers, the methods do have considerable "antecedently established credibility," in that established "disciplinary tradition" supports their reliability in application to new cases—that is, they are "projectible" (2010, 137). Further, there are, of course, computational techniques related to fitting numerical models generally, such as Monte Carlo methods, that can be applied whatever the subject. In the ISB case, though, it is often an open question whether and what engineering modeling methods can be applied or how they might be adapted. As I noted, we often saw modelers experimenting with the application of methods from the engineering domain in which they were trained, such as wave-smoothing techniques from telecommunications to smooth noisy biological data. Still, with respect to the credibility of model-building methods, even though these are drawn from a discipline other than that in which they are used, much of what Winsberg argues about *how* they gain credibility does apply.

He calls techniques and assumptions made in applying various methods "self-vindicating," by which he means "whenever they produce results that fit well into the web of our previously accepted data, our observations, the results of our paper-and-pencil analysis, and our physical intuitions; when they make specific predictions or produce engineering accomplishments—their credibility as reliable techniques or reasonable assumptions grows" (Winsberg 2010, 122). His is a thoroughly pragmatic stance: methods are vindicated by the fruits they bear, which is the case across the history of development of scientific methods. So, too, the techniques and assumptions of the methods that are transferred from engineering systems and adapted to biological systems gain this kind of pragmatic credibility and become projectible as they develop an interdisciplinary history in biosystems modeling. Further, productive strategies used to solve frequently encountered problems gain traction as reliable parts of the practice for both the individual and the community, as do the novel methods for building models of a specific type (e.g., G10's two-step method for modeling in the lignin domain) developed within these emerging epistemic cultures, which have not yet become established tradition.

Winsberg's other claim is that well-established theory in physics-based modeling plays an important role in helping to mitigate some of the arbitrary features of the model, and enhance its credibility. In lieu of that source of credibility, I consider what aspects of the model-building process in ISB might serve to confer credibility on the model, especially as these relate to the role of the pathway representation in the process.

First, the scope and range of the data integrated into the model cover data for all related systems. Initially, the data are split into two sets, one the modeler uses to build the model and the other the modeler uses to run cross-validation tests after the model is fitted. The integration of data develops in interaction with building out the pathway structure, which lays out the causal sequence of connections among the elements of the system. In this interactive process, the modeler can take different pieces of the network and simulate their behavior in various combinations and configurations. The modeler can run unlimited simulations (recall G10 ran ten thousand to examine just one piece of the pathway). These simulations enable the modeler to consider whether adding or deleting pieces of the pathway are biologically reasonable moves.

Second, the simulation process provides the modeler with significant ability to control and manipulate the model's behavior. The modeler can stop and start the simulation in every state, which enables her to track the system variables (nodes in the pathway) that generate specific behaviors and to determine detailed measures of significant variables. The modeler can also track every time point of the state of the simulation, which enables her to change the time at which some process kicks in, among other modifications. Such manipulations enable the modeler to interrogate the dynamics of the system and develop a sense of how the pathway as constructed could generate the experimental data, as well as a sense of what changes in the pathway might be productive, again, consistent with their biological reasonableness.

Third, the mutually constraining nature of the data, parameters, and pathway in the model-fitting process helps to mitigate some of the arbitrariness of model-building. The notion of fit is complex. It does not mean that the model provides a point-by-point replication of the data for all variables. Rather, at least for the kind of modeling we have studied, it means that the model replicates trends in the experimental data for most major variables. "Fit" is often construed as a matching process in which there is a satisfactory match between the data generated by the final model and the experimental data, usually determined by comparing graphs of each. However, from studying the practices of the modelers in our labs, we have come to understand fit as a dynamic and interactive process among the three elements that works to enable the modeler to home in on a satisfactory representation for both pathway and model.

There are three changeable components—pathway structure, experimental data, parameter values—that become increasingly constrained by their interactions in the highly recursive fitting process. To estimate unknown parameters, the modeler uses fit with experimental data as an anchor. For each change in parameter, the way the output of the model maps to the experimental results changes. Only parameter values that improve fit, or keep it at its current state, are retained. Although it cannot be done for all parameters, the modeler screens parameters for their biological plausibility to the extent possible. Each replication of experimental results in a simulation infuses more, and disparate, data into the model and changes the parameter structure. During the process, fit is used to add or delete components of the pathway network. As we saw, inferences about how to build out or modify the causal network derive largely from the behavior of the simulations. These simulations enable the modeler to infer whether her conjectures as to network structure are on the right track by running the model with and without various pieces, which requires changes in parameters. Equally important, simulation enables modelers to infer missing network structure, which they check for biological plausibility, such as G10's addition of reverse fluxes and channelization of the S and G monomers. And, although we were told such outcomes are rare, simulation has the potential to point even to the possibility that an element thought to be outside an established pathway could be affecting the behavior of the system, as in the inference about the element "X" made by G10.

Through the three-way locking-in process, the model gains complexity. The fitting process is, of course, not without risk of introducing unwarranted elements into the model, especially as there are often some parameters that can only be fit by Monte Carlo simulation. Even a well-fitted model is underdetermined. However, there are ways to further enhance its credibility. Once a satisfactory fit is obtained, for instance, the modeler performs cross-validation tests on the model (or a small ensemble if the fit is not unique) with additional data and diagnostic tests, such as how it responds to perturbations, which, if passed, add to its credibility.

At the end of these constructing, fitting, and validating processes, the modeler can build sufficient warrant to believe, provisionally, that a robust and stable model does enact the behaviors of a generic system, that is, it exemplifies the behaviors of the class of biological systems. As repeatedly expressed by the modelers in our labs, though, they aim to build models that not only replicate the available data, but also provide substantive predictions. As with the in vitro models built in BME, ISB modelers transfer predictions about behavior from the in silico models they build to the target systems using analogical inference. ISB modelers deem predictions that derive from a stable and robust model that exemplifies the known behaviors of the target system sufficiently credible to warrant investigation by experimentalists. If these predictions are verified, it further enhances the credibility not only of the model, but also of the model-building methods.

5.4 Summary: "Getting a Grip" with/on In Silico Simulation Modeling

For some time, scientists have been using computational simulation to gain epistemic access to the behaviors of complex dynamical systems, from colliding galaxies to climate systems. Only recently, though, has it been used to investigate biological systems. This has been due, in part, to the methodological problem of how to build computational models of these systems in the absence of the resources provided by a theoretical basis and, in part, to the technological and methodological problems of how to collect sufficient data of the right kind (time series) or to get around the lack of data with appropriate algorithmic strategies. Although ISB is a diverse field, the modeling practices in labs we have been studying are representative of a major area that draws conceptual and methodological resources from engineering fields, including electrical engineering, control engineering, systems engineering, and telecommunications engineering, to build models of complex biological systems. Although the long-term objective expressed by researchers in the field is to develop a systems-level theory by which to understand and predict behaviors of complex biological systems, our labs expressed more modest aims. The lab G director is the one who expressed their more immediately obtainable aim with the particularly apt phrase, "*getting a grip*" on systems behavior—that is, an understanding sufficient for predictions that at least can enable manipulation and control. As we have seen, modelers also need to get a grip on the challenge of building models of large-scale biological systems in the face of numerous constraints. We have examined in detail some of the practices modelers in lab G have been developing to manage the complexity of this challenge.

Instead of articulating theories into informative computational models, researchers in this area of ISB need to compose their models by collecting the needed dynamical and structural information themselves from a variety of sources, including their own simulations, in an iterative and incremental fashion. Our cognitive-ethnographic investigations on how they build models provide valuable insights into the processes that are unlikely to be found in examining only the published literature, as has been the case with most of the philosophical accounts of physics-based computational modeling, or even archival material, to the extent it exists. We have been able to detail how they build models from the ground up by piecing together in nest-like fashion principles from molecular biology, experimental results, information gathered from literature surveys and databases, canonical frameworks, and computational algorithms to create representations of biological systems in data-poor environments.

Simulation, which is the central methodology for experimentation in computational modeling, is often seen as the end phase of the research. Our in situ examination of the model-building processes brings to the fore the key roles of simulation in building the model itself. Simulation is a means through which the modeler develops the biological pathway and comes to learn and assemble the relevant ontological features of a system. The modeler continues to adapt the pathway network in conjunction with simulation throughout the model-building process until pathway, experimental data, and parameter fit coalesce into a stable and robust model (or small set of models). The pathway representation is shaped by issues of available parameters and parameter estimation tools. Likewise, pathways are tailored to fit the capacities of the mathematical frameworks and whatever mathematical tools the modeler can bring to bear. At the same time, these frameworks determine the extent of the parameter fixing problem. The modeler keeps all these elements in dialogue during the model-building process. In this regard, simulations play an important functional role in how modelers learn how to assemble information and to construct a computational model that gives the right kind of representation. Modelers assemble the needed information in the course of an exploratory process that involves preliminary simulations, both computational and pen-and-paper, and subsequent refinements and revisions. This process enables the modeler to build up her own understanding of the dynamics and relevancies of particular pathway elements. Building this understanding can require the modeler to make judgments about adding elements to the pathway not discussed in the literature, such as hypotheses about elements that must be playing a role or about feedback relations that are not documented but are required by the model, as we saw with G10.

In sum, we have analyzed ways in which the processes of incremental and iterative model-building and their attendant processes of simulation are the means through which modelers come to understand their model and their biological systems. This, in turn allows them to make better judgements about what to include or exclude and which tools and techniques will help and which, not. As per the nest analogy, simulation provides them with the means to work out the best, or most stable, way to pack the pieces together.[15] There, thus, is an important cognitive dimension to simulation in that the iterative back-and-forth interaction between the modeler's mental model and the computational model, which we characterize as "coupling" between these parts of a D-cog reasoning system, is an essential part of the ISB problem-solving practice that builds models of complex systems that lack a basis in theory. The affordances of simulation as a cognitive resource in the ways we have delineated make building representations (pathway and model) of such complex systems without a theoretical basis possible.

Modelers uniformly use the expression "*a feeling for the model*" to characterize the understanding they develop of the behavior of the model over the course of the building process. Simulation is a major source of this feeling. The modeler comes to understand the model's dynamics through numerous iterations of simulation under various conditions and uses the feeling to guide the direction to develop the model. From our observations and interviews, other aspects of this *"getting a feeling"* include a growing insight and understanding into the target in vivo system and a growing belief in the credibility of the simulation model as a dynamic enactment of the behavior of the system. A further aspect that needs to be considered is the affective dimension. The "coupling" developed between the researcher's mental model and the in silico artifact model creates an intimate connection, the importance of which should not be discounted.

"Feeling" is intimate language. In computational modeling it is directed toward an artifact with dynamic behaviors with which the modeler is interacting intimately ("coupling") as she creates it.[16] Some readers are likely to recall Barbara McClintock's expression of a "feeling for the organism," which she deemed critical to her biological research, especially to her discovery of genetic transposition, which was dismissed at the time but, many years later, was awarded the Nobel Prize. What Evelyn Fox Keller, a scientist herself, has said in her penetrating biographical analysis of McClintock, applies equally here: "Good science cannot proceed without a deep emotional investment on the part of the scientist. It is that emotional investment that provides the motivating force for the endless hours of intense, often grueling, labor" (Keller 1983, 198). We have witnessed such emotional investment across the labs we investigated, and have analyzed how expressions of it, too, are tied, importantly and intimately, to epistemic achievement. We have examined this investment in depth in the BME labs in particular (see, especially, Osbeck et al. 2011, chapter 3). Affective engagement with the objects of one's research does not taint scientific knowledge, rather it makes it possible. The computational modelers in the ISB labs themselves recognize their "feeling" develops only through the hard, slow work of building out the model, which is necessary for them to develop insight into and understanding of the model, as well as the target system. The lab G director shows his recognition of the importance of this work when he strongly encourages his modelers to invest considerable time in exploring and playing around with their models. He does the same in the biosystems modeling classes he teaches.

6 The Bimodal Model-Building Strategy

As ISB develops as a science, it continues to face a classical methodological problem that has been present since its inception: how to manage and integrate wet-lab experimentation with model-building. As I have discussed in chapter 5, the enterprise of ISB is to model large-scale biological systems using modern computational and modeling techniques. In that undertaking, modelers are dependent on bioscientists for data to build models, for experimentation to validate models, and for biological expertise to understand the possibilities and limitations of model-building. ISB is far from settled on what are the best methodological practices or on what are the best modes of research organization to further its model-building practices (see, e.g., Calvert 2010; Calvert and Fujimura 2011; Nersessian and Newstetter 2013; O'Malley and Dupre 2005). As a field in development, it has some flexibility to experiment with respect to its methodological practices. The labs we have investigated belong to the area within ISB in which the lack of data of sufficient quantity or quality are among the foremost problems. Contrary to the widespread 'omics rhetoric, these researchers rarely have easy or sufficient access to high-throughput, time-series data. In chapter 5, we saw how this data problem runs through the range of constraints within which ISB modelers have to work. There I cast the process of managing the complexity of the model-building task as "adaptive problem-solving." Adaptive problem-solving ranges from specific practices of individual researchers to strategies by lab directors to organize their labs in the configurations they consider most effective to facilitate model-building. In this chapter I examine how the lab C director has organized her lab primarily around what we have called the "bimodal strategy": modelers are trained to conduct the wet-lab experiments they need to build their models.

Lab G was organized to comprise modelers, with backgrounds in engineering or applied mathematics, who collaborate with experimentalists external to the lab. The adaptive problem-solving in lab G relies totally on what highly mathematically and computationally skilled modelers can do to build a computational model. What we call the "unimodal strategy" is the typical organization of ISB model-building. As we saw, in unimodal model-building, adaptative problem-solving takes the form of such practices as reducing the scope of the problem; taking advantage of the affordances of simulation to build out pathway networks and the model, as well as to validate the model; and using and developing sophisticated algorithmic techniques to fit parameters in the absence of data.

Most often the modelers are collaborating with bioscientists in different labs, but there are some labs that are organized to have both within them. With this organization, the hope is that collaboration might run more smoothly, with data produced in a timely manner. The lab G director was skeptical. He thought this organization unlikely to work unless you had many experimentalists per modeler, and large amounts of funding to support them. He pointed to the well-known lab of Douglas Lauffenburger at MIT, which has around fifty members to accommodate modelers and experimentalists in productive collaboration as a prime exemplar of such an effective lab organization. Further, he contended, the biological systems investigated would be limited by the constraints of the experimental setup. We did not investigate this kind of lab. When we learned that lab C had a wet lab, we anticipated it to be that kind of lab, so we were surprised to discover its novel bimodal strategy. We decided to continue with it because it seemed a potentially important attempt at hybridization in ISB, despite the fact that it meant our project would not be studying pure experimentalists.

The lab C director was trained in combined research labs, where she was the only modeler in her PhD lab and one of several modelers and experimentalists in her postdoc lab. As I detail in the next section, she had unusual training as a "hybrid" researcher in ISB, having learned, sequentially, first to do modeling and then to do wet-lab experimentation. To organize her own lab C, she decided on an unusual form of adaptive problem-solving in which modelers were to be trained concurrently to conduct their own biological experimentation as part of the model-building process. The "coupling" of modeling and experimentation builds a kind of distributed cognitive-cultural system with some epistemic affordances and limitations different from those of lab G. Although both labs build mesoscopic models, lab C, as we will see, builds models more "bottom up" by accumulating parts and dynamics of systems through the interaction of modeling and experimentation to solve problems. In contrast, lab G uses the averaging power of power laws and the parameter flexibility of the models it builds to close "downward" on a satisfactory representation.

6.1 Lab C: Redox Systems Biology

Lab C had been in existence less than three years when we entered. Its lab director was, then, an assistant professor, who, as with the lab G director, described for us an original and serpentine route to becoming a systems biologist, which was still not an established field when she began, even though she was much younger than he. By the time we met her, she had developed into the experienced kind of hybrid researcher the BME program was envisioning. However, even from the perspective of the senior faculty in that program, as well as from the perspective of her developing field of ISB, her vision to build a lab that trained graduate students to perform research that required they simultaneously learn to build systems models and to conduct wet-lab experiments to investigate complex biological processes was a high-risk undertaking.[1] In her own training, she had learned to do these sequentially, and she felt strongly that path had held her back: *"I tell my students to never do this because you should always do these things in parallel. It kind of delayed my graduation date because I ran into all the learning curve issues that [my] early graduate students face, only here, I was 4.5 years in and starting from scratch."*

As the director recounted her learning trajectory, she had set out in college to be a biomedical researcher, having decided not to be a doctor, and so began as a biology major. Because she had tested out of freshman biology through her high school AP exams, her college education started with the required physics (thermodynamics), calculus, and chemistry courses. The thermodynamics course was full of premed students, which most of the biology majors were at her institution. She found what she experienced as a *"cutthroat environment"* of premed culture, where *"it's a big game of 'if I didn't get an A on the first test, I drop the class.'"* This was off-putting, so she decided to change her major. During a summer internship at a medical institution with a physicist who conducted research on protein structures, she discovered you *"don't have to do a bio major to do bio research."* In what she described as *"flipping through the course schedule . . . in a kind of process of elimination,"* she discovered that the nuclear engineering department had a radiological sciences track, which required cellular physiology and included other biology selections a student could take as part of their degree requirements. She called it a *"general engineering degree"* that has a curriculum *"very close to today's BME majors."* With that major she was able to take enough biology courses for a biomedical engineering minor (there was no major yet). For her undergraduate research, she joined a lab that was building computational models to investigate the effects of X-rays passing through soft tissues and did a modeling thesis.

She decided to focus on computational modeling with a bioengineering PhD. In the lab she selected, she *"wound up with a professor that wasn't a modeler at all. . . . He's a jack of all trades."* Her supervisor conducted research on muscle physiology, mainly with NMR spectroscopy. She described herself as *"the only modeler in the group—everyone else was an experimentalist."* Her research focused on building models of the temporal dynamics of supply and demand of phosphorous (P-31) metabolites in muscle contraction based on the data the lab collected. She described her modeling (ODE) work as requiring *"a lot of applied mathematics"* that used *"third party software."* The final step in her development toward being a bimodal researcher came when she found that to test her model, she *"had utilized all the literature possible"* and that *"there were certain things we couldn't measure with a magnet [by NMR]."* She stated that she realized that the only way to put her model *"through stringent tests to see if some of the things that were emerging as properties of the system actually happen was if I did some of the experiments myself."* So, at the point where she should have been graduating, she began a whole new line of wet-lab research with mouse muscles. Although she characterized this as a setback (quoted above), she also saw it as a *"opportunity,"* since her adviser was an expert experimentalist and she *"hadn't taken advantage of that aspect of his skillset."*

She conducted postdoctoral research in a lab that specialized in modeling protein signaling networks and *"where a lot of mathematical and modeling advances were being made."* Signal transduction, basically, is a process that communicates a message to a cell to grow, divide, alter metabolism, and so forth, in which proteins are the main actors. That lab comprised both modelers and experimentalists who worked in collaboration. She decided, in the context of this lab, to work only as a modeler on an immunology project using T cells, in part because *"it would be kinda cool, because I didn't know anything about it."* To figure out how to model such systems, she had to learn a lot of immunology and develop a new, *"more of a statistical,"* modeling method based on partial least squares, both of which would prove important for the research projects in her own lab. Her two model-building experiences with quite different biological systems also made her aware that there was a *"disconnect"* between metabolic pathway and signaling pathway research, even though in a biological system metabolic and signaling processes are integrated in vivo. She felt her training offered *"a way of integrating those things together"*—a way of building models to *"understand the way all these things are regulated."* The more she investigated, the more she both *"realized, wow, no one studies this because it's just too complicated"* and that this is *"all the more reason why you need to do these computational approaches in parallel with the experimental while you were making progress."* So, at the outset of establishing a research program, she saw the *"parallel"* approach as both a necessary and a justified risk, because the novelty of her project meant that the kind and quantity of data required for the model to do the *"integrating"* could not only be found through searching the literature and databases, but also would need to be determined and collected as the models were being built. She stressed in her interviews, as well as several of her research presentations we attended, that her lab's wet-lab research was in the service of model-building.

That said, every presentation the director made of her lab, including the numerous visual representations she made of the lab's research, put the biological dimension of the research at the center, rather than the model-building, whereas the director of lab G always gave the model pride of place. This difference in emphasis, which reflects how they conceptualized their research, as well as organized their labs, can be seen quite clearly in the representation each drew when we asked them to "draw us a picture of your lab, that is, the problem-space of the lab's research." Unfortunately, the lab G director finished his first and showed it to the director of lab C, so she chose the same format. But it turned out to be a good thing, because it made for a striking contrast. His depiction (figure 6.1a) placed the model and methods for model-building at the center of the lab's activities, while hers (figure 6.1b) placed the biological context there. Hers also depicted the technologies the lab was developing to carry out their experimental

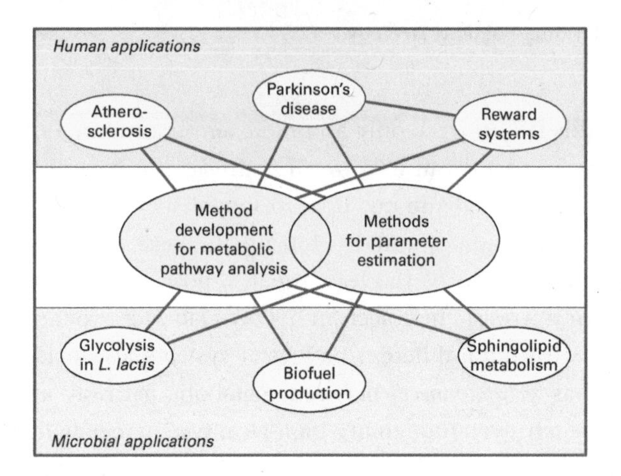

Figure 6.1a
The lab G director's schema of his research problem space. He placed the lab's goals to develop methods for model-building at the center, with various kinds of biological systems feeding into those developments. Lab G's experimental collaborators were in university, medical school, and industry experimental labs locally and worldwide.

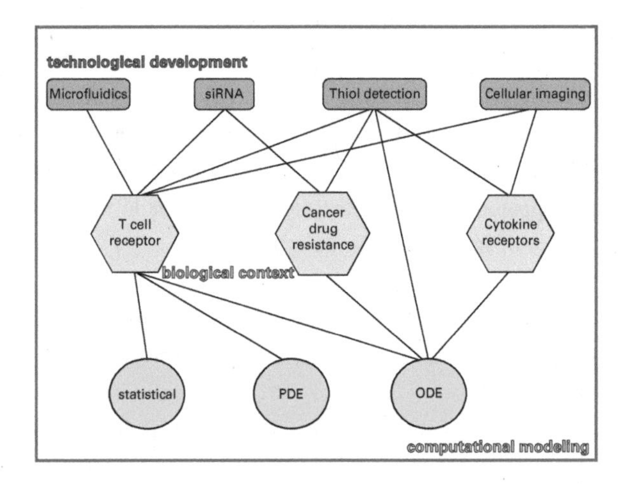

Figure 6.1b
The lab C director's schema of her research problem space. She placed the "biological context" of the lab's research at the center, while noting that they use various kinds of model-building formats and develop experimental technologies. Lab C had experimental and engineering collaborators in different fields within her own university, including chemists and electrical engineers, and at a nearby medical school. They both listed the names of researchers associated with each of the systems or methods, which I have removed.

research, as well as the different kinds of models they built, most using off-the-shelf packages. In addition, their recommendations for reading to help orient us on their research also underscored this difference between the labs. The lab G director recommended a graduate text on modeling biochemical systems using the BST framework and the lab C director, an undergraduate immunology text.

As mentioned in chapter 1, these lab directors were not collaborators, though they had significant interaction, and occasionally served on the dissertation committees of each other's students. The students in these labs did not interact much. Our research project introduced the only occasion we know of in which they presented their research to one another in a joint lab meeting. Since we were studying both labs, the directors thought it might be a good idea for their students and their research projects to be introduced to one another, and so they arranged two joint lab meetings for these purposes. The students did, of course, attend one another's presentations of research in departmental community forums and, often, dissertation defenses. In working with our group, the lab directors developed and taught together a graduate-level introductory systems biology modeling class.

As I noted in chapter 5, the lab G director stated there was a *"philosophical divide"* between ISB modelers who typically take the unimodal approach of his lab, with collaborators at a distance, and those who work more closely connected to experimental research, and he especially noted the divide with respect to the bimodal approach of lab C. Importantly, lab G and lab C have differences in their epistemic agendas. As we saw in chapter 5, the lab G director was quite emphatic about the importance of his lab's aim to advance and enrich mathematical theory and mathematical analysis to further investigation of complex biological systems. Sophisticated biosystems modeling requires rigorous and novel mathematical analyses that can capture a wide range of nonlinear behaviors within a tractable formalism that keeps complexity under control. To produce these kinds of models and the tools for mathematical analysis, however, requires high levels of applied mathematical and computational skills that, in the lab G director's view, cannot be achieved in combination with doing one's own experiments. As he saw it, the trade-offs are, *"If you do the experiment yourself, you know what the data are like; you know how reliable they are. You know the kind of assumptions you made in order to produce the data. . . . So, you get a better idea about the whole context. On the other hand, life is complicated, and to do good modeling is a full-time job;*

to do experiments is a full-time job. And if you don't want to do two full-time jobs, then something will suffer from it." From his perspective, lab C's modeling *"suffers"* in the respect that *"the models that are being developed . . . are . . . by and large, off-the-shelf type modeling approaches that are, not always, but that are often rather simplistic. . . . Our models are at least going into much more depth."* But, as we saw, parameter-fitting was a significant, overarching problem in lab G research, because adequate data and data of the right kind for the system they are modeling are often missing. Lab C modelers rarely experienced these problems. Further, as we saw, the need for unimodal modelers to collaborate with experimentalists is fraught with difficulties that increase the complexity of the model-building task. As we will see in this chapter, lab C's epistemic agenda tracked more closely with contributing to medical research and molecular biology through getting a grip on targeted biological systems.

The *"philosophical divide"* the lab G director referred to might well reflect deeper attitudes in the field as a whole. As O'Malley and Dupré (2005) have argued, there are, in organization and in research strategy, divisions over the practices and aims of systems biology, even though they might not always be debated openly. Some systems biologists are relatively pragmatic and aim to use modeling as a tool to further develop molecular biology, while largely pursuing that field's theoretical agenda; others have a strong systems-theoretical agenda to advance the role of mathematics in systems analysis and to promote the development of a mathematical theory of biological systems. In the case of lab C and lab G, these divisions are acknowledged explicitly and are expressed in the form of laboratory organization that each director has chosen, and in the kinds of model-building practices they favor.

As mentioned earlier, lab C research is driven by a specific theoretical agenda: to determine whether and how particular biochemical systems play a key part in the regulation of cell signaling[2] and metabolic processes. The director sees this issue as critical to the advancement of molecular biology and physiology: *"So they are almost like two different camps in cell biology—all enzyme-based people cared about was how it got into the cell, all the metabolites that are involved, and didn't really think all that much about what was controlling the expression of these proteins or how the muscle was responding to any other cues from its environment or anything . . . and all these signaling people were interested in is what's connected to what and how this receptor is, you know, causing changes in gene expression, but they don't think about basic things like the energy supply in the cell. . . . So, there was this disconnect between the two camps."* She claimed that she *"was thinking about this in a really different way,"* likely because of her unusual hybrid training, which also provided her with *"a way of kind of integrating those things together."* Thus, as with the other lab directors we have discussed, she staked her claim to be conducting innovative, frontier research. But she also noted that the problem was recognized, if not addressed, in the field: *"It's not like there are no other people out in the world thinking this, otherwise I wouldn't have gotten these ideas. But I realized no one was trying to take this [integrative] approach."* She framed her overarching research goal as to make a contribution to molecular biology, but distinguished her lab as importantly different from those of molecular biologists: *"What we really specialize in—what can we do that no one else can do—is put it in terms of the context with respect to the rest of the network."*

This goal likely explains the director's clear preferences for working with more tractable and experimentally accessible systems than those of lab G. The models lab C builds tend to be smaller-scale, more mechanistic, models that help to demonstrate how specific mechanisms operate within the system. The modelers' preference for Michaelis-Menten or simple mass-action models of interactions engages with common representational techniques of molecular biologists.[3] Unlike in lab G, we heard few complaints in lab C about parameter estimation, because wet-lab experimentation by the modelers serves to keep the unknown parameters of their models mostly under control. As the director noted, with the bimodal approach, when faced with *"this issue of having more parameters than we're capable of fitting with the data, we have to say, 'ok, what data do I need to collect just to fit the model to these parameters?'"* Her goal was to train the modelers in her lab to be able to design and *"run the experiments under these conditions, do the analysis, and then plug these values into the model."*

The overarching epistemic goal of lab C is to understand cellular oxidation (a metabolic process) in regulating immunological and cancer cell signal transduction.[4] Such understanding has the potential to be used to develop interventions for numerous diseases associated with the oxidation state of cells (atherosclerosis, HIV, Parkinson's disease, lupus, cancer, and so forth) and can, especially, be utilized in personalized medicine. In particular, the lab conducts research on the impact of the redox environment on proteins. "Redox" is an abbreviation of reduction-oxidation, which is a chemical reaction that changes the oxidation state of atoms. Under normal physiological conditions, cells maintain a reduced oxidation environment. However, oxidizing molecules and free radicals are produced by cells as part of physiological processes, or they can enter them, and can react with cellular components, including DNA, cell membranes, and proteins. Such reactions have been implicated in several diseases. Cells use enzymes to counteract these oxidants and proteins to mitigate the effects of oxidation. The communities that investigate redox and oxidative stress have only recently begun to appreciate the need to understand the interplay of these processes in order to determine the mechanisms of disease and to conduct therapeutic interventions.

Lab C's research focuses on the impacts of oxidants on proteins, which are part of signaling pathways. As we saw above, the lab is a pioneer in seeking to integrate these phenomena, as they are in the in vivo biological system. As she explained it, "*Oxidative stress is coupled with metabolism. . . . I realized that there is a way that the byproducts of oxidative stress, which are these reactive oxygen species, can bind to signaling proteins and affect the way they operate. For me, that's the kind of missing link, 'cause oxidative stress is controlled by metabolite levels. . . . I saw this as a really different perspective on the traditional signaling cascade.*" Modeling provides the means to study the dynamics of these metabolic and signaling processes in an integrated manner. However, given the novelty of this biological problem, the modelers need also to conduct experiments under specific conditions to obtain much of the data they need to build out the integrated pathway and find the parameters required to fit the model. In the period of our investigation, the lab's specific modeling problems within their redox agenda were both generated from the interests of the lab director and brought to her by experimental researchers outside of the lab.

Lab C is located in a new building that was designed to facilitate collaboration among labs. The lab spaces are largely open, with dividing walls surrounded by a wide corridor in which the expensive technologies in common use are housed. Lab C comprises a large wet lab where experiments are conducted, in which there is a walled-off cell culturing room and a dedicated space for conducting western blot assays, and a "grad cave" with cubicles where the graduate students do their modeling work on laptops and store their stuff. The wet lab has the typical accoutrement of a molecular biology lab, which includes pipettes, centrifuge, test tubes, a biohazards waste bin, a cryogenic freezer for the immortalized cancer cells they purchase in bulk, incubators, and wall pegs for hanging the clean white lab coats all members don when entering the space. Importantly, the wet lab was the center of social as well as experimental activity in lab C. As one researcher reported, *"In doing experiments you just sort of gain an understanding of how the lab [lab C] runs. . . . I'm not even talking about just the technical skills that you gain—I'm talking about more of the social aspect of the lab. . . . A lot of stuff happens in the [wet] lab area not in the desk area where I do my computing."*

Cells of various kinds figure in the experimental research of the lab members. The immortalized cells line they purchase are, primarily, HeLa (cervical adenocarcinoma), JurKat (acute T-cell leukemia), and Caco-2 (epithelial colorectal adenocarcinoma). Given the sensitivity and cost of primary cell types, such as T cells and neutrophils, they use the lab members own cells from freshly donated blood drawn in the health center when these are needed, which they can maintain for a short period. There was considerable joking among the researchers around their donations and the characteristics of their donated cells. All these types of cell lines provide what the director called the "*model-systems*" of the research, because they are used "*in substitution of what may actually be occurring in normal [in vivo] cells.*" She explained that in biological research, such cell lines are called "model-systems" because the processes of maintaining them alter them in some ways, and, further, some cell lines, for instance cancer cell lines, are not "*even normal to begin with.*" However, all in vitro cell types still provide a "*fairly good representation of what's happening in the real cell.*"

The lab absorbed the considerable expense to purchase its own Bio-Plex machine for immunoassays, which sits on a dedicated counter space, because the researchers use it so frequently for studies of temporal dynamics on the primary cells, which are difficult to acquire and age rapidly in vitro. The director thought the investment worthwhile, given that it can run eight different time points on one sample. As she commented with considerable enthusiasm when she demonstrated to us how it works, "*My machine can do these things simultaneously and then it's like 'Wow you got eight different measurements with one single time point, you got sixty-four measurements with this one input!' So, it's a way of generating the data that can supply our models.*" The last statement is indicative of the way this lab differs, significantly, from a customary molecular biology lab, even though it has the look of one: the data are collected by modelers who need it to "*supply,*" or "*feed*" (a commonly used expression), their models.

The grad cave space has several whiteboards the students use for work, leaving reminders, playing games, and joking with one another. Unlike lab

G, the students were in the lab space much of the time, even the one dedicated modeler, and "the lab" had the feel of an active community. They often ate lunch together and reported gathering for social activities outside the lab. (Lab G researchers also reported such gatherings.) There were weekly research meetings with one or two presenters to update lab members on research or to troubleshoot experimental or modeling problems. There were also weekly "journal club" meetings in which the lab members discussed pertinent papers in the literature, selected by the members on a rotating basis. The lab director was frequently in the wet lab, where she conducted her own experimental research as well as supervised the student projects. So, she was usually on-site when students ran into difficulties or had questions. None of the students had conducted experimental research prior to entering the lab, nor had any developed biosystems models. Not all of the students were bimodal modelers when we entered, but by the time we ended our investigation all had followed that path. The process of developing skills in both modeling and experimentation had become central to what made one a part of the cognitive-cultural system of lab C.

When we arrived, the lab consisted of the director, three PhD students, nine undergraduates (all in the new BME major), and a research technologist, who had an MS in biology and who carried out the responsibilities of lab manager as well as conducted some experimental research in collaboration with the director and grad students. Although the lab membership expanded while we were conducting our research, because our time and resources were limited, we focused on "founding" members. The lab members were a diverse group internationally, spanning four continents. The grad students had undergraduate degrees in electrical engineering (C10), materials science and engineering (C9), and biotechnology (C7). Because there was no PhD degree in ISB, their current degree programs were located in electrical engineering, biomedical engineering, and bioinformatics, respectively, while the lab and the director were in the newly formed BME department. Interestingly, just as with the director, the students all characterized their degrees as *"general engineering"* in their initial interviews. C10, for instance, contrasted her degree program with electrical engineering programs in the United States, such as her current department, by saying, *"They learned us to learn, not to learn something."* I suspect that the fact that the broad-based engineering programs they came from had required less rigorous education in the high-powered applied mathematical skills than we saw lab G members possessed might have attracted them to the research agenda of lab C and predisposed them to be more open toward learning biological experimentation methods.

The research technologist's undergraduate and MS degrees were in biology, and she had worked for several years in the biotech industry before joining lab C. We interviewed her and followed her research, as we did with the PhD students. She provided an interesting case as the only person trained as a biologist and, initially, solely engaged in experimentation. Near the end of our research, she transitioned to a PhD student, and we did continue to follow her progress for a case study on learning even after we had stopped intensive data collection. She intended at first to do only experimental work, but ended up building models as well, after she took the introductory biosystems modeling class we helped the lab G and C directors develop for the PhD program. I say more about her experience in chapter 7.

C10 and C9 were bimodal researchers and had been in the lab two and three years, respectively. C7 had just joined, did only modeling, and maintained he would not be doing experimentation, but, as our study ended, he, too, began experimental research because his modeling project needed data that were not available in the literature. To carry out her research, C10 needed to collect high-throughput data, and so her first project was to collaborate with members of an engineering lab to design and fabricate a microfluidics lab-on-a-chip device that she could use to collect sufficient experimental data for her modeling project. Since our primary interest was to understand the epistemic affordances and limitations of the bimodal strategy for research, as well as the challenges it poses for learning, we developed detailed, longitudinal case studies of the practices of C10 and C9, and a briefer one of C7.

A primary research contribution of all of the undergraduate researchers was to conduct western blot assays, which identify specific proteins and measure their amounts in a sample, for the graduate students to which they were assigned. The students assigned to C10 and C9 assisted in other aspects of their research projects as well. One undergraduate, who had joined when the lab started, had his own research project supervised by the lab director. The director had collaborations with researchers (engineers, chemists, biochemists, and medical researchers) located within her department and at a nearby medical school, some of whom we were able to interview.

As with lab G, lab C researchers identified functionally in accord with their epistemic practices rather than as systems biologists. To us, they identified primarily as modelers. When we asked the bimodal researchers, including the director, how they identified themselves to other researchers, they all responded that it depended on the person with whom they were talking: sometimes as a modeler, sometimes as both modeler and experimentalist. The students noted how their engineering education provided them with the skills for model-building and technology design. They expressed confidence in their ability to use and modify the third-party software, as well as do limited de novo coding when needed, but they also noted, explicitly, their technical skills were not at the level of the modelers in lab G—whom they called "*theoretical modelers*." The director and all the graduate students noted the importance of concepts, theory, and methods from control engineering, in particular, in their model-building practices. Control engineering, itself, could be called a "general engineering" area, since it is an interdisciplinary mix of various engineering fields, including electrical, mechanical, telecommunications, computer engineering, and product engineering.

All the graduate students said that learning to conduct experiments presented a significant challenge. Their research required specialized in-depth knowledge of a specific system, so they, too, felt that taking a number of biology courses, beyond those that their degrees required, would not be useful. Biological concepts and experimental techniques were the subject of most journal club sessions. With respect to experimentation, the standardization of molecular biology techniques and the availability of prepared assay components from vendors were significant factors that contributed to their ability to carry out experiments. They often commented that experimentation mainly comprised skills one needed to practice repeatedly, such as "*pipetting to make sure it was accurate*" and "*following a recipe*." The latter comment would elicit vehement objections from the research technologist, who had the uncanny ability to hear these and other comments she considered disparaging of the complexity and sophistication of biological research even when uttered quietly across the room from her.

As in the BME labs, anthropomorphic language figured prominently in discussions about the cells and their behavior in this lab. Lab members talked about seeing things from the "*perspective of the cell*" and of the need to keep cells "*happy*," especially so that they do not "*commit suicide*" (apoptosis). They often took the perspective of the cells when discussing their behavior, using phrases such as *"if I were a cell."* The researchers exhibited the same kind of affective engagement with their cells that we saw in the BME labs. A vivid example is the explanation one researcher offered about why they need to stimulate the cells in an interview that turned into a three-way discussion with another researcher, who concurred: *"You have to stimulate them . . . so they are happy—and so it basically, because it is stimulated—it tells her [the other researcher], 'oh, I am useful, so I cannot commit suicide because somebody needs me.'"*[5] We interpreted such language as an expression of how their intensive interaction with the cells developed cognitive partnerships with them, as in the BME labs.

To examine the bimodal strategy, I first take a brief look at C10's design of the microfluidic device to demonstrate that an important dimension of lab C's ability to conduct their own experiments is the members' ability, as engineers, to create technologies for experimentation, as were depicted by the director in figure 6.1. Once the microfluidic device is built, it changes experimental biological practice and, in this instance, makes it possible for an engineer to more easily collect her own *"gold-standard"* time-series data, since it replaces a complex series of experimental manipulations. The data from the device allows the researcher to build detailed models that can make more accurate predictions than those built on scant data. Such predictions often lead to novel experimental manipulations, which create still more data for modeling, thus generating a positive feedback spiral in the direction of a more accurate model representation and deeper understanding of the system.

We followed the iterative and incremental design process for this device, as it was constructed and tested in experimentation with cells, through to its completion. We were not able to follow her research through to the phase where C10 conducted the experiments needed in the course of her modeling project. In this case study, we detailed the nearly three-year process through which C10 designed and fabricated a high-throughput device to automate the experimentation she would need to conduct to build a model of T-cell senescence. In addition to our coding of it as a part of the research in the lab, we recoded our data for the project, separately, as a case of engineering design (for a detailed analysis, see Aurigemma et al. 2013). The rest of the chapter focuses on an in-depth examination of how C9 used the bimodal strategy to couple experimentation and model-building in her investigation of the differential sensitivities of cancer cells to a chemotherapeutic drug, and its implications for our research themes.

6.2 "You Need Very Precise Stimulation at Very Precise Time Points": Turning Experiments into Devices

When we were initially told about the research to design a *microfluidic device* so the lab could carry out their own experimentation on T-cell senescence, we, of course, thought it might be something along the lines of the simulation devices in the BME lab, where we first heard the terminology of "device." However, once we understood that this was a high-throughput data collection technology and not an in vitro simulation device, we asked the director about the term. She explained that engineers use "device," generically, to mean *"the man-made object that's being used to manipulate cells and change them in some fashion. In our case, we are breaking them open at some point. . . . We're able to treat them, break them open, and collect all the outlet proteins that come out of it [microfluidic device]."* Thus, in vitro simulation, as carried out in labs A and D, is just one way in which a device can manipulate and change cells.

The lab director explained that they wanted to understand the possible role of redox processes in T-cell senescence (the aging of the cell that leads to an inability to replicate), especially with respect to a clinical application. Clinicians were starting to develop procedures to boost a person's immune system by harvesting their T cells, multiplying them in vitro, and then returning them to the patient. Although they had some success with the procedure, the rapid aging of the cells in vitro presented a considerable obstacle. The lab director suspected redox processes were the culprit, and C10's project was to investigate these processes in T-cell signaling. However, there were little data available on their behavior, so lab C needed to collect their own in order to build a model of this system. They needed primary T cells for the research, and the fact that they could obtain only small quantities of these, coupled with the fact that T cells age rapidly, presented problems for the research. They needed to be able to collect multiple data time points rapidly—more than were feasible by hand. High-throughput data collection technologies are designed to overcome these problems. Engineers develop what they call "lab-on-a chip high-throughput devices" (hereafter, LOC) to bring together in a single apparatus—and replace—a complex series of experimental activities that would need to be executed by researchers in biology labs. This technology has been a major contributor to the development of computational systems biology, since vast amounts of time-series data can be collected quickly and efficiently from a population of cells (thus, the designation "high-throughput"). These LOC devices improve the accuracy and quality of measurements, as well. C10 decided to build a specific microfluidic LOC device to conduct her experimental research on T cells.

Initially, C10 came to a microfluidics lab in an electrical engineering department to do research on LOC design for a MS degree. Microfluidics engineering builds devices to precisely control and manipulate fluids that are geometrically constrained to a small scale. As C10 explained, "*When I arrived, I didn't know anything about microfluidics, about biology, and about research.*" When we met her two years later, after she had completed that degree, she had developed expertise in microfluidics and was learning T-cell biology. She had also transitioned to being a member of lab C, as well as of her original lab—indeed, she served as the bridge in a collaboration between them. She came to know lab C because the director had become interested in LOC devices. To develop computational models of cell signaling requires time-series data that are difficult to collect in benchtop experimentation, because signaling events happen rapidly (sometimes within twenty seconds of stimulating a cell). The LOC device can automate the stimulation of the cell and the collection of cell samples at different time points simultaneously. This automation improves data collection, particularly for early signaling events that occur right after stimulation, and signaling events that occur in quick succession, thus providing cleaner and richer data for modeling.

The problem C10 wanted to investigate using the device was to quantify senescence in T cells, specifically, to determine which biomarkers change in correlation with age. This research required primary T cells, and those she planned to use were collected freshly from human donors, which were available in limited quantities. These cells also immediately begin to age rapidly, and so can be used for experiments for only a few days. One of the advantages of the LOC is that C10 would need only a limited number of cells to collect a significant amount of data as compared to benchtop methods. To investigate senescence with this approach, the T cells need to be stimulated (mixed well with a reagent), which causes different proteins to form in the cells (as a result of the signaling process), and then measured at many time points, ranging from twenty seconds to twenty minutes after stimulation. With the data from the LOC, these measurements can be done at both the population level (a certain number of cells) and at the single-cell level. The measurement of proteins, itself, is not done in the LOC, but separately, with biological instruments. The device freezes the cells' internal state at different time points by quenching the biochemical reactions in the cell. This is done by lysing the stimulated cells (adding a reagent that breaks open the cells, which creates population samples) and fixing them (adding formalin, which creates single-cell samples). Proteins, whose internal states are frozen by the LOC at different time points in the signaling process, can then be measured.

The LOC design process, in general, requires the designer to translate the goals and actions of an experimentalist executing a complex lab routine into mechanical procedures that can be accomplished by the device, which is only a few centimeters in size. In this case, the device C10 was building needed to automate three processes: (1) stimulate the cells (by mixing with stimulant); (2) freeze the cells internal state by lysing (by mixing with lysis buffer) half of the samples and fixing (by mixing with formalin) the other half; and (3) do this at precisely the right moments (as defined by the desired time points). In the initial stage of the design, she only considered lysing and added fixing toward the end of the design. One of her early design decisions was to have a modular design, one module for the mixing process and another for the freezing process. The two modules would be connected by tubing of different lengths, so that the liquid (stimulated cells in media) in each tube would take different amounts of time to reach the second module, where the biochemical reactions in the cells are then quenched. The varying tube lengths thus function as an analogue for different time points, turning time into space. C10 built the device in PDMS, which is a 3D CAD (computer-aided design) software program, using soft lithography.

Although I will not detail it here, the device design involved complex iterative and incremental processes through which C10 distributed her cognition across various kinds of representations she constructed as she created what would become the final LOC, which she would use to build a computational model of T-cell senescence (see Aurigemma et al. 2013 for a detailed analysis of these processes). These processes involved numerous interactions among components of a D-cog system that comprised C10's mental models, computational models that simulated design possibilities for various geometries for the modules, sketches, fabricated device prototypes, different cell lines, and visualizations, which included those created by tagging cells, computational visualizations, and sketches, as well as numerous problem-solving sessions with lab mates and experimental collaborators in the microfluidic lab and lab C. The LOC device needed to accommodate and integrate engineering and biological constraints. C10 encountered numerous problems, and each time she would develop a computational model of the LOC (in MATLAB) to simulate the effects of design possibilities. The most difficult problem was to develop a solution for the geometrical configuration of the pressure drop chamber (PDC).

The PDC was designed, initially, as rectilinear channels folded in a rectangular zigzag pattern, which she thought a good solution to the engineering constraint that the PDC needed to be long and thin, but fit into the footprint of the device. When the device was tested with fluids, it worked perfectly. But, when she tested with JurKat cells, which they used in the design process because they are plentiful and longer lasting, they became stuck in the corners and were *"getting stressed."* As C10 stated her frustration, *"If you don't have cells, it's almost perfect. You put cells, nothing works anymore."* She tried various zigzag configurations with fewer turns, but the problem became even worse when she tried with primary T cells, which she discovered were larger than the JurKat T cells. The final design solution came at a lab meeting where she once again discussed the problem with the cells still getting stuck. C11 hit upon the idea of circles and then, echoing one another rapidly, the lab members proposed it could be *"like a spiral,"* which they elaborated could be drawn from the center, *"the way a seashell is made . . . like nature."*[6] C10 balked at the idea because it would be *"painful to draw one—drawing a circle in AutoCAD is painful,"* to which they responded *"but you only have to draw it once."* In the end, the spiral design solved the problem. C10 fabricated the final version of the LOC device and was ready to begin collecting data to build and test models of T-cell signaling processes when we finished our data collection.

This C10 case shows how the lab could use the affordances of their engineering skills to obtain the experimental data they needed to carry out, and manage the complexity of, the bimodal model-building strategy. In the next section I develop a case in which we followed how a researcher, C9, used the bimodal strategy of coupling experimentation and model-building to manage the complexity of modeling a biological system about which she and the lab director had formulated a novel hypothesis.

6.3 "As I'm Building the Model, Things [about Experiments] Are Popping Up in My Head": Investigating Cancer Cell Drug Sensitivities

C9's research provides an example of a distributed cognitive-cultural system with epistemic affordances different from those we have considered thus far in ISB. An examination of the path C9's research took over the course of her PhD is central to understanding how the bimodal strategy works as a form of adaptive problem-solving, which aims to manage the complexity of biosystems model-building. The bimodal strategy directed and determined C9's investigative possibilities. Through this strategy she was able to leverage affordances of both in silico simulation and wet-lab experimentation as an effective means to handle her complex problem-solving task. As is likely common in bimodal research, her problem-solving process took a circuitous path, driven by how the dynamics of her interactive methodological system, in particular, generated novel relevant phenomena. C9 was the first graduate student to enter the lab, and when we began our research, she was in her fourth year. However, until that time she had been building models with data obtained from the literature and from a large, unused data set provided by an experimentalist whom she called *"a mentor"* to both her and the lab director. So, we were able to follow her use of the bimodal strategy from start to finish.[7]

By the time we first interviewed her, C9 clearly saw herself as distinct in terms of the kind of research she undertook and the kind of researcher she was. She was investigating a problem about cancer drug sensitivities and thought *"it was very interesting because no one had approached it that way before."* She clearly saw herself and her lab as on the frontiers of research. In her undergraduate education, she had not taken any *"hard core biology"* courses, but only chemistry *"with sprinklings of biology."* She had also taken applied math and done some modeling with MATLAB software as part of her engineering degree. She stated that what had drawn her to biosystems engineering was the *"interesting"* idea of *"using math to describe biology."* She thought her lack of intensive training in molecular biology had allowed her to begin model-building without the typical experimentalist biases against modeling: *"So coming in I might not have had those biases, you know, that some experimentalists might have—so I had, maybe I was more of an open canvas for accepting modeling."* At the same time, she also distinguished herself from what she called *"theoretical modelers"* (of the lab G type): *"We don't just come up with ideas and then just shoot them out there and wait for people do to them [wet-lab experiments]."*

The overall aim of C9's research was to try to explain different sensitivities in cancer cell lines to the chemotherapy drug doxorubicin (Dox). A clinical researcher at a medical school had brought this intriguing problem to the attention of the lab director. The director hypothesized that the sensitivities are somehow related to signaling functions of ROS, such as hydrogen peroxide within cells. She ventured this hypothesis on the basis of two plausible assumptions, which would inform C9's research. The first assumption is that this signaling system is sensitive to drugs like Dox, which generate hydrogen peroxide, and the second is that this signaling system modulates pathways relevant to cell apoptosis (self-initiated cell death) and proliferation. C9's research goal was to figure out whether redox systems play a role in Dox metabolism and Dox-mediated cell signaling and what that role might be. They hoped that this research would contribute to understanding the mechanism behind the differential sensitivity of cancer cells to Dox and, thereby, make a contribution to personalized cancer therapy. C9 saw this research as very much a joint project with the lab director, and often shifted between *"my"* and *"our"* when talking about it.

Over the course of her research, C9 constructed four models (which we have labeled Model 1 to Model 4). These were constructed consecutively and form the problem-solving tasks around which her research was organized. Although C9 carried out all the model-building and experimentation for her project, as the research progressed, she had extensive discussions about how to interpret what she was finding and how to proceed with the director and a senior biochemist from outside the lab who, as mentioned above, had become an informal mentor to the director and later joined C9's dissertation committee. The lab G director was also on her committee, although they did not have as much interaction. She was well into the work of building Model 2 when we arrived, and she had recently defended her dissertation proposal. So, I present abbreviated descriptions of Model 1 and 2 from her retrospective accounts, and focus on the latter models, which required her to do wet-lab experimentation, and for which we have concurrent data in the form of progress interviews, field observations, presentations, journal club discussions, and dissertation.

6.3.1 Phase 1: From Local Simulation to Global Simulation

C9's research began with the task of modeling a specific pathway thought to be an important instance of how the redox environment of a cell (the balance of oxidants and antioxidants, or of oxidized and reduced chemical agents) affects signaling processes within the cell. The particular signaling pathway with which her research began is that of the activation pathway NF-κB (nuclear factor kappa B), which is a transcription factor that regulates genes responsible for immunity and is involved in programmed cell death or apoptosis. C9 framed the "*working hypothesis*" of the research in the following way: "*So, our working hypothesis has always been that, some cells are preferentially resistant to Dox because Dox does something that leads to signal transduction with the cell that leads to, you know, anti-apoptotic transcriptions or something like that. And we know in the literature, also, that there are certain points in the NF-κB pathway that are ROS-regulated. So, then it didn't take too much to say, 'ok if you have this drug that induces ROS, it is a possibility that the ROS that's induced can affect this pathway within this cell that might lead it to be pro-survival.'*"

They suspected the NF-κB transcription factor might be relevant in this respect to the response of cancer cells to redox environments. She claimed this line of thinking was a new "*perspective*" on NF-κB, the pathway for which was well-known, because it had never been approached in terms of "*the underlying mechanisms that control*" calcium fluxes that influence NF-κB. C9 spent her first year constructing an accurate topology of the NF-κB pathway by searching through the literature and looking up or determining rate constants and chemical concentrations, building an ODE model on the basis of these, and then testing the model simulations against published data. By the end of the first year, she and the director reasoned that they had a "*pretty good model*" that simulated the interaction of the products of ROS processing with NF-κB and the regulation of these processes.

She reported that her presentations of this model at conferences drew reactions that took two forms. On the one hand she received encouragement for the basic concept the model seemed to illustrate: NF-κB is redox regulated. On the other hand, she encountered resistance to the fact that the model represented such a small fraction of the in vivo physiological process, and so the conditions used to build the model were "*very far from what occurs physiologically.*" At this point their biochemist mentor encouraged them to shift their attention from the small NF-κB model to the whole system of redox regulation itself—that is, from the entry of ROS, such as hydrogen peroxide, into the cell through to the processes by which ROS are processed and cycled. She could then situate the NF-κB model within this larger global system and would have a more accurate and realistic understanding of the smaller system through simulating, in particular, the environmental factors that influence the NF-κB model's various inputs or control points. The biochemist had been interested in how redox buffering and regulation could explain more general disease phenomena, and had collected relevant experimental data on cardiovascular disease, which he gave them for the general model.

The process of building Model 2 required C9 to build out the biological pathway from the existing literature and the biochemist's data. The pathway diagram from which she built the simulation model appears in her dissertation defense presentation, announced as the "*first ever comprehensive account of the mammalian antioxidant system*." Model 2 was an ODE model that used a simple model of enzyme kinetics (Michaelis-Menten) commonly assumed by modelers of metabolic systems (as we saw in lab G) to describe the changes in concentration of cellular redox buffering components. It contained four branches or pathways of H_2O_2 elimination. The model follows the entry of H_2O_2 into the cell and the processes of redox buffering that eliminate the incoming hydrogen peroxide. Her simulations with Model 2 replicated the basic dynamic data for key proteins in the network, glutathione (GSH) and thioredoxin (NADPH). The pathway/model does not mention NF-κB, but its modular structure would allow for Model 1 to be incorporated, as was her original intention in building the global model.

Although the model reasonably accounts for the structure of the system and its participating elements, the move to a "whole cell" perspective and a general model of redox buffering multiplied the number of components, which in turn multiplied the number of rate constants and concentrations she needed to unearth. In all, there were twenty-two kinetic parameters, and C9 spent nearly two years foraging through the literature for these parameter values. In this process she encountered the kinds of problems I discussed in the general description of lab G modeling practices. She was just finishing up this process when we met her. C9 was able to draw on successful simulations with Model 2 to make a number of inferences, including a sensitivity analysis of parameter responses and an analysis of the relative burdens of different proteins in peroxide removal. In the former instance, she analyzed membrane permeability as the factor that produced the greatest network response in the model. In the latter instance, she found the protein thioredoxin (Trx1) does much greater antioxidative work than the protein glutathione (GSH). In possession of a global model of how a cell deals with oxidative stress, C9 was now in a position to move on to the problem of how cells might respond to different parameter configurations or, more specifically, how DOX might alter redox environments.

6.3.2 Phase 2: Building the Dox Bioactivation Network Model

In building Model 2, C9's task had been more or less to assemble the information in the form of a dynamic model, since the structural features of the system had already been well-established in the literature. In the next phase of her research, C9 faced the situation of needing to derive, experimentally, unknown features of the system. Once she began wet-lab experimentation, the literature took on a different role: she mainly went to the literature when something unexpected came up. In the process of experimentation, she discovered that some of her expectations, based on the literature and her previous models, about how the system should behave were wrong and that the essential mechanics of some of the components of the system were in fact unknown. She had to localize and isolate these inaccuracies and unknowns. This part of her discovery process started with Model 2, the global model of ROS. As she explained her problem-solving strategy up to that point, *"So, one of the main [questions] that came up was how does the cell . . . deal with these increases in oxidative stress? No one really knew about that so . . . we need to answer that question before we can move on to how is this drug able to alter the redox environment—to explain how something modulates that redox environment—why or how that works."* C9's plan, after she had built Model 2, was to move on to formulate a mechanism that could confirm their hypothesis that it was redox regulation of Dox that up-regulated NF-κB activity, thus helping insensitive cells survive the drug. She planned to model this mechanism by perturbing the values of hydrogen peroxide entering the Model 2 system, given that Dox was known to raise its levels, and then feed the outputs into the redox-sensitive points of the pathway for Model 1. But, to *"get [that] model up and running,"* she would need to do *"a lot of [wet-lab] experimentation . . . to go in and get some of the rate constants and concentrations."*

Knowledge of how Dox functions to cause toxicity was in fact limited, although it was thought to intercalate DNA and thereby lead to apoptosis. One of its main known side effects is, though, the production of ROS, which made it a candidate for C9's analysis. From the clinical researcher, who had introduced them to the problem, she knew that different cell lines have different responses or sensitivities to Dox. C9's goal was to explain these sensitivities: *"If you ask in the field how exactly does doxorubicin work, they would not be able to give you a set answer. Specifically, particularly since in different cell types you have different responses and no one understands why that is. So hopefully, by the time I am done with my dissertation, we can shed some light on whether or not the ROS generation portion of the doxorubicin story is something that leads to differences in the sensitivity of cells to—to this drug."*

The clinical researcher provided two patient cell lines, EU1 and EU3, for her wet-lab experiments. EU1 is Dox-insensitive and was retrieved from a patient who had not responded to treatment. EU3's patient, however, had had a good response that led to remission. As mentioned previously, when she began the wet-lab experiments, C9 planned to combine her global model (Model 2) with the NF-κB pathway (Model 1). But then a significant problem emerged. It happened early in the course of putting together the findings from her initial experimental research to draft a paper she planned to submit to an experimental journal of her research linking Dox to the NF-κB pathway.[8] As she posed the problem, *"The issue that I think we are having or, I don't even know if it's an issue, not really sure yet 'cause we, we don't know what's going on, is that, the cell lines are what would be logically expected based on literature and what not, with regards to the cell that's insensitive to Dox should have more NF-κB activation—and that's what our model is sort of predicting, but experimentally, we are seeing kind of the opposite and we're not really sure how reconcile that."*

This observation from her initial experimentation led C9 to an extended novel investigation in which she went back and forth between modeling and experimentation. Her experiments with Dox were producing data that were the opposite of what her model was giving and of what would be expected from the existing experimental literature. It is unclear (from our records) on which model she based her expectations, but they were most likely based on a combination of the Model 1 and Model 2, by extrapolating from different levels of ROS input to NF-κB levels. Those levels, according to her models and the experimental literature, were expected to be more active or up-regulated in the case of the insensitive EU1 cells, and down-regulated in the case of the sensitive EU3 cells. Her wet-lab experiments had, however, found more up-regulation in the EU3 cells. The extra ROS generated by Dox should regulate NF-κB, but at this point she was beginning to doubt whether NF-κB was even *"a link in the story"*—or at least a link in the way she had anticipated. As with all the research undertaken in the bioengineering labs, C9 often encountered impasses or failures. As she recounted, *"Some days I come into work and I don't feel like doing anything. . . . I mean, if your models aren't working, experiments are failing, you need that extra push to keep you going."* What helped to motivate her were *"the camaraderie"* with the other researchers who were also *"struggling"* and also thinking about *"the patients that could benefit from it."* She also noted that, now, *"talking about it"* in our interviews *"reinforces what you're doing, so, I can go back and feel motivated."*

To isolate the conflict between her model output and her experimental data, she conducted further experiments with the cell lines, while at the same time she planned to use Model 2 to *"try and explain what we are seeing experimentally."* Her experimental study of ROS in those cell lines did confirm something expected; namely, there was more ROS in EU1 cells due to Dox, which ruled out that the problem was something in the general ROS or NF-κB mechanisms of the cells. She then found a paper that associated the generation of the toxic form of Dox with a ROS-reducing enzyme, NADPH, which she and the lab director thought to be an important experimental result. Her further experiments showed that the NADPH levels were low in EU3 and high in EU1. This suggested that both toxicity and the extra ROS emerged from the mechanisms by which Dox is activated into its toxic or reduced form. As a result, the lab director decided that an additional step was required in their problem-solving process: C9 needed to model the production of ROS by Dox, that is, to build a new Model 3, rather than simply input the estimated amount of ROS produced straight into the global Model 2, as she had planned. The new model-building would serve to open up an area of Model 2, otherwise black-boxed, in order to look for differences between EU1 and EU3. Instead of combining *"all these reactions into one single arrow [black box] and then just have an estimate of what the culmination of all of these reactions would be—we realized that there were areas where there are differences between the EU1 and EU3 cells particularly with their NADHP."*

Once again, she was turned away from her intentions to follow the downstream aspects of Model 2 and return to the role of NF-κB. Instead, she focused on the processes involved in the reception of Dox into the cell and the production of ROS. To build Model 3 she relied on a two-pronged strategy. In the first part, she constructed *"hypothetical mechanisms"* in close conjunction with her NADPH and other wet-lab experiments on different interactions needed to *"build out the model."* For instance, she investigated the role of superoxide (SOD), O_2, and CPR (an enzyme that catalyzes redox reactions) when feeding Dox into the different cell lines and compared the results. As she went along, she simulated the hypothetical mechanisms she constructed and checked the simulation outputs against controlled wet-lab experiments on the different interactions. Finally, she ran simulations in interaction with experiments in order to fix parameters. The model contained ten parameters and ten initial conditions (concentrations) for what she called *"a relatively simple network."* Only two parameters needed to be estimated. Through this interactive process she built Model 3 as she constructed a bimodular pathway structure for Dox metabolism (figure 6.2).

In the second part of her strategy, once a model was up and running, C9 tested it by running simulations to perturb the model and performing corresponding controlled wet-lab experiments, which she cast as having *"physically experimented on the model."* For instance, she conducted computational simulation experiments to rule out other possible causes, such as

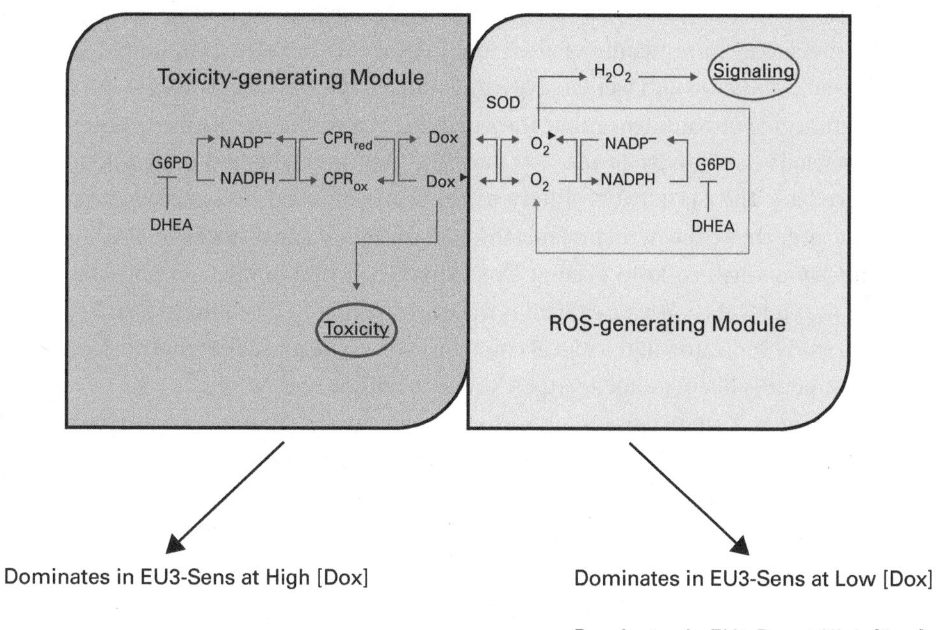

Figure 6.2
C9's proposed bimodular pathway of Dox metabolism in leukemia cells for Model 3.

efflux of Dox from the cell at different rates in the different cell lines. She also "*experimentally perturb[ed]*" the model to test it outside standard clinical ranges by simultaneously inhibiting the key elements of the cycles that had been determined by the model, namely the enzyme G6pd (important to the production of NADPH) and superoxide (SOD). In parallel with these simulations, C9 conducted wet-lab experiments on the two cell lines using chemical agents to inhibit these elements physically and determine the outcomes.

In accord with her experiments, the Model 3 simulations predicted that, in the case of low NADPH, redox cycling takes place. When NADPH is high, however, it absorbs the oxidative species that would otherwise be reduced by toxic Dox (Dox˙ in figure 6.2) leaving toxic Dox to go free. The reductive conversion/toxic Dox production model is represented by the pathway module on the left in figure 6.2 and the redox-recycling model is represented by the module on the right. C9 stated that the agreement between her model and her experiments "*is again proving that our model has stood these different perturbations and interventions and it's still predicting what's happening. So, it further validates the fact that we might have actually gotten it right.*" Because of the successful outcome of the close interaction between computational model simulation and wet-lab experimentation, she was able to assert with a high degree of confidence that they had discovered the mechanism behind the relative sensitivity of the EU1 and EU3 cell lines. She had been able to trace back the cause of sensitivity to the levels of G6pd possessed by each line and, thus, to whether or not the line could replenish stocks of NADPH quickly enough to keep cycling Dox. This result had immediate potential clinical relevance, because G6pd is measured regularly by clinicians, and so the study indicated that its level could be used as a signal as to whether Dox treatment is likely to kill a patient's cancer cells or not.

There is a final twist in this story, however. C9 and the lab director wrote a paper on this research and sent it first to an experimental journal that rejected it because the reviewers thought a two-cell-line study would likely not be generalizable, and therefore would "lack impact." They then sent it to a well-known computational biology journal. In this case, they were surprised when the reviewers complained that the levels of Dox that they were using to study their cells were higher than those used clinically, even though she and the director maintained to us that they had seen numerous experimental papers that used their levels. So, C9 went back and used lower values with the cell lines. These experiments produced radically different behavior, which—somewhat to their surprise—the model reproduced. She considered the fact that her model was able to reproduce her experimental observations as a powerful validation of her discovery of a new mechanism.

6.3.3 Phase 3: Wrapping Up: More Surprises, More Discoveries, Model 4

Even though her research thus far had led to the discovery of a novel mechanism to explain Dox sensitivity, with, of course, some help from the literature, she still found herself dealing with unexpected and unknown behaviors in the final phase of her research that again required interactive experimental and modeling work. In this phase, she had planned to return to her original problem, namely, the redox regulation of transcription factor NF-κB. The questions of how EU1 cells handled both the extra oxidative stress these cells generated under Dox treatment and how they survived whatever toxic Dox they generated still remained. The answers, they thought all along, had to lie in the redox regulation of the NF-κB pathway. She declared she felt she had now "*gone full circle*," and had finally done "*the preliminary stuff I needed to do in order to answer this question.*" Of course, that she had needed to build those models and do those specific experiments had not been clear to her at the outset, but had only emerged as she tried to formulate and tackle pieces of the problem.

C9 had already built a validated computational model for NF-κB (Model 1) based on the literature, so the issue now was to establish a connection between it and the other models and make whatever modifications might be required. To do this, she started a new line of experimentation. First, she established that Dox treatment in EU1 cells created higher levels of hydrogen peroxide. Second, she showed that Dox is correlated in these cells with increased NF-κB production. When she introduced antioxidants at the same time, NF-κB went down again, which established the relation between Dox-induced ROS and NF-κB levels. She continued in this vein to establish that adding antioxidants to cut the production of NF-κB pushed cell survival rates down. C9 told us that she thought this level of experimental detail or "*fine resolution*" is necessary to convince other researchers of these causal relations. It enabled her to show that the causes she hypothesized in her models were robust: "*There's nothing written in stone about the steps you take. You need to sort of say to yourself, 'ok, how fine of a resolution am I comfortable with, or how fine of a resolution do I actually need . . . to get other people to believe this is actually what's happening?'"*

Once she established these relations experimentally, it was time to get into the *"nitty gritty"* details of the NF-κB pathway itself. Returning to Model 1, the question was which pathway elements would be modulated by the increase in ROS; in other words, what are the potential points in the NF-κB network that govern redox regulation (see figure 6.3, left pathway). With the model as the basis, she planned to work through wet-lab modulations of the different components using a specific antioxidant (NAC: N-acetyl cystine). She intended, initially, to use an experimental procedure of soaking up oxidants to see whether she could confirm model predictions about which pathway modulations are due to excess oxidants caused by Dox and which are not.

When she went through elements of the NF-κB pathway, such as NEMO (an essential NF-κB modulation gene), with a more detailed wet-lab examination adding Dox, she discovered one protein, IKK-β, whose s-glutathionylation levels changed when Dox was added. S-glutathionylation modification is caused by oxidation of the protein and thus provides an indication of the sensitivity of that protein to oxidative agents. She isolated the IKK complex (which binds IKK-β) as a ROS-sensitive component, circled in the NF-κB pathway above, and confirmed this in the wet lab by adding NAC and seeing IKK-β levels drop accordingly. However, when she tested with this antioxidant more expansively, she discovered that the effects of IKK-β varied non-monotonically with the levels of NAC in the system. This was unexpected and non-intuitive for her and the lab director, and put them again in the situation of having discovered complexity that they had not anticipated. Once again they would need to draw on C9's interactive experimentation and modeling strategy to untangle the knot. They decided C9 needed to build out Model 1 by adding a model of the IKK complex s-glutathionylation (Model 4) to see if they could explain the wet-lab findings. C9 built this model by working quite closely to the chemical details, targeting the *"mechanisms by which a protein such as IKK-β can be s-glutathionylated in the presence of an ROS promoting agent like Dox."*

C9's literature search yielded ten potential biochemical reaction mechanisms by which IKK-β s-glutathionylation could occur. She used both the conditions under which those mechanisms were observed in the literature in comparison to the mechanism she postulated, and what conditions were

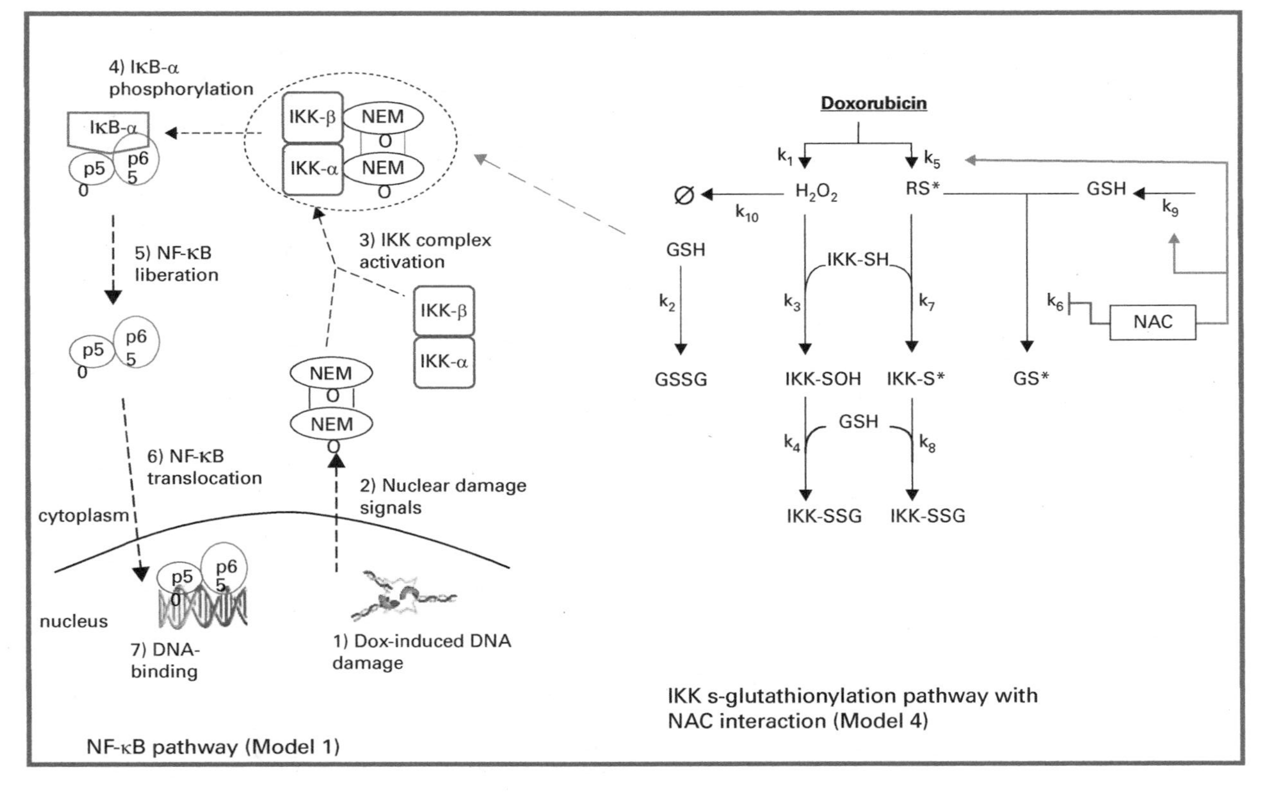

Figure 6.3
The left pathway is the one C9 built of NF-κB pathway-Y (Model 1). The right pathway is for the IKK s-glutathionylation system with the action of NAC built in, where IKK-β is the active component of the IKK complex, circled on the left pathway. On the right pathway, the values of fixed parameters k_5, k_6, and k_9 were used to help reveal the action of NAC. The values of IKK predicted by the right pathway in response to Dox and antioxidation could be used to modulate IKK in the NF-κB pathway.

possible under the treatment conditions she employed in the wet-lab experiments, to narrow the ten down to three possible mechanisms. She used simulations of different concentrations of NAC to build the models of these candidate mechanisms and to fit them to the experimental conditions in order to choose the best of them. The fitting process revealed certain modified parameters that suggested NAC has both pro- and antioxidant effects (see figure 6.3, right pathway). Through more literature searches, simulations, and parameter fitting, she was able to postulate a mechanism by which NAC operates on the glutathionylation process. Her model reproduced the nonlinear response of the IKK-β protein. Above certain concentrations it has pro-oxidant effects. In clinical terms, her findings indicate that it might be counterproductive to use NAC to treat cancer in certain instances.

When C9 began what she thought would be her final examination of the NF-κB pathway, she had not intended to build a model to describe NAC effects, but found herself compelled to in her attempt to explain the counterintuitive effects she saw in her experiments. She separated the use of a model in this instance from other uses she had made in the first two phases, where she had needed the model at the outset of her investigation to develop a basic understanding of the phenomena. Here, she had presumed she could do without it and rely on a black-boxed relation (an "*arrow*") for the input of antioxidants into the system, but had found herself short: *"It was doing an experiment, seeing that it was crazy, and having to build a model. . . . With the other two I started with the model and used the model to try and inform experiment. And in this, I did the experiment not expecting to build a model, but had to build the model to explain what I saw experimentally. So that's the relationship there."*

She found this final model-building process quite difficult, and it was the one time she expressed the wish that she had been able to work with the director of lab G to build a better model. But time was short to complete her dissertation within the allotted time. The model applies a relatively straightforward set of functions to the relationship between NAC and the rates ('k's in Model 4, figure 6.3) of various protein syntheses in the ROS network. The process of building Model 4 was tightly reasoned on the basis of chemical reactions she found in the literature. The model was able to capture their nonlinear interaction and to replicate the nonlinear effects observed in her wet-lab experiments. In all, although not an exhaustive representation, C9 thought *"it's a good enough approximation to get things done."* As she explained what she felt she had achieved in this part of her research, *"Since the experiments were good and done, I think we accomplished what we needed there, but . . . [in] the generation of the model—I wouldn't necessarily call it a full-blown model, it was more of a mathematical framework to try and understand what was going on."* This framework could provide a basis for another graduate student or postdoc to work from. She also felt, *"in hindsight,"* that if she had collaborated more with the lab G director, she could have gotten more *"feedback from him to tweak the model and make it a little more robust and applicable. . . . But I feel I've done enough in my doctoral career."* So, although it might have been possible to negotiate more time to her degree, given its interdisciplinary nature, she also felt she had spent enough time as a student.

This work proved to be the final research of the dissertation, and the focus of her third paper. However, as she discussed in her final post-defense interview, the primary aim of discovering how manipulation of NF-κB affects cell viability through ROS signaling remained an open and unresolved problem. When looked at as a whole, she saw the result of her PhD research as reinforcing the idea that *"redox mechanisms do play a role in chemotherapy administration and more attention should be paid to those mechanisms."* She characterized her research as having opened up the discussion without having come to precise conclusions that could be used clinically. In acknowledging the limitations of their research into Dox and NF-κB, she noted, *"I would say [it is] even more far removed from a clinical setting because we only looked at one anti-oxidant, a particular range, and we didn't even look at the effect of altering NF-kB signal transduction on viability. All we said is this does alter NF-kB. There is a huge question that's left to be answered that is: once you have found out that it alters NF-kB what does that mean for viability?"*

In the end, her research never managed to make the final connection between the NF-κB pathway and the insensitivity of EU1 cells. Overall, her findings definitely were novel and important, but as with most PhD research, time constraints, and the feeling of having *"done enough,"* at this point shared by the student and the committee, in the end led C9 to not continue on to the *"viability"* question. She settled (or satisficed) for having shown that Dox did interact with certain parts of the system without having determined the mechanism by which it produces its ultimate effects. She had made substantial progress, though, and based on her research, a future member of the lab, or someone else in the field at large, could pick up the research from where she left it. The complexity of the biological system was greater than C9 and the lab director had anticipated. As the director would say, "*we started out doing this project, and realized we kind of had to back up a few steps.*"

6.4 Epistemic Affordances of Coupling Simulation and Experimentation

The foregoing analysis of C9's investigative pathway tells the tale of an intensive research process that ranged over different systems and took turns and detours on the way, meeting unexpected obstacles while making valuable discoveries. Our reconstruction and analysis of her model-building process offers important insights into the ways in which the bimodal strategy can provide researchers in ISB another means to manage the complexity of modeling biological systems through creating a kind of distributed cognitive-cultural system different from the typical one, which requires collaboration between modelers and experimentalists. C9 built a D-cog system that, through their intensive interaction, coupled experimentation, computational modeling, and mental modeling in her model-based reasoning processes. The epistemic affordances of this system enabled her to undertake the challenge of building models of a system for which there was little understanding, and which required new experimental information and experimental testing to succeed.[9] As a result of being able to collect the data she needed as she was building the model, she was able to find most of the parameters required to fit it and to test its simulation outputs, so there were considerably fewer arbitrary features in the models.

The system-nature of her research practices also informed and shaped the ways in which she was able to make progress on her epistemic project. Chief among the affordances of the bimodal strategy is the ability to do her own experiments when and how she determined, which facilitated, significantly, her ability to fit and test her models. As we saw, she was able to design experiments to target the specific data she needed. As she described her process, "*I like the idea that as I'm building my model things are popping up in my head: 'oh wow this would be a good experiment'. I plan out the experiment myself and then go into the lab and I do it.*" Further, those processes were much more efficient when compared with modelers who have to rely on experimental collaboration alone—often significantly delayed, as we saw in lab G. For one thing, the bimodal strategy solved the considerable problems pure modelers face when trying to convey their experimental needs to pure experimentalists: *"I personally think [my approach] is better only because—I could tell someone what I needed, but—I think, not really understanding the modeling aspect, they can't accurately come up with a good enough experimental protocol to get what it is I need."* Importantly, doing both modeling and experimentation enables the researcher to *"understand the links between them."*

In collaborative relationships, experimenters and modelers interact with respect to a model, but as we have witnessed, and been told numerous times, there are typically many inefficiencies involved in such relationships. On the one hand, experimentalists often do not understand what kinds of experiments and data modelers require, because they do not understand model-building. On the other, modelers do not understand the nature of wet-lab experimentation sufficiently to know how to frame a request appropriate to the affordances and constraints of experimentation or, more precisely, those of their specific collaborators. Experimental collaborators usually conduct experiments to test hypotheses derived from the models only after the modelers are finished, not during the building process. A failed collaboration within lab C of C7 and C11 bears out this point. C11 tried to conduct the experiments C7 needed in the course of building his model, but the lab director ended up having to take over the experimental work to get at what the modeler needed. Only much later, while she was taking the biosystems modeling course, did C11 realize what was needed by C7. As she formulated it, *"You know, I wish I had taken this class two years ago . . . and it would have been very helpful for me to understand what kind of data he needed, to understand what kind of questions he should be asking of me. . . . [I] didn't have an insight to what exactly—what kind of data would be useful for him."* She also noted that C7 had begun conducting his own experiments, so he had *"figured out the same thing from his end. It's easier if you have more knowledge on the other side."* Chapter 7 discuss insights from our experiences in these labs on ways to facilitate collaboration in bioengineering sciences and other interdisciplinary practices.

C9, of course, was not confined by such problems. As a result, she was able to coordinate her modeling and experimental activities efficiently and, most importantly, to make sure that her experimentation was well-adapted to obtain the information she required, within the constraints of her experimental abilities. This coordination gave her the ability to run experimentation and modeling (computational and mental) as an effective coupled system. Here, as with the coupling between model and modeler I discussed in chapter 5, the overall effect of the back-and-forth exchange between these components of the D-cog system is to extend human inferential powers so as to facilitate the modeler in building out the pathway while building out the model and improving its parameter fit, as well as to provide model validation. Such a coupled system has all the epistemic affordances of a modeler-model coupled system we discussed in chapter 5. However, adding wet-lab experimentation to the coupling provides significant new epistemic affordances in that it enables the modeler to discover and extract relevant and needed experimental information from a complex jumble of biochemical interactions, as well as to control and direct the information flow in the course of building models.

Further, wet-lab biological experimentation, in general, has important epistemic affordances that "google biology" lacks. Experimental engagement with target systems provides another, different kind of epistemic access to the target system from that possessed by those who do only modeling. For one thing, in developing the ability to design and run biological experiments, C9 was able to develop her own sense of what is "reasonable" biologically, which we saw was a constant question for lab G modelers. Her embodied engagement with the biological entities informed her mental models and, undoubtedly, gave her an understanding different from unimodal modelers, for whom these are abstract entities, known only through literature, pathway logic, and simulation.[10] C9, too, talked about getting a "feeling" for the behavior of her models, and, as with the lab G modelers, the system properties and dynamic behaviors of the biological system were only available to her through the model simulations. However, her "feeling for the system" was not based solely on the model and literature, but also on her engagement with the phenomena as she designed and conducted experiments. As we saw with lab G, aspects of developing a "feeling"—for the model or system—in the context of research include a growing insight into and understanding of the target system, belief in the credibility of the model, and affective engagement in the research.

6.4.1 "I Did Them at the Same Time": The Coupled System

To discern the epistemic affordances of this kind of coupled system, I start from an overview of C9's modeling practices. As with other model-based reasoning, once she built an initial computational model, it became the main cognitive apparatus through which C9 interpreted the experimental data and advanced her research by her choices as to what to investigate, experiment on, and model further. She relied on the behavior of the model itself to inform her on the properties and functions of its parts, particularly to reveal the existence of missing parts. The model provides information only available at the system level, such as the relationships between indirectly related variables. As she ran simulations and produced visualizations of the resulting variable and parameter relations, each model provided information otherwise inaccessible to wet-lab experimentation. For instance, the model provided measures not possible to determine in experiment, and simulations helped to reveal which pathways bear most of the burden in a system or which points are relevant to control a network. Determining network control points enabled C9 to black-box mechanisms that did not appear to affect control, as well as to target her experiments. In comparing experimenting in building the model (simulation) and in the wet lab, she pointed out that she could "*tease apart*" things in experiments with the model that she could not in the wet lab and claimed, "*You have access to everything in the model . . . in that sense, I can actually see deeper into the experiment in my model than I can in the real experiment.*" However, she had more "*trust*" in what she could "*see*" in the wet-lab experiment, which "*obviously is showing you what actually happens.*" Of course, what she could "*see*" there is what "*happens*" in the context of an experimental—in vitro—situation.

As we saw, in many instances C9 conducted experiments while in the process of building and running simulations with her models. She called her research process "*iteration*" between them. When we asked which was driving her results—models or experiments—she replied that she was doing them nearly simultaneously: "*You can say that I did my model first, but I don't . . . I don't see it as I finished my model and then I did my experiments. I see it as kind of like I did them at the same time.*" She referred to their interaction as "*synergistic.*" We detailed the aspects of this coupled interaction by looking at the reciprocal roles model-building played in directing her experimentation, and conversely the role experimentation played in her model-building. Figure 6.4 provides our schematic of the way her model-building and experiment interacted.

C9's usual strategy was, in the first place, to construct a diagram of the topology of the biological pathway based on her extensive review of

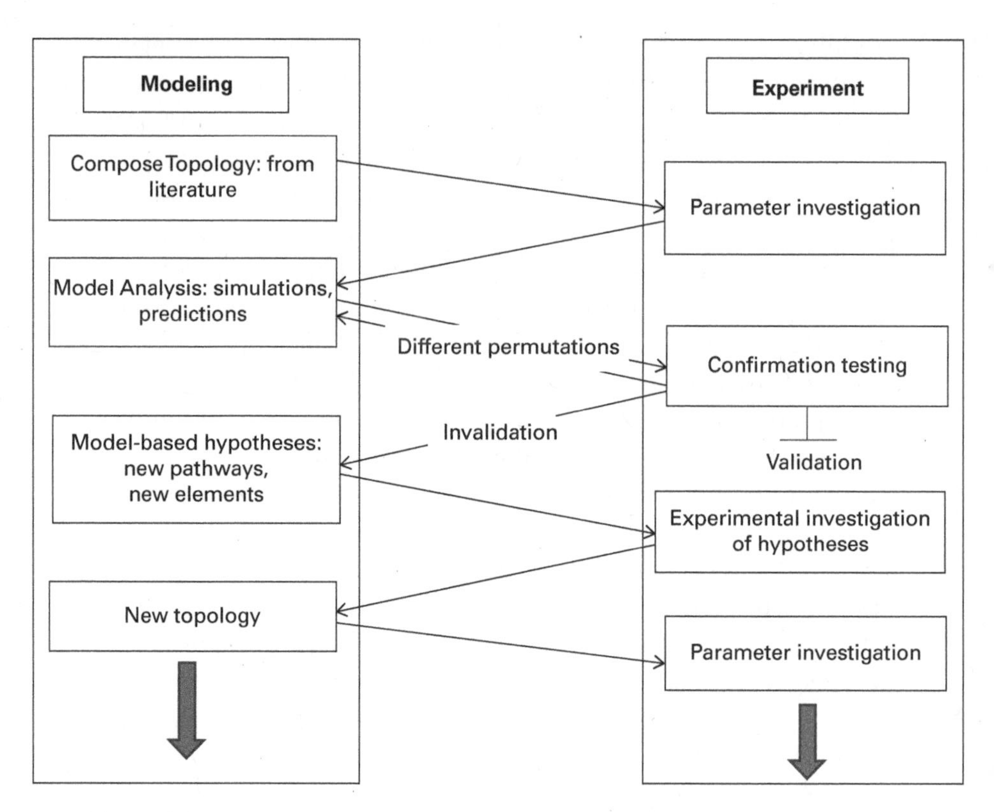

Figure 6.4
The diagram illustrates the coupling of modeling and experimental procedures in the course of C9's model-building as an iterative interactive process of discovering the correct topology and parameter values for the networks on which she was conducting research. The process ends when validation is achieved. Experimental results that depart from the model simulations ("invalidation"), require the generation of new hypotheses, which are derived from the existing models or their modification/elaboration.

the literature. This process is much the same as what we saw in lab G, that is, she would trace and piece together parts of the network from different sources in the literature and add or alter components on the basis of simulations of the preliminary model. In our interviews on how she was building the pathways, she describes an iterative and incremental interactive process that contributes to building her model-based understanding of the system. She noted the specific value of the visual nature of these pathway representations, that are themselves "*models*" that are "*just visual now. They don't have any math affiliated with them. I have to visually see what's happening before I can start writing the equations."* As I have discussed, the pathway representation provides both a conceptual and a visual model of a causal chain of interactions within the biological system that lead to the observed data. C9 further explained her process of *"visually seeing"* to us as, *"I have a notepad where I do a lot of my calculations, thinking, sketching, different things. I have notes and sometimes, like right now with the NK-κB model, I have to come up with some of the connections. I have to hypothesize for some of those because it's not known in the literature. So, I'm going back and forth drawing different things."* These processes of diagramming, sketching, and calculating on paper while thinking and making inferences contribute to coupling her mental models with the pathway model and the computational model.

As with lab G modelers, to build the topology, she also needed to collect as much parameter information from the literature as she could. However, after she had gathered and pieced together whatever information she could find through those searches, there was an important difference: her next move was to rely on her own wet-lab experimentation to obtain pathway and parameter information that was not available in the literature. Once she put together this information in the form of an ODE computational model, she could begin the process of running preliminary simulations to get a feel for the model's behavioral dynamics and to build out the pathway, as well as to produce predictions that she could investigate and possibly verify experimentally. The bimodal strategy enabled her to run simulations and experiments nearly simultaneously, as we saw in the drug Dox (Model 3) case. She could try to replicate interventions and perturbations she made in the models in the physical systems as, for instance, she did with the aid of known blocking or suppressing agents.

The systematic and organized manner in which she coupled experimentation and model-building had a range of epistemic affordances. A prime affordance was the ability to evaluate, experimentally, parts of the model as she was building it. When simulations did not replicate the experimental data, she could rely on the elements of the model she was relatively confident about and could probe those parts she was not. In this way, her models and their limitations provided resources from which she could make inferences as to how to direct her wet-lab experimental activity, as she tried to track down solutions in the form of new pathways and new functional metabolic elements that could fill out the structure of the models. The salience of these elements would not have been apparent on the basis of a program of experimental investigation alone, but required the system-level perspective provided by the model to be identified. For instance, when building Model 1, she discovered the role of cysteine by virtue of the model: "*I sort of reached that IKK . . . needed to be disulfide bonded in order to be active. I reached that through a conjecture because I was like 'something is missing.. When I was drawing my pictures, I was like 'okay, there is a big gap here. How is it that IKK has to be oxidized to be active?' And I was like 'There must be a redox sensitive cysteine.'*" She noted at another time that, "*our model can help us tease out that information . . . like when the model helped us tease out that thioredoxin dependent proteins are basically reacting more speedily with hydrogen peroxide than like glutathione dependent proteins. That's something we couldn't have gotten at without the mathematics.*"

In other instances, when things turned out unexpectedly with her experiments, she could build a model to figure out the problem. When, for example, she discovered that her experiments provided results that were in opposition to her model-based (Model 1) expectations that NF-kB levels would be up-regulated in EU1 cells and down-regulated in the EU3 cells, she built a new Model 3 of the production of ROS by Dox. As we showed in the case analysis, that model informed which questions to ask and experiments to run, and which "*arrow*" in the pathway to pursue as the avenue to best resolve the discrepancy between her models and the data. She again needed to build a model (Model 4) when she discovered, contrary to her expectations, the non-monotonic behavior of IKK-β with respect to NAC levels. She talked repeatedly about how she used the models as explanatory tools, saying, for instance, in this case, "*We have to go to the model to explain what we are seeing because we can't do it experimentally . . . and going to the model and kind of pairing it with the literature evidence, we were able to come up with an explanation that we believe sort of encompasses the changes we saw. . . . I couldn't explain what I saw without the model. It just wasn't possible for me to do that.*"

The repeated coordinated interaction of experimentation and simulation created a coupling between these processes and her mental models. Through this coupled system, C9 could construct the topology of these complex systems by going back and forth to gather data, measure parameters, and make new hypotheses. Her simulations produced novel behavior that would tell her to "*pay more attention to this [novel model behavior]*" on the "*experimental side.*" Experimentation provided a source of ongoing epistemic validation for the model and reinforced its role as a platform for hypotheses. It also informed the mental models through which she reasoned about how to improve her computational models, what specific experiments to undertake, and the dynamics of the phenomena. Further, as Wimsatt has perceptively and correctly noted, "*processes of validation often shade into processes of discovery*—since both involve a winnowing of the generalizable and the reliable from the specific and artifactual" (Wimsatt 2007, 56; emphasis added). The discovery dimension of coupling is highly significant in C9's case. By running controlled experiments and simulations side by side, C9 was able to discover reliable pathway structures, novel biochemical mechanisms, and parameter values with precision, as well as to build robust, validated models.

Through coupling experimentation and modeling, C9 was able to manage the complexity of the problem-solving tasks she faced, because the strategy provides an efficient and effective means by which to explore and constrain the large possibility space of her tasks. As we saw in chapter 5, ISB modelers, because of limited data or data of the wrong kind, face large possibility spaces that are difficult to search through because the systems are nonlinear, and so it is much more difficult to determine what pathway structure or parameter hypotheses might be in the neighborhood of a solution. They often have to test a large number of alternatives. The bimodal strategy, however, provides advantages for narrowing down the search through a parameter space. For example, C9 was able to diagnose and localize uncertainties in her models and to discover relevant sources of information to improve them by simulating and comparing parts of them against controlled experiments. Such localization afforded her an ability to posit and to test, experimentally, tractable hypotheses about uncertain mechanisms in order to extract relevant information she lacked.[11]

In particular, C9 used perturbations of the model and experimental perturbations of the biological system to derive strengths and weaknesses of the models she had built from the available literature. She could isolate parts of the models, simulate those interactions, and check them against experiments that isolated these relations physically in order to establish the accuracy of those parts or to measure their parameter values. When, for instance, her experiments showed differences in NF-κB activation for EU1 and EU3 from what she had expected from her model and the literature, she was able to trace the problem to "*a single arrow*" in the pathway through a combination of wet-lab experimentation and model simulation. She could also use her models to identify sources of biological information necessary to improve the model. In general, as the result of a coordinated probing of the model by means of perturbation and controlled simulations and of probing of the physical biological systems with targeted wet-lab experimentation, C9 was able to establish, robustly, the parts of the model that were accurate, as well as localize the problematic parts or parts that were relevant to further development of the model. Once localized, she could conduct experiments to run through various model-based hypotheses about the interactions of those components, as she did when she formulated a mechanism for the IKK-β s-glutathione system and the antioxidant NAC. Her further wet-lab perturbations to the biological system, such as by changing chemical inputs and using other chemicals to suppress specific reactions, could be checked against the dynamics predicted by the model and used to fix pertinent model constraints. This gave her, in particular, the ability to fine-tune the specific parameters of the reactions and interactions she was experimenting on without having to do the kind of large-scale algorithmic fitting we saw in lab G. For instance, rather than having to model each of the ten possible mechanisms for NAC and trying to infer a best fit, C9 used controlled experimentation of their mechanisms to narrow the model-building task down to three.

This case analysis provides an exemplar of how the bimodal strategy creates a coupled problem-solving system that serves to manage the complexity of model-building in ISB. C9 relied on information embodied in her models, derived from experiments and culled from the literature, to triangulate the location of inaccuracies and missing elements. This bimodal strategy enabled her to build confidence in the parts of the models as she built them through wet-lab testing and perturbation, which allowed her to localize problems and locate and reduce uncertainties. She was able to narrow down avenues for solving model-building problems to a limited set of possibilities, for which she could make hypotheses and test them through controlled experimentation. The bimodal strategy, in particular, enabled her to avoid the challenge faced by unimodal modelers that, given limited experimental data, they have to sort through pathway structure alternatives and parameter values algorithmically, knowing these often will need to be refit. However, this strategy does have limitations, which include significant ones with respect to what and how biological systems can be modeled.

These limitations underlie what the lab G director called a *"philosophical divide"* about how best to build models, as well as train modelers, in ISB.

6.4.2 Limitations of the Bimodal Strategy for ISB

Although some kind of interaction between experimentation and simulation is required to model complex biological systems, the three-way coupling (modeler–model–experiment) of the bimodal strategy is atypical in the current landscape of ISB practice. For the most part, interaction does not extend beyond that of an experimentalist supplying initial data and then data to verify or falsify the model's predictions, as we saw with the G12 case. Such a "collaborative" strategy is often fraught with difficulties, as we have witnessed and heard described by lab G researchers. In some instances, it can even lead to the abandonment of a line of research. It is not easy for a modeler-experimentalist team to overcome a range of barriers created by specialization and create an efficient and effective problem-solving system. The bimodal modeler does have an advantage in this respect. However, one should not take away from the C9 case that a bimodal strategy is necessarily the best, or most effective, epistemic route to building models ISB. There are trade-offs associated with each approach. Further, as we saw, the bimodal approach does not mitigate, completely, the epistemic interdependence I noted to be inherent in ISB. First, modelers using the method still rely on the experimental literature to build and validate models, and, second, since they have not been trained as biologists, they still need to rely on the expertise of bioscience advisers or of collaborators who might be on their project, even though they conduct all the experiments themselves. They are, of course, better equipped to take advantage of the expertise and to collaborate than dedicated modelers. All the experimentalists we interviewed in connection with research in lab C or Lab G were aware of the practices of both labs and uniformly expressed their opinion that the bimodal strategy was the better approach, since it was likely to enhance collaboration between the fields, and expressed the hope for this *"new breed of students"* to become dominant in ISB.

Importantly among the trade-offs, there are limits to the scale of the system a modeler can manage reliably by a bimodal strategy. The strategy worked well in C9's case because she was dealing with relatively small-scale systems, with a manageable number of unknowns. This fact enabled her to contain her experimental work and direct her model-building to keep them within reasonable constraints. Because she had sufficient data to develop detailed models of the biochemical interactions, the need to use mathematical averaging techniques and fitting algorithms was reduced, along with the potential for error these bring with them. So, she did not require the level of mathematical and computational sophistication we saw with lab G to build useful models of her system. Even so, she did not get as far on her problem as she had initially thought she would, and as would be needed to solve the clinical problem. She, herself, felt she could have benefited from more sophisticated modeling skills. By contrast, as we have seen, a dedicated modeler can handle a significantly larger system either because there are sufficient data available or experimentalists with whom to collaborate or because they possess a high level of mathematical and computational skills that they can use to try to get around the lack of data. As we have seen, the rhetoric surrounding ISB promotes its promise to build models of large-scale systems, and so favors a unimodal approach.

Another way to get around the lack of experimental data for large-scale model-building, and to provide it in a timely manner approaching the efficiency of bimodal strategy, might be to have a lab organization that comprises both dedicated modelers and dedicated experimentalists. This was the kind of lab in which the lab C director did her postdoctoral research. The lab G director felt this strategy would require supporting a huge lab financially, which is not usually practicable. He stated that the ideal ratio is, as he has *"said it for twenty years, you need ten experimentalists for every one modeler."* This is because experimentation and modeling work on different, asynchronous, time scales. Wet-lab experiments needed for specific data to build the model or to check the outcome of a model simulation can take many months. We often saw lab G modelers waiting around for data from collaborators to continue their modeling work, while in the meantime they worked on algorithm development. On the flip side, building a robust, stable model takes a long time, too, and, as experimentalists told us, they had often *"moved on to working on something else"* by the time the modeler gets back to them with hypotheses to test, and they are unwilling to redirect their time and money.

As to modelers doing their own experiments, the lab G director expressed worries about the risk of experimental bias or the unwitting manipulation or interpretation of results to fit the model: *"If you produce the data to validate your model, implicitly or explicitly . . . there is a lot of room for interpretation."* He felt the best way to avoid this risk was to have someone else do the experimental validation research, particularly at some arm's length (in different labs). On the other hand, his concerns contrast with what the lab C bimodal modelers told us. For instance, C7, who had considerable experience building models from data collected by others before he started to conduct experiments, claimed, *"I feel like I have more confidence in the data if I'm doing it on my own part rather than having someone else do it."* Of course, confidence does not mitigate the possibility of bias. Another plus for bimodality, though, is as C9 told us and as was confirmed by experimentalists, it is often difficult for a modeler to convey—or for an experimentalist to determine—what experiments will satisfy their needs.

Perhaps most importantly, for the question as to whether the bimodal approach will become more common, is that to become skilled in both experimentation and model-building takes significant time and can involve compromise in the level of skill developed, unless, perhaps, one opts for intensive sequential training, either by taking a longer route to the PhD or in postdoctoral research. The lab C director, however, felt she had *"lost time"* by training sequentially, but clearly, she was more accomplished modeler and experimentalist than the students in her lab would become in the customary five-year PhD research period. The lab G director thought the sequential approach to training was best if one wanted to adopt the strategy. Backing up the lab G director, the postdoc who was then training in his lab to be a modeler after getting an experimental PhD told us he would not organize his own lab to have graduate students use the bimodal strategy: *"What I get from my personal experience is that I lose time going from one side to the other. . . . Both fields have their own problems, you know. . . . A student needs to have his brain focused—and performing experiments is an art, just like modeling."* Modelers in ISB tend to come from applied mathematical or engineering backgrounds and have little if any knowledge of experimentation. But they also need to be trained in the *"art"* of systems biology model-building, which requires that they not only learn how to apply and further develop their mathematical and computational skills to model biological systems, but also develop new skills for how to search for the data in literature and databases (conduct "google biology"), as well as for how to develop a basic grasp of the nature of the system. It is no easy task to learn these skills, especially given that today one might be modeling a yeast system and tomorrow a cancer-producing system. And, we would add from our investigations, modelers need to invest in developing collaboration skills much more than is currently done. Researchers who adopt the bimodal strategy need to develop all of these skills, but in addition they need to learn experimental design and benchtop skills.

As noted, with respect to bimodal training, the *"philosophical divide"* extends to whether to learn experimental and modeling practices concurrently or sequentially and, in the latter instance, whether to do so in the context of a PhD, or first do a modeling PhD and then learn experimentation as a postdoc or vice versa. Concurrent training risks the consequence, as the lab G director noted, of ending with *"modeling lite and experimenting lite,"* but sequential bimodal training puts the researcher in the position, as the lab C director noted, of basically *"starting over"* after several years of education. As we saw, C9 did express some regret, in our final, post-PhD defense interview, that she had not been able to develop her modeling skills more, especially that she had not taken sufficient advantage of working with the lab G director. She felt that having those more sophisticated skills might, in particular, have helped her to work better with the complex nonlinearity of the system in the NAC case. But she also said she would not have given up the ability to couple experimentation with her model-building to rely on collaboration instead.

6.5 Summary: Getting a Grip with/on Bimodal Model-Building

The bimodal strategy can be understood as a particular response to features of the problem-solving contexts of systems biology: namely its complex problems, lack of theoretical starting points, lack of data, and other constraints. With the bimodal strategy, experimentation and simulation are closely coupled in the model-building process, not only to validate model-building steps, but also to provide an effective means of limiting search spaces and triangulating on a good representation. The bimodal strategy has both advantages and limitations when compared to the unimodal strategies we have studied. It does offer unique epistemic affordances.

In the central phases of her research, where C9 needed to use the bimodal strategy, she gave much more weight epistemically to wet-lab experimentation than to simulation. The uncertainty over the structure and properties of her systems prevented her from building a model by starting from accepted assumptions and filling in numerical details of an established modeling framework. She was never in a position where she could rely on a model to conduct only simulation experiments on her systems. The limitations of the models, on the other hand, proved to be resources that directed her experimental activity to track down solutions in the form of new pathways elements and new functional metabolic elements that could fill out the structure of the models. The interaction of experimentation and simulation provided continual epistemic validation for the model-building steps along the way and reinforcement of the model's role as a platform from which to both build understanding of network dynamics and make hypotheses about the behavior of the biological target system.

As we have seen throughout this book, methodological innovations enable scientists to create or enhance their native cognitive powers by building distributed cognitive-cultural systems appropriate to their complex problem-solving tasks. Most often, philosophers and cognitive scientists, when they address cognition and complexity in science, have focused on heuristics that help narrow down search spaces for complex problems (e.g., Newell and Simon 1972; Wimsatt 2007). Cognition is interpreted in terms of its constraints, which heuristics help overcome. We have seen, for instance in the case of mesoscopic modeling, ISB model-building methods can, indeed, be developed with cognitive limitations in mind. The bimodal strategy, however, is not just a means to employ heuristics to deal with cognitive limitations. The bimodal strategy enables the modeler to actively distribute her cognition and build her cognitive powers through coupling simulation and experiment. In the C9 case, running simulations provided a method by which she could calculate and visualize dominant network patterns, which, in turn, helped her develop and simulate mental models of the system dynamics. She was able to infer from the understanding obtained through these simulations how to direct experimentation, manipulate the biological materials, and interpret the results, thus turning experimentation into a sharper investigative tool that could help her efficiently and intelligently search through the space of network and parameter possibilities. The whole problem-solving system served to augment her cognitive capacities to investigate the complex biological systems.

Direct experimental engagement with the biological systems also served to reduce the risk of error. In particular, she developed sophisticated strategies to localize inaccuracies in her model, and thus localize quite precisely where wet-lab experimentation was required. She would do this by running sets of controlled simulation experiments (fixing particular sets of parameters and varying others) against controlled biological experiments that replicated the model controls physically. In this way she could work through the model to pick out specific problematic relationships and investigate them further. This fine-tuned operation gave her a specific ability to handle the complexity of her problem-solving task, not available to unimodal modelers. Unimodal modelers are usually forced to explore complex large parameter and structure spaces algorithmically or with Monte Carlo methods to discover approximately accurate structure and parameter values that fit the whole model to the data. C9 could cut down the space significantly with well-targeted experimentation on specific relationships. Once she localized errors to specific structural hypotheses or parameter values, she could conduct wet-lab experiments to run through sets of hypotheses about the interactions among those components and measure their parameter values. When it came to parameter-fixing, C9 declared that she had always at most two to three parameters to fix. This compares with the upward of twenty that most lab G members report.

In sum, the bimodal strategy tightly integrates the two modes—modeling and experimentation—into a system to generate and validate information about complex biological systems. This *coupled methodological system* provides a novel means to manage the complexity and uncertainty of biological systems, but its challenges and limitations make it not yet a widespread methodological choice in ISB. It is not evident whether bimodality will be the choice of a *"new breed"* going forward; however, one thing is clear from our investigations: modelers and experimentalists need to become more conversant with one another's epistemic practices, norms, and values than is the current situation.

7 Interdisciplinarities in Action

> The moment you cease observing, pack your bags, and leave the field, you will get a remarkably clear insight about the one critical activity you should have observed—but didn't.
>
> The moment you turn off the tape recorder, say goodbye, and leave the interview, it will become immediately clear to you what perfect question you should have asked to tie the whole thing together—but didn't.
>
> The moment you begin analysis it will become perfectly clear to you that you're missing the most important pieces of information and without those pieces of information there is absolutely no hope of making sense of what you have.
>
> Know, then, this: The complete analysis isn't . . .
>
> Analysis brings moments of terror that nothing sensible will emerge and times of exhilaration from the certainty of having discovered the absolute truth. In between there are long periods of hard work, deep thinking, and weight-lifting volumes of material.
>
> —From Halcom's[1] Iron Laws of Evaluation Research
>
> (Patton 2002, 431)

The ethnographic research and analysis presented in this book makes no claims to being exhaustive. No less a master of ethnography then Clifford Geertz acknowledged, "I have never gotten anywhere near to the bottom of anything I have written about. . . . The more deeply one goes, the less complete it is" (Geertz 1983, 58). After twenty years of wrestling with ethnographic data as I have tried to fathom what there is to be learned about the epistemic practices of the bioengineering sciences and the nature of interdisciplinary research, I concur with the sentiment, but also with his claim to have made progress at least by "a refinement of debate" (1983, 58). I do claim to have made progress—and furthered debate—on the question I posed at the outset:

"How can we understand and account for the epistemic accomplishments of science given that scientists are limited beings and the natural world is vastly complex?" I framed one, significant, approach to answering that question this way: to seek to understand how cognition and culture are integrated in the modeling practices scientists create to investigate the world. That framing stemmed from years of investigating scientific epistemic practices as evidenced in historical and contemporary scientific research and from engaging with literatures that take science as their object of study, namely, philosophy, history, sociology, anthropology, and cognitive science. Once we began our preliminary investigations, the centrality of a specific kind of model-building in the research of each lab immediately became evident, and as we further investigated, its role as an integrative practice became evident as well. We focused our research, then, on how researchers create and use this kind of material culture through which they think and reason about complex biological systems that are otherwise inaccessible to them. Our research has taken an in-depth look at problem-solving practices in two different fields of bioengineering sciences, as well as two different subspecialties in each. The detailed case studies and analyses presented here provide important insights into the epistemic project of twenty-first-century biological engineering, which aims to get a grip on complex biological systems by making use of material, conceptual, methodological, and technological resources of engineering to both formulate and work toward the solution of biological problems of significant human consequence. Strikingly, across all labs, our interviewees stated that "helping people" was the primary reason for their choice of bioengineering, which offered them an exciting opportunity to be pioneers in developing approaches to doing so. Keeping that reason in mind also motivated them to persist through times when nothing seemed to be working out in their research.

At the end of each chapter, I have provided a high-level summary of its major points, which, now that you have reached this point, it might be good to review. In section 7.1 I follow these up with a summary of summaries. There I briefly remind the reader of three highlights in particular: distributed cognition as an analytical framework for investigating cognitive-cultural integration, the integrative epistemic practice of distributed model-based reasoning, and how we have seen epistemic warrant for specific models, as well as the modeling practices in general, is developed by the pioneering students and their visionary directors who have served as our protagonists throughout.

In section 7.2 I take a look at interdisciplinarities in action, more broadly considered. I like to conclude my books with a forward look to ongoing research and further implications—yet to be fully worked out—of the research I have discussed. In the case at hand, I look at the current epistemic situation of interdisciplinary practice, and then go on to point out what insights our own ethnographic investigations have thus far yielded, with a view to how such practice might be facilitated. Despite the fact that there have been major successes, clearly there is more research to be done to devise means to facilitate more effective ways to conduct twenty-first-century interdisciplinary research than is widely acknowledged to be the current state of affairs across the sciences, humanities, and arts. Thus, there are important insights to be gleaned from the in situ study we conducted of different kinds of interdisciplinary practices.

A major implication of those insights is how to foster learning in interdisciplinary research communities. What philosophers of science need to realize is that education is part of the *epistemic infrastructure* of scientific research. This issue comes to the fore in emerging fields, which lack established curricula or texts. Although scientists, administrators, and grant agencies recognize the need to build such infrastructure, scientists often cannot step back and see the learning requirements of their emerging practices on their own. This is where philosophers and cognitive scientists can help, and this is why we have been investing a great deal of time and energy to do so. I have given some indication of our educational research in BME in chapter 4. Here I focus in particular on five specific "interdisciplinary epistemic virtues" that we ended up distinguishing, and discuss some of the ways we have cultivated these through targeted educational experiences in BME and ISB. The chapter ends with a look at the responses from some of the ISB researchers who took part in brief experiences we devised to improve their abilities to collaborate, which show that even small, targeted interventions based on studying a kind of interdisciplinarity-in-action can have significant payoffs. These researchers told us with a lively sense of joyful amazement how these experiences had helped them gain new, realistic insights into what moved and occupied their counterparts from the other side of the disciplinary divide.

7.1 Highlights

7.1.1 Distributed Cognition as an Analytical Framework for Cognitive-Cultural Integration

In chapter 1, I argued that the D-cog framework, with suitable extension, provides both a method—cognitive ethnography—and a conceptual and analytical framework to investigate cognitive-cultural integration in scientific practice. In particular, it provides a means to analyze problem-solving in science as a *system phenomenon* in which scientists, as embodied agents, extend and enhance their natural cognitive capacities by building material, social, and cultural environments for problem-solving. A major scientific environment is the research lab, which I cast as a distributed cognitive-cultural system with epistemic goals. The extensions needed to accommodate science are, first, to include in the analysis an account of pertinent cognitive contributions of the individual, embodied mind/brain and, second, to include in the analysis an account of the epistemic aspect of scientific practice. There are multiple dimensions to each, and I chose to focus, respectively, on the role of mental modeling in reasoning processes and on the issues of how bioengineering scientists build models and develop warrant for their models, as well as for their innovative modeling practices.

In each lab we determined the most important cognitive-cultural resources for problem-solving, which include concepts, methods, materials, and epistemic norms and values. These resources are put to use in building models, which serve a dual function as cognitive artifacts needed for problem-solving and as material culture of the specific kind of epistemic community. As we have seen in this analysis, models are the loci of cognitive-cultural integration in the research of these labs. In the various bioengineering sciences labs we investigated, the researchers overwhelmingly came from engineering or applied mathematics backgrounds, and in their research they transferred and adapted engineering resources and epistemic norms and values to formulate and address problems that would enable them to get a grip on complex biological systems. We called the labs *adaptive problem spaces* in which researchers learn to adapt problems, concepts, and methods to manage the complexity of the biological systems of interest. In such adaptive processes, researchers learn the affordances and limitations of these resources for addressing their problems.

The multidimensional notion of interlocking models emerged from our coding process as an important cross-cutting thematic category, specific to cognitive-cultural integration. Interlocking models is a system-level interpretative notion that, in the first instance, captures dimensions of modeling practices that cut across many of our coding categories. In particular, it provides a means to specify in what ways the simulation models in each lab serve as hubs in which various dimensions of the cognitive-cultural system are fitted together as the models are developed and used. As with transportation systems, where many service lines interlock at central stations, models serve to interlock many dimensions of practice. Among the ways in which models serve as hubs, considered in this book, are how they provide the locus for interlocking problem-solving, learning, lab history, and mentoring; for interlocking facets of interdisciplinarity, including concepts, methods, materials, and epistemic norms and values; for interlocking configurations of devices in experimental situations (model-systems); and for interlocking researcher mental models and artifact models in simulative model-based reasoning (dynamic representational coupling).

In each chapter I present a detailed analysis of the epistemic practices of the research lab and specific cases of problem-solving. These analyses are based on many iterations of our research group's rigorous coding, thematic analyses, detailed case analyses, and triangulation of information from a range of data we collected to arrive at our interpretations. Following these thick descriptions, I go on to analyze the theoretical implications of our findings about the practices with attention to specific philosophical notions and theories and show how the concrete details are indispensable for developing the abstract accounts ("productive interplay"). Each chapter provides a different, though in some instances overlapping, focus on the concrete and the abstract based on data from a specific lab, but in many cases, it would have been possible to address similar issues or themes with a different lab. Section 7.2 will look at some important commonalities with respect to interdisciplinary research. In general, the overall analysis presented in this book provides, I believe, a demonstration of the fruitfulness of the cognitive ethnographic approach for the descriptive and normative projects of philosophy of science and for the advancement of the interdisciplinary project to establish a cognitive science of science.

7.1.2 Distributed Model-Based Reasoning

A major epistemic practice of biological engineering fields is to build models as the means to understand or control complex dynamic biological systems. These are systems for which there are no general theories of the biological phenomena under investigation, and so models are built from the ground up with the aid of engineering concepts and methods. Modeling the dynamics of such systems, whether in vitro or in silico, comprises cycles of design, construction, evaluation, experimentation, and redesign, that is, cycles of *building* to discover. In building models, researchers have to manage the complexity of not only the biological phenomena, but also of a variety of conditions pertaining to the kind of interdisciplinary practice, for instance, the lack of time-series data or effective collaboration for ISB or the need to fuse biological and engineered materials while maintaining biological functionality for BME. As we have seen, the course of interdisciplinary model-building is never smooth. Impasses, obstacles, and failures are ubiquitous. Cell cultures die in the midst of an experiment, in vitro model-systems fail to behave as anticipated, in silico models prove difficult to fit, borrowed concepts or methods fail to provide insights or even prove to be obstacles, and so forth. What we witnessed was the remarkable resilience and creativity with which the researchers addressed these challenges, largely through looking at them as opportunities to learn—an attitude that is inculcated into student researchers from their very first days in their lab. What is also remarkable is that, although it often took a few years for a student's research project to solidify, once determined, no one needed to abandon their model in the face of setbacks. Rather they found productive ways to modify it or their interpretation of its behavior.

Model-building not only addresses the research problem, but is also what drives the creation and evolution of the distributed cognitive-cultural systems of each lab. To understand how, in each case, requires numerous dimensions of the dynamics of processes through which models are built to be examined. Importantly, in collecting field observations and interviews around these processes, we were able to log the various methods, steps, and iterations of building; probe the decisions and judgments behind developing and altering that specific model; examine how and what kind of inferences experimental simulations with them enable; track the formation and changes in problems and goals; and note interactions among researchers relevant to the process. Our data provided a wealth of insights about the nature of the epistemic practices that would never have made it into the historical records—not even those of the likes of a Michael Faraday, who kept extensive, elaborate, and annotated records of his research practices.

In both fields, their research problems and goals require researchers to build models as epistemic tools through which to probe and learn about the behaviors of selected system components (endothelial cells of the cardiovascular system) or of the system as a whole (lignin production system in plants). In both instances, the models provide dynamic simulations of biological phenomena under experimental conditions that can be manipulated and controlled by the researcher. In the BME case, these hybrid in vitro simulation models comprise biological and engineered materials that enable researchers to isolate and experimentally control entities and processes of interest. In the ISB case, the models synthesize as much of the available data as possible to provide computational (in silico) simulations of system-level behaviors, such as intracellular signaling and metabolic processes, that enable researchers to perform experimental manipulations under real-world and myriad counterfactual conditions the researcher might consider relevant to the problem.

A central epistemic aim in both cases is to build a model that will provide the basis for inference about the target system, that is, to *build an analogical source*. Noting that these models are to function as analogical sources brings to the fore an overlooked aspect of analogical reasoning that is central to its use in creative scientific problem-solving. Often there is no ready-to-hand analogy to retrieve and map to the target problem. Instead, the analogical source needs to be created specifically to be mapped to the target. Such built analogies—at least of the sort I have examined thus far in bioengineering and, earlier, in physics (Nersessian 1992a,b, 2008)—are hybrid constructions that merge selected features and constraints from both source(s) (often multiple) and target domains. Building is a bootstrapping process in which a model is developed toward becoming a viable analogical source (think of the computational dish model) or refined in the direction of providing a better one (think of the construct–flow-loop model-system). In some instances, such as with the BME in vitro models, analogies need to be built in a nested manner; for example, the flow loop provides an analogy with blood flow shear forces; the construct, with the blood vessel wall; the animal cells and tissues, with human cells and tissues; and the flow-loop–construct model-system, with shear forces in the arteries. Once developed, models provide structural, functional, or behavioral analogue systems through which researchers can reason, about both the model and, potentially, the real-world system. In these pioneering labs, most of the reasoning we observed focuses on the model, especially its current capabilities and limitations and how to make it a better analogical source, which of course requires the researcher to think not only about the biological target but also about a range of resources available for building, including the constraints of the materials and methods. Only near the end of a building project—when the model is deemed satisfactory to the purposes at hand—is it possible to transfer inferences as hypotheses about the target system or claim to have provided scientific understanding about that system.

Our analyses show that model-based reasoning is, itself, a system phenomenon. Put in another way, in building models, researchers *distribute cognitive processes* across material artifacts, a process we cast as building distributed model-based reasoning systems. Typically, distributing cognition to artifacts has been cast as *off-loading* certain cognitive functions or processes to the artifact, such as memory. However, our metaphors of interlocking and coupling suggest a different relation between mental and material resources—that of *incorporation* into a D-cog system. We have argued, in particular, that the repeated back-and-forth information exchange between mental and real-world models during the building processes gradually incorporates them into an inferential system that enhances or expands the capacity of the researcher for simulative model-based reasoning. Two of the instances considered in this book are how the simulation capabilities of in vitro and in silico models enhance the reasoners' ability to imagine and probe counterfactual scenarios (thought experimenting) and how the dynamic visualization capabilities of the in silico models can drastically alter the problem-space. Of course, some processes are off-loaded within the system in that certain cognitive functions are performed by the researchers and others by the artifact, so the metaphors are not incompatible, but our emphasis on incorporation captures the system nature better. Finally, in the case of in silico models, we also considered the affordances and limitations of a D-cog system that incorporates wet-lab experimentation into the model-building process, with the bimodal strategy.

7.1.3 Building Epistemic Warrant

Science is an epistemic practice. As such, its methods and claims need justification, including a specification of their scope and limitations. As we saw, the bioengineering labs are by and large methodological pioneers in the application of engineering, mathematical, and computational concepts and methods to the investigation of complex biological systems. What this means is that they need to provide evidence and arguments for the credibility of both their models and their modeling practices. Our examinations of issues around credibility with each kind of modeling practice again demonstrate the value of an ethnographic approach for traditional philosophical concerns. In most cases, we were able to collect data on the assessments and decisions researchers were making about the makeup and performance of their models while they were in the process of building them and assessing their credibility. We were also able to probe them further about these in interviews. With respect to potential epistemic claims, researchers in all the labs evaluate models in relation to their function as analogical sources, which our analyses show requires an assessment as to whether the model *exemplifies the relevant features* of the biological system with respect to the problem and that nothing germane has been left out, or, if something has, a determination has been made as to ways the inferences from the model are limited. That a model exemplifies the relevant features of the target system provides assurance that the model has the potential to produce candidate inferences to transfer to the target system as hypotheses. Thus, analogy and exemplification are bound together in model-based reasoning. Although I have not demonstrated this here, I contend this relationship is important not only for the cases I examine, but with respect to modeling more widely.

With respect to the BME case, in designing an in vitro model, researchers aim to exemplify relevant biological entities and processes, subject to the constraints both of biology and of the engineering methods and materials used to construct it. They begin research by focusing on what they expect to be salient features of the phenomena, while bracketing potentially irrelevant features or those deemed too complex or not feasible to address at the outset. For instance, the tissue engineering lab deemed the endothelial cells that line the artery to be the relevant entities to study with respect to the problem of determining the effects of mechanical forces on the cardiovascular system on the basis that they are in closest contact as the blood flows through the lumen. The lab built the flow loop to exemplify laminar shear stress forces to a first-order approximation at normal blood flow rates, but also included the capacity to produce abnormal rates and turbulent forces that could be used should higher-order effects be determined to be relevant as their investigations moved along. The latter point underscores that part of a research project with in vitro models is to determine more specifically what are the relevant features. As we saw in the chapter 2 analysis of the design and evolution of the in vitro model-systems in both labs, the researchers are able to articulate the ways in which their model systems do or do not exemplify specific features of the in vitro phenomena that are considered relevant at that time, as well as to provide assessments of why specific features had been selected and in what way their simulation outcomes are delimited by those choices. But we also saw that even when the researchers deemed features relevant to their problems, building a model with those features might have to await developments in engineering methods and materials. Envisioning such further developments is also part of their research agenda. For instance, although the model-system made up of the flow loop and endothelial cells on slides could provide valuable information about morphology and proliferation, the researchers recognized it does not exemplify the functional behavior of the cells in the blood vessel wall because, at the very least, in vivo they interact with smooth muscle cells embedded in the wall tissue. But to build a blood vessel wall model—the construct—to exemplify that interaction and other features of the blood vessel wall required developments in tissue engineering capabilities. However, the negative analogy between the cells and the blood vessel wall opened a research opportunity to develop a better model-system that would exemplify more relevant features. Thus, in vitro models are built toward exemplifying relevant features, which themselves are further specified in the course of the research. A better, potentially more productive, analogy improves or enhances the relevant features the model exemplifies. Because experiments cannot be performed directly on human targets, the inferences drawn from the models are evaluated with respect to whatever data on the target systems are available. For instance, the response of the endothelial cells in the constructs to mechanical stimulation can be compared with genetic markers of the cells in the in vivo system, and so provide confirmation of stimulation effects. Additionally, the construct-baboon model-system marks progress in the direction of specifying the requirements for an arterial tissue graft for humans by experimental evaluation of the behavior of the construct seeded with EPCs in an animal model-system.

With respect to the ISB case, researchers aim to build robust and stable in silico models that exemplify the behaviors of complex biological systems. Here, too, model-building is an iterative and incremental process in which, at each phase, researchers assess how well the model data exemplify (replicate) the available data in the experimental literature, usually by means of comparing output graphs. As the model gains complexity, it develops the capacity to enact known and potential system behaviors. Once a stable and robust model, or convergent ensemble of models, is produced, it has the potential to provide an analogical source from which the researcher can derive experimentally verifiable hypotheses. For instance, we saw that in modeling the lignin system, G10 was able to make inferences about how to modify lignin production to create a better biofuel by knocking out specific genes and even to infer a missing element in the established lignin pathway. His experimental collaborators were able not only to verify these modification hypotheses, but also to determine, in a highly significant collaborative discovery, that missing element. Such discoveries establish that an in silico model not only exemplifies known features but also has the capacity to predict, and so exemplify, heretofore unknown features. In vivo experimental verifications of hypotheses that derive from a model confer credibility on it as an analogical source, as well as on the methods that produced it.

As to the methods, the researchers in all labs transfer and adapt engineering, mathematical, or computational methods largely developed in the context of building or modeling human-made systems to investigate biological systems. Unlike with established methods in a discipline, bioengineering researchers in emerging fields need to build credibility for this transfer and adaptation. As the methods gain credibility and develop an interdisciplinary history in biosystems modeling, they become projectible for future research. As we have seen, the main criteria are pragmatic, centered on success. Do the models built with them provide significant (or at least useful), verifiable information that enables the researchers to make progress on the research agenda to understand or control the behavior of the biological systems? Verification, to the extent and means possible, reflects back on the credibility of both the model and the methods.

In lab A, for instance, the in vitro flow-loop–cells-on-slides model-system provided useful information about the changes in cell morphology for a population because of controlled experimentation with shear forces, which can be compared against various in vivo changes detected in normal and diseased arteries. The success of this method in providing support for the hypothesis that pathological forces cause disease provides epistemic warrant for the continued use of the method of flow-loop studies. Limitations on the kind of information the method can provide can open new avenues of research. For example, limitations of flow-loop simulations with only cells led to the development of the construct, with which diverse cells and tissues could be subjected to flow-loop studies. Experimental outcomes on the functional behaviors of the cells—for example, that A7's preconditioned EPCs produce thrombomodulin in the construct-baboon model-system simulation—can be compared to in vivo cell behaviors. Verification of the outcome, in this case, establishes that the in vitro simulation methods both provide new understanding of the ways in which EPCs can become mature endothelial cells (by stimulation with mechanical forces) and make progress in controlling the EPCs behaviors toward the vascular graft application goal.

The success criterion also is central with respect to the methods used to build computational simulation models of biological systems. Some methods used in modeling biosystems are, of course, long-established computational methods, tested in a range of fields, such as Monte Carlo sampling. But many are being imported for the first time to use with biological systems; these are usually drawn from systems and control engineering, but can also be related to the specific engineering background of the researcher, for instance, the lab G researcher using wave-smoothing techniques from telecommunications engineering. We also saw that the lab G researchers continually innovate in algorithm development to build and fit models, such as the two-step procedure developed by G10 to build models of lignin production. All these methods gain credibility as they produce stable and robust models that are informative about the behavior of biological systems, which means that they replicate known data, as well as predict new, experimentally verifiable behaviors. These behaviors range from useful new information, such as what genes to target in the lignin system to produce a better biofuel, to highly novel and significant discoveries, for which G10's prediction of a missing element in the long-established lignin pathway provides an exemplar. Methodological innovation, as with model-building, is a bootstrapping process.

7.2 The Epistemic Situation of Interdisciplinary Practice

Interdisciplinarity is widely cast as a hallmark of frontier twenty-first-century research in the sciences and engineering. Interdisciplinary research is customarily characterized as "integrative" and "innovative," yet difficult to achieve. The obstacles lie in the complexity of the problems posed, the need to develop novel investigative practices, and the fact that interdisciplinary collaboration is fraught with difficulties that increase with the distances between the collaborating disciplines. Although a broad range of empirical methods is used to investigate these dimensions, studies of the dynamic processes of interdisciplinarity practices—that is, how interdisciplinarity is enacted in situations of scientific research and the challenges posed for researchers—are scant.[2] Further, although detailed taxonomies of different kinds of interdisciplinarity have been elaborated in the abstract since at least 1972 (see, e.g., Klein 2010), richly nuanced accounts of interdisciplinary practices, too, are needed when it comes to thinking about how to promote learning or how to facilitate a specific kind of research. Ethnography has long been a method used by anthropologists to study and interpret cultural practices situated in naturalistic settings. Most importantly for understanding challenges of interdisciplinary practice, ethnographic research enables one to examine both the insider ("emic") perspective of the participants and to develop the ethnographer outsider ("etic") interpretation of practices of interest. As the research presented in this book demonstrates, cognitive ethnography is particularly well-suited to examining the conceptual, reasoning, and learning dimensions of interdisciplinary problem-solving, where differing and often incompatible epistemic practices, values, and norms are in play. The method is perhaps uniquely suited to investigating the processes of integration in epistemic practices because it enables collecting fine-grained data as researchers attempt to solve interdisciplinary problems within a complex context of cognitive, social, material, and cultural resources and constraints. Cognitive ethnography provides nuanced findings about specific interdisciplinary practices—how they come to be as well as how they are used—that not only enhance our understanding of interdisciplinarity but also can help faculty and policy makers figure out how best to facilitate research, especially as they develop programs to educate the twenty-first-century scientist. Although valuable in themselves, findings from cognitive ethnography can also be used to enrich or validate findings from more theoretical or global methods of studying interdisciplinarity, such as bibliometric analyses of patterns of interaction and influence (see, e.g., Roessner et al. 2013).

It is widely agreed that the chief characteristic of interdisciplinary research is *integration*.[3] Integration is held to be what promotes creativity and innovation. What is needed, though, is both a more nuanced understanding of what "integration" means in the problem-solving practices of quite different interdisciplinary epistemic communities and of the specific challenges encountered in trying to achieve it. Cognitive ethnography has enabled us to examine in fine detail how the researchers determined how to reconceive a complex biological system with the engineering and computational resources at hand so as to be able to solve—or at least make progress on—the target problem. In both fields, problem-solving requires adapting concepts, methods, or materials from engineering to manage the complexity of the biological problem. We, thus, characterized these labs as *adaptive problem spaces*, in which different forms of adaptation emerge specific to the nature of the problems and goals and the requisite resources for problem-solving in the field or subdiscipline (Nersessian 2006; Nersessian and Newstetter 2013). In general, adaptation of the complex interdisciplinary systems within these spaces is a process of continually reconfiguring the components from which these are built, as the system gains experience (see, e.g., chapter 4). Research in these adaptive problem spaces requires that the individuals themselves achieve a measure of interdisciplinary integration—in how they think and how they act. The nature of the integration depends on the requirements of the kind of interdisciplinary problem-solving, which, as we will see, differs for BME and ISB.

As one can imagine, long-term investigations provide a wealth of data to mine, and our findings are rich and varied. We do not claim to have captured all the nuances of the range of interdisciplinary practices in either BME or ISB, but we have been able to formulate some significant insights. An important goal of ethnographic research of multiple sites is to assess *transferability*: to ascertain what abstracted insights might be in common across sites and possibly extended to the broader field, and which ones are unique to a site. Many of our findings of the challenges of integrating engineering and biology in BME transferred robustly across the two labs. The ISB labs differed in various aspects of the in silico model-building process. However, our major insights about the methods for and challenges of integrating biology, engineering, and computation in ISB problem-solving practices do transfer. We have presented our findings to audiences of researchers outside of our studies in each field and have done sufficient broad sampling of each of the fields to feel confident that our research provides significant insights relevant to the practices and challenges of interdisciplinary research and training across the fields.

I begin with a discussion of what I have been calling "interdisciplinary epistemic virtues" that facilitate interdisciplinary problem-solving. We derived these virtues from our assessments of the challenges and requirements for successful problem-solving faced by the researchers in the different kinds of interdisciplinarity I have discussed in the preceding chapters. We first determined and described the challenges and requirements as they arose in our coding process, and then considered them in light of analyses of pertinent notions in the literature on interdisciplinarity, where possible. I put this section first, even though the analysis of epistemic virtues came near the end of our investigation, so that I can use the notions in the subsequent discussion of interdisciplinarity in each field.[4] Sections 7.2.2 and 7.2.3 provide a brief overview, focused on the kind of interdisciplinarity, of actual epistemic practices and challenges in BME problem-solving and in ISB, respectively. Section 7.2.4 focuses on some of the challenges of collaboration in ISB, strategies we proposed to help mitigate them, and the enthusiastic insights of the students we tried them out on.

7.2.1 Five Interdisciplinary Virtues Distinguished

As part of the educational contribution of our research, we have sought to distill from our findings some overarching learning requirements for effective interdisciplinary research. We focused, in particular, on determining characteristics that could usefully be cultivated in the course of graduate education. To that end, we first determined the characteristics from our intensive coding of the in situ studies of interdisciplinary practices on the basis of what we found either to be present and effective in the practices of the labs, or to be lacking, and so posing an impediment. We then, where possible, related our codes to concepts in the theoretical literature on interdisciplinarity, while further elaborating both these preexisting concepts and the new notions uncovered in the course of our own research. Overall, we determined five highly significant characteristics that lead to effective interdisciplinary problem-solving:

1. Cognitive flexibility
2. Methodological versatility
3. Resilience in the face of impasse
4. Interactional expertise
5. Epistemic awareness

The case studies in this book provide numerous examples of these characteristics in problem-solving or, in some cases, the consequences of their absence. It is important to understand at the outset that although these characteristics are attributed, customarily, to individuals, on my analysis, they can be features of distributed problem-solving systems as well.

By *cognitive flexibility* I mean the ability to see or understand a problem from different perspectives, which facilitates the kind of adaptation needed to transform a complex problem into one that can be solved. It also promotes collaboration. Strictly speaking, developmental psychologists mean by cognitive flexibility an executive function that develops as the prefrontal cortex matures, not therefore through learning. However, in educational fields, the term is being used broadly in relation to learning, and that is how we use it as well (see, e.g., Spiro et al. 1994; Spiro et al. 1992). We have seen instances of cognitive flexibility in each lab. For instance, in the tissue engineering lab we saw researchers framing the interactions between cells and blood flow from the perspectives of mechanics and of biological properties and functions. In the combined computational/wet lab we saw researchers looking at interactions between cells and a therapeutic cancer drug using the perspectives of systems engineering and of ROS biology to build out and model pathways. We also saw how introduction of the in silico dish model into the D-cog system of lab D provided a different perspective on bursting phenomena in the in vitro dish.[5]

Methodological versatility is having multiple methods in the tool kit with which researchers can tackle a problem. Instances of such versatility we have seen include the ability to draw from computational model-building methods in several engineering fields (labs C and G); or to have facility with mechanical engineering design methods and wet-lab methods for culturing cells and engineering tissues (lab A); or to use both computational simulation and wet-lab experimental methods (lab C). We have also seen the advantage of having multiple methods in collaborative D-cog systems, such as the capacity to use neural engineering signal processing methods, software development, and computational modeling (lab D).

In pioneering interdisciplinary science research, failure, obstacles, and impasses are ever-present, as we have had ample opportunity to observe, so *resilience* is needed to find a way through them and even to see failure as an opportunity for learning. As we saw, cell cultures die, parameter fittings do not work out, collaborators do not respond, and so forth. In the lab D case, in particular, we saw their repeated failure to stop the dish from bursting for over a year, and then, after quieting it, they were unable to sustain a stimulus pattern from which it could learn ("drift" problem). Each researcher demonstrated resilience in trying different approaches to get around the problem. In particular, one member introduced a method novel to the lab (computational modeling of an in vitro model), which not only provided insights to move the research forward, but also created a more cohesive and resilient collaborative research system able to overcome significant obstacles, and, ultimately, jointly solve the problem.

Further, interdisciplinary researchers need to develop interactional skills for collaboration. *Interactional expertise* is a notion introduced first by Harry Collins and Robert Evans (Collins and Evans 2002) to characterize the nature of the expertise required of sociologists doing fieldwork. It marks a distinction between the development of conceptual understanding of the practices of collaborators, which enables each to engage linguistically with the practices, and the ability to perform the practice (contributory expertise). Collins, Evans, and Michael Gorman (2007) extended the notion to interdisciplinary collaboration more widely and stressed that, beyond language, interactional expertise is "tacit knowledge-laden and context specific" (661).[6] Again, our research shows that all researchers coming from either the engineering side or the biosciences/medicine side of biological engineering start with little understanding of the other side, and, where collaboration is required, research is slowed down, if not impeded. But we also saw in the BME case that such expertise can be cultivated when it is attended to explicitly in the systematic development of a curriculum. But, as I show in the ISB case (section 7.2.3), it can also be cultivated through limited informal, targeted interventions. Both kinds of approaches seek to promote individual learning, but aim, also, to create more cohesive and effective problem-solving systems.

We introduced the notion of *epistemic awareness* to call out epistemic norms and values in problem-solving (Nersessian 2017; Osbeck and Nersessian 2017). The notion comprises a metacognitive awareness that one's epistemic identity and epistemic norms and values play an important role in research, and that what constitutes good scientific research can be different from one discipline to another. Epistemic awareness, then, is the ability to reflect on the epistemic dimensions of one's own discipline and research practices as well as on those of the collaborators in the problem-solving system. We introduced this notion, in particular, because we witnessed in all the labs, including the ones requiring hybridization, that researchers coming from engineering and computational sciences had little awareness of the epistemic norms and values of biological research. The bioscience collaborators of the computational labs we interviewed also demonstrated a lack of awareness of those at work in modeling. Finding remedies for these problems remains a major challenge in the developing fields of biological engineering, but the first step is for researchers to become conscious of the need for such awareness. Again, our research has shown such awareness can be cultivated with explicit attention.

From an epistemological perspective, these five characteristics can be cast as epistemic virtues for the conduct of good interdisciplinary research, that is, as *interdisciplinary virtues*. According to Linda Zagzebski, an *epistemic virtue* is "a deep and enduring acquired excellence" motivated by and reliably successful at achieving intellectual ends (Zagzebski 1996, 137). Aristotle first introduced the notion that there are intellectual as well as moral virtues, and Zagzebski asserts along with him that virtues are acquired by practicing them. Interdisciplinary epistemic virtues, too, have sociocultural dimensions that can enhance the possibility of achieving intellectual ends. For instance, cultivating them can promote the development of effective collaborative communities of researchers.

The set we determined is doubtless incomplete, but we found these specific ones to be both central for creative and effective interdisciplinary problem-solving and open to being acquired in principle in both BME and ISB, when appropriate means are devised to cultivate them. We also found that when characteristics (3) and (4) are lacking, this significantly increases the complexity of problem-solving in ISB. In particular, both are interrelated with cognitive flexibility in interdisciplinary research. Generally, the skills associated with these characteristics are not easily acquired on one's own. Further, the challenges of cultivating these characteristics differ with respect to the context, in particular, the current state and aims of the field. As I discuss in the following sections, as part of our educational research, we investigated ways to cultivate such characteristics by means of experiences targeted to the research requirements for BME and ISB, respectively. A brief discussion of our efforts provides an opportunity to highlight, in a different way, the differences between these fields in the kind of interdisciplinarity practiced—and desired.

7.2.2 BME Problem-Solving: Hybridization

The overarching problems BME poses are directed toward how to use engineering design methods and principles to understand basic biological phenomena in order to control disease processes or create interventions for specific medical disorders. The problems investigated in the tissue engineering lab aimed at understanding the biological influences of the mechanical forces of blood flow in arteries, with an eye to determining the requirements to construct living implants that can perform normal functions of arteries. In the neural engineering lab, the problems focused on understanding the network behavior of neurons, in particular by teaching a cultured neuronal network to learn from feedback from its "body." A potential application would be to develop brain-controlled prosthetic devices that neurons can learn to use. In the years of our investigation, both labs were focused on the basic research, especially on how to develop in vitro simulation models as epistemic tools to investigate complex biological processes. In chapters 2 through 4, we examined in depth the epistemic practices of in vitro research in each lab. Here I focus on salient interdisciplinary features labs of these kinds have in common.

BME researchers develop programs of in vitro research that build physical simulation models to investigate selected aspects of complex biological systems because the problems the field poses require a level of control that either is impossible to achieve in animal research or would be unethical to conduct. These simulation devices are hybrid artifacts in which cells or cellular systems interface with nonliving materials in model-based simulations that are run under various experimental conditions. In each BME lab, more than one device was central to the research program. Because the devices are created to address the specific research problems of a lab, they are usually built in-house through several iterations. The devices participate in experimental research in various configurations of hybrid model-systems. The daily challenges of building devices and model-systems require the researchers to determine the relevant, selective interlocking of biological and engineering concepts, methods, and materials for the problem at hand. The ongoing processes of building simulation models create emergent hybrid problem-solving systems—with artifactual *and* mental components—within the adaptive problem spaces of BME. We, thus, coded the chief characteristic of interdisciplinary integration as we saw it enacted in the BME labs as *hybridization* to capture the processes of combining distinct elements into an inseparable whole.

One way in which problem-solving with in vitro simulation model-systems in BME requires cognitive flexibility is that researchers need to be able to transform a complex biological problem, such as neuronal network learning or pathologies in the cardiovascular system, into one that can potentially be addressed with conceptual and methodological resources from engineering. To build in vitro model-systems, work with them, and interpret and evaluate experimental outcomes requires the availability of a range of methods within the lab, not only from engineering, but also from biology—for instance, cell culturing or gene profiling—as well as the ability to use biological instrumentation, such as the confocal or two-photon microscope. The iterative and incremental processes of building a model to exemplify the relevant features of the biological system entails trial and error. The death or contamination of a cell culture can result in months of work being wasted. Impasses or failures of various sorts are a frequent occurrence in a context where *"no one has done this before"* is an oft-repeated sentiment, so researchers need to develop resilience to step back and evaluate the situation to figure out whether to persist in a direction or how to start in a new one. Although the research projects we saw in the labs were not collaborative, researchers still needed to develop interactive skills to take advantage of the expertise of others in the group and to participate in problem-solving sessions with researchers with different kinds of engineering backgrounds in the lab. Further, even hybrid biomedical engineers need to develop the skills to interact productively with people trained in medical and biosciences fields as they venture into careers in academia, industry, or policy. We called individuals with the ability to interact productively within interdisciplinary contexts *boundary agents*.

The BME labs we investigated were located in an adaptive problem space in which interdisciplinarity is explicit, reflective, and intentional. Their kind of in vitro research program was initiated by engineers who either could not recruit bioscience collaborators or who found such collaborations inherently difficult because bioscientists lack the requisite quantitative and engineering knowledge to facilitate collaboration. These researchers aimed to create a model of interdisciplinary research different from the standard, "team science" model of two or more researchers from different disciplines in collaboration. Thus, their learning aim is to create interdisciplinary integration not only with respect to concepts, methods, and objects of research, but also at the level of the individual researchers.[7] The educational program in which the BME researchers are embedded aims to design a kind of researcher who might, themselves, be considered emergent hybrid systems—bio-medical-engineering researchers who are not only self-sufficient in problem-solving with hybrid in vitro models, but are also able to collaborate fluidly with disciplinary colleagues in any of the three fields. The direction of emergence would depend on the subfield, such as tissue engineering or neural engineering. In developing the problem-driven learning environments I discussed in chapter 4, we aimed to begin to cultivate all of the interdisciplinary virtues we had identified students would need to be effective biomedical engineers and equip them to develop these further in the context of their subfields as they advanced in their research projects and chosen fields of employment. We aimed to equip them to become both an integrative biomedical engineer and a potential boundary agent.

7.2.3 ISB Problem-Solving: Synthesis

ISB is a young field, though it shares objectives with an older systems biology philosophy. The overarching goal of the field is to develop analyses of complex nonlinear biological phenomena at the system level. The traditional biological approach of well-controlled experimentation focused on characterizing select components or processes is seen as necessary, but not sufficient, to investigate how higher-level functionality emerges from myriad interactions at lower levels. The confluence of new kinds of data production and collection (high-throughput) technologies, computational resources (e.g., high-performance computing and novel parameterization algorithms), and the development of curated biological databases and Internet search engines for seeking biological literature has made it possible to bring quantitative and computational methods to bear on the problem of developing an integrative analysis of the behavior of complex biological systems at all levels, from intracellular interactions to ecosystem processes.

Finding solutions to the problems posed by the field creates an essential *epistemic interdependence* (MacLeod and Nersessian 2016; Andersen and Wagenknecht 2013) among the participating fields, which is likely to remain, given the complexity and sophistication of the research required from each field. These fields comprise various engineering fields, computational sciences (including applied mathematics), and biological sciences. ISB at present does not have a unified vision of what a researcher needs to learn/know to be an effective problem-solver. In a general sense, the adaptive problem space of ISB is integrative in that, to formulate and solve problems, researchers draw from engineering and mathematical concepts, engineering modeling methods, computational algorithms and methods from applied mathematics and computer science, engineering technologies, and knowledge, concepts, and data, and, in some instances, experimental methods (bimodal strategy), from molecular biology. What is striking is that the various possible configurations for research in this adaptive problem space are numerous and continue to emerge. Our labs provide a subset of possibilities. Still, we have gained important insights about interdisciplinary engineering-oriented research in this field.

Building computational simulation models of complex biological systems is the main epistemic practice in ISB. A major issue we witnessed in both our labs is that without effective collaborations, lack of biological knowledge and insufficient or inadequate data increase the complexity of the modeling work. Every problem requires modelers to adapt or tailor methodological strategies to transform it into one they have the potential to solve. Modelers are required, themselves, to search through the available biological literature and databases to build out the metabolic and signaling pathway diagrams of the system under investigation sufficiently to inform the modeling process. Usually, modelers start from a small piece of a pathway provided by a collaborator or found in the literature and then fill it out by making "*guesses*" about "*what is reasonable*" to add/alter in conjunction with running simulations, with and without pieces, as they build the model. They need to predict what effects a modification of the biological pathway representation will have and locate where a modification is needed to solve the problem. They try to check their guesses with their collaborators, but often find them unresponsive. Further, what resources are to be used to build the model are largely at the discretion of the modeler. Systems biology lacks the established domain theories that, in physics-based sciences, provide representational resources and methods to build reliable in silico simulation models. As we illustrated in the G10 and C9 case studies, every model is a strategic adaptation to a set of constraints, ranging from those of the complexity of the biological problem to the fact that simulation experiments and real-world experiments take place on vastly different time scales and, further, to the human cognitive constraints and to the challenges of collaboration. Most of these constraints cannot be eliminated, but our interviews with modelers and experimentalists did lead us to insights into limited learning interventions that might prove useful for enhancing collaboration, as I discuss in section 7.2.4.

Interdisciplinary "integration," then, in the ISB context largely means infusing experimental data gathered from a range of disparate sources into the in silico simulation models as they are built with systems engineering concepts and methods that combine these elements into a *dynamic synthesis*. Ideally, the output of a stable and robust model or small ensemble provides understanding into the system-level phenomena, or at least insight with respect to control of selected features, as well as novel hypotheses to guide biological experimentation. "Integration" at the conceptual level also means a kind of dynamic synthesis, as one researcher noted: *"The tasks of this new frontier require thinking beyond linear chains of causes and effects—thinking in terms of integrated functional entities; thinking in systems, networks, and models."*

We have sometimes characterized this kind of interdisciplinary field loosely as a "transdiscipline," but definitions of "transdisciplinary" in taxonomies are vague and often contradictory and do not quite capture the nuances of ISB practices. The kind of interdisciplinary integration we witnessed at the level of participating fields has features of what Peter Galison (1997) calls "intercalation," where fields keep separate identities and practices, though it is possible for practices within one field or the other to be transformed in significant ways in their interactions. But the need to work partly in the field of the other is not captured well by his analysis, and the kinds of adaptive processes we found did not fit well into his, now customary, characterization of interdisciplinary spaces as "trading zones." "Symbiosis" is perhaps a better characterization than "trade" of the relationship among engineers, computational scientists, and bioscientists in the adaptive problem spaces of ISB research.

As we have seen, problem-solving in ISB requires the cognitive flexibility to manage the complexity of a wide range of constraints that influence the model-building process for a specific problem. In addition, the same modeler needs to be able to function in a highly adaptive manner to work on a wide range of biological systems. The lab G director maintains that modelers have the ability to tackle a range of biological problems because they have the "*flexibility to recognize shared features of control/regulation across disparate domains*," which comes from experience with engineering systems. But that engineering understanding of control/regulation needs to be adapted to biological systems in order to transform intractable problems into potentially solvable ones. Problem adaptation is an iterative and incremental process in which researchers search through and adapt strategies for representing the problem and avenues for solving it given the governing constraints. The researchers we investigated do not generally follow specific methodological norms but pursue whatever strategy looks like it will enable them to get a handle on the specific problem they have, usually tailored to the kind or quality of data for the system. The situation puts a premium on having on hand a range of methods and strategies, as well as innovation and creativity in methodological approaches. Modelers need to be able to use a range of heuristics and to experiment with multiple methods drawn from their backgrounds or the experience of the lab director. In pursuing diverse methodological options, individual modelers can contribute understanding about the value of different methods to the wider field. Because of the wide range of options to build and to fit models and their attitude of "seeing what works," modelers need to anticipate failure and impasses and to look at these as resources for developing insight into the problem or direction of solution.

In ISB, methodological choices extend from specific low-level decisions by individual researchers about how to represent a reaction mathematically to much higher-level decisions by lab directors, such as how to organize their labs, whether to collaborate externally or integrate internally the various requirements for model-building, and even to the manner in which they choose to conceptualize the goals and aims of systems biology. As we saw with lab C, for instance, a possible adaptation is to develop hybrid modeler-experimentalists, that is, bimodal modelers. However, in the present situation, wherein specialized modelers (mainly engineers, applied mathematicians, and physicists) and specialized experimentalists (mainly molecular biologists and biochemists) are the dominant participants, the standard methodological choice is collaboration among specialists. As we have seen in chapter 5 and as I discuss further in section 7.2.4, with little knowledge of one another's fields, collaboration is fraught with difficulties. In the present state of the field, our research indicates the onus is on the modeler to be the boundary agent, who is required to step into the biological arena to build her models. Because the domain is continually shifting, however, our modelers all maintain that deep knowledge of a specific biological field would not be helpful. Thus, collaboration with experimentalists with deep knowledge of the biology of the problem at hand is critical to the objective of system-level analysis. From the situations we investigated, and from reports from the wider field about its current state, it is clear that most researchers on either the modeler or experimentalist sides of the "collaboration" lack the interactional expertise and epistemic awareness needed for collaboration to be effective, and feel frustrated by this.

In the next section I look at specific challenges that arose with respect to collaboration for researchers in lab G (the "unimodal modeling" approach), and strategies we used to start to mitigate them. I focus on this first, because collaboration, and not the bimodal strategy, is the dominant approach to model-building in the present state of ISB. This situation creates an interdependence that was expressed perceptively by a senior bioscientist who was just in the process of establishing a collaboration with lab G: "*Team science is the only way it's gonna work these days. Its gonna get hard to write a single investigator R01 [NIH grant] these days and expect to get it funded because everyone is now realizing the interconnectedness of everything. And for me to sit here and think that I can have all the expertise in my tiny little brain to do everything with all these approaches that I don't understand at all is ridiculous. So, we really are trying hard to put together a research team. . . . So, you know, we gather data, we talk with [lab G director] about how these data need to be put together, and what kind of inferences can he help us generate out of them. . . . You really need to have an interaction with people. . . . You're gonna be much more on one side or the other. So, you need the other half of your brain [bioscientist] to be in another person [modeler]*"

The push for the "*team science*" collaborative approach is the major direction in interdisciplinary science more broadly, especially as promoted by funding agencies (e.g., NRC 2015). However, productive "*interaction with people"* in the other field is difficult to achieve. It proved to be the case even with this researcher who expressed the need so clearly. Thus, it is possible that the kinds of strategies we developed can be applied more widely in team science. Collaboration needs to be attended to explicitly, which leads to the second reason I focus on it, which is that the vivid responses of the student researchers to the experiences we devised show how even just a little attention to building interactional expertise contributes to cultivating the other epistemic virtues.

7.2.4 Challenges and Strategies for Collaboration in ISB

Across interdisciplinary fields generally, the dilemma is couched as whether to educate researchers as specialists or polymaths to meet their problem-solving demands. Our investigations have led us to see the response to the "specialists or polymaths dilemma" as lying in compromises that are adapted to the specific situation of a research approach. Cognitive ethnography provides a unique means to investigate the details of these compromises and adaptations as they are made during the problem-solving process or diagnosed in response to problem-solving difficulties. In ISB problem-solving, modelers and experimental collaborators both have the objective to produce a computational simulation model that should be biologically informative, especially with respect to providing experimental guidance. Our focus has been on modelers, but our analyses also have been directed toward their interdependence with experimentalists. Although the requirements for effective model-building in these contexts lie more toward the specialist end of the spectrum for both, we found that effective collaboration requires more than cursory acquaintance with the collaborating field. Yet what we witnessed in the labs we investigated (and have been told by numerous other researchers is the current state of the field more globally) is that modelers have little understanding of the possibilities and constraints of experimental practices, and experimentalists have little understanding of the nature and requirements of model-building—and, I would add, neither has an understanding of the epistemic norms and values of the other.

Our strategy was to determine from the nature and challenges of the model-building practices we witnessed, and they discussed in interviews, what are some learning requirements, at a metalevel, for effective research. In analyzing the challenges of collaboration, we found it useful to form an understanding not only of needs, but also of what each side of the collaboration viewed as the deficiencies of the other with respect to collaboration.[8] Here I present a sampling of how some of the researchers, from each side, expressed their needs and perceptions of collaborators that were important for our choices about how to facilitate collaboration.

Our studies have identified several principal reasons for collaborative difficulties that result in significant challenges for the modeler. The modeler's primary need is for sufficient, high-quality data appropriate for the problem at hand. The collaboration usually starts with the experimentalist, who has become aware that modeling might help them get useful information out of the data they collected. As one modeler cast the interaction, the experimentalist approaches the modeler with *"You're a modeler and I do 'systems biology.' So, model these data for me."* The quotes around the term systems biology indicate the modeler recognizes there is a possible difference in understanding of just what that means or entails. But, from the modeler's perspective, *"the biologists produce the data they want. But those data are not actually what we want when we do parameter estimation—so there might be some gap between these two, between us. But even so, they don't produce enough data—they don't measure the concentration for example. And they have few kinetic data."* Models usually have specific parameter requirements, such as kinetic concentration and rate data for ODE models. However, modelers are usually not aware that to measure these kinds of data can be difficult, expensive, and time-consuming. As the experimental collaborator with the modeler just quoted told us, *"The data they want from us is something that is not simple to generate. So, if they want kinetic rate for an enzyme, we have to purify that enzyme. Then we have to create all the conditions to measure it in vitro. That's not a simple undertaking. That's probably six months of work. . . . The second problem is, yeah, if we are going to . . . spend six months generating what they want, then we would like—we would need—to have something that's going to come out of it."* As she viewed the situation, modelers, in general, are *"not taking it to the step where it's useful for the biologist."* Further complicating the situation, the modeler in this instance did not realize that, for her collaborator, as a vascular biologist, to produce the kind of data she needed was not something she ordinarily did since it was not of value for her own research project. And, further, sometimes, as another experimentalist noted, *"they ask things that are not biologically possible."*

The difference in the time scales of modeling and experimentation also creates an issue. On the one hand, as we saw with G10, the modeler can wait around months for data. But, on the other hand, it often takes several years to build a productive model, and by the time the modeler comes with hypotheses for which they want the collaborator to conduct experiments, their experimental research program has moved on. As another experimentalist told us, such new experiments, *"would be time consuming, and [cost] money and effort. Sometimes we already passed that point."*

On the other hand, modelers often expressed the view that experimentalists mainly do not understand the capabilities of models and the power of mathematical techniques to derive network structure and derive valuable predictions even from limited data through approximation. Further, they claim experimentalists fail to see the value and inferential power of the literature synthesis the model builds from years of data that experimentalists are no longer interested in looking at. Modelers contend that experimentalists often see them as just reproducing *"old"* data—or producing models that are *"tautologies"* that can offer no insight. Indeed, as an experimentalist told us, the modelers she was interacting with are *"trying to model something published fifteen years ago—well what are you going to do with that?"* and modelers are modeling *"for the sake of the model,"* not of the experimentalist. Further, modelers often note how experimentalists are skeptical about models: *"They think of it [model] as something that's—just hooked up to—to, you know, match figures, . . . So, for them, it's like you're using your data and then plugging in some numbers to fit the output of your model to that, and then they would not possess a lot of faith in those models or what they predict."* One reason for lack of trust on the side of the experimentalists is that they view modelers as not understanding the complexities of biological research and its impact on the data they are using: *"We know how complicated the system is . . . one change in experimental condition can totally change the result."* The modelers did often present a naive view of experimental research, for instance, *"biology is memory"* or *"it's not that difficult—like a recipe, when you cook."* Experimentalists also cast modelers as not valuing accuracy: *"They are not really interested in actual numbers . . . it's more like getting a sense than accurate."* At the same time modelers cast experimentalists as not understanding the importance of system dynamics and the reasons for why they model trends rather than exact numbers: *"All they care is up/down—they don't care dynamics,"* as well as not understanding the model and its capabilities in general: *"They treat it as a black box. . . . They will not get deep into the model's detail because that's maybe too complicated for them."* When there is significant misunderstanding and productive interactions are lacking, each side ends up with a caricature of the other. In general, the lack of understanding and the frustration on both sides can lead to each positioning the other in unflattering and unproductive ways, which in turn impedes collaboration and building trust (see, also, Andersen, 2010, 2016; Andersen and Wagenknecht 2013). Further, we found that often each side positions the other as a service provider rather than a collaborator. The experimentalist requests the modeler to *"model my data,"* and in turn the modeler, as we frequently heard, *"order[s] my experiments"* from them.

It is clear that experimentalists do not understand much about how models are built and are not comfortable with what they do know of modeler practices, such as using data gleaned from a variety of experimental conditions, modeling trends in the data rather than exact data points, and making other abstractions. This lack of understanding often results in a lack of interest, not responding to queries in a timely manner even when they have requested the model-building, or, even, as we saw, being unwilling to part with unpublished data that the modeler needs in order to proceed with their request. On the other hand, modelers usually have no understanding of the experimental practices that have led to the data. We found that none of the modelers in lab G had even taken a biology class with a lab (except, of course, the bimodal postdoc)—and few in lab C. Their general attitude was that they could easily pick up any part of the biological knowledge they would need because it was *"horizontally organized,"* unlike mathematics and engineering, which are *"vertical"* in structure and require progressive learning. Whether this might be the case with biological subject matter or not, it is not the case for sophisticated experimental practices, which require coordinating multiple kinds of biological, skilled-based, and technological knowledge that takes years of experience to acquire.

As I noted earlier, there is not an agreed-upon approach for how researchers in ISB should be educated, as there was in the BME program. There are institutions, for instance, that are working to develop full modeling curricula for biologists, on top of their biological education, in response to the widely recognized collaboration problem. The educational context of the modeling labs we investigated required the ISB graduate students to earn their degrees in an engineering major, bioinformatics, or BME (only C9 in our study) since there is no ISB degree. This meant they needed to cover a range of required courses in those fields. Neither students nor faculty seemed interested in extending the time to graduation. We decided that the most effective thing we could do in the learning dimension of our research on graduate education would be to propose or develop minimalist learning interventions that would facilitate smoother collaboration. Our strategy was to determine from the nature and challenges of the problem-solving practices we witnessed, and they discussed in interviews, what the important learning requirements for collaborative research in ISB are, as currently practiced. Then, because each side stressed the limited time available to spend on work that was not strictly modeling or experimenting, we needed to determine how such learning might be achieved using a "small interventions, big payoff" approach. In general, the issues for ISB epistemic practices around the theme of managing complexity that I identified in chapter 5, create significant demands for all participants in the cognitive-cultural systems of ISB research. Our analyses identified three of the interrelated characteristics discussed in section 7.2.1—cognitive flexibility, interactional expertise, and epistemic awareness—as most important to focus on for effective collaboration in this context..

At least in the current state of ISB (and quite possibly a necessary feature of this kind of research), the full "hybrid" curriculum is not desired as modelers and experimentalists need deep training in one discipline, sufficient to be solely a computational scientist (engineer, applied mathematician) or an experimentalist (bioscientist, medical researcher). But to realize the full potential of ISB requires some degree of penetration of each kind of researcher into the field of the other. At a minimum this means that modelers need to learn to adapt what they know to complex biological problems across a range of areas, as well as learn to know what biological information they need and how to seek and evaluate it, and that experimentalists need to learn enough about the nature and potential of modeling biological systems to produce the kind of data needed, and both sides need to know, as one experimentalist put it, *"the right kinds of questions"* to ask about how each can contribute to the model-building process in order to further a collaboration. Within our project we experimented with two learning interventions that aimed to develop, especially, the three interrelated characteristics early in the student's research career in order to mitigate some of the struggle of collaboration.

Because the model is central in ISB problem-solving, the engineers/modelers are taking the lead in moving the field forward. As we saw, modelers do more than just feed biological data into a model and provide predictive outcomes to experimentalists. They have to understand how to search the literature to find relevant data and build out the biological pathway, both of which require discernment and judgment about biological phenomena, about what it is feasible to do in experimentation, and about the reliability and relevance of the data, as well as the ability to discuss problems with experimentalists as they build the models. On the other side, sophisticated biological experimentation requires equally specialized training, but to be able to collaborate effectively with modelers, experimentalists need to understand the basics about how a model is built so as to, at the very least, devise experiments to produce the kind of data modelers need to construct and to test, experimentally, informative models.

We undertook two interventions with the newer researchers in our labs that proved quite successful, which I will discuss briefly. On the modeling side, as we saw, modelers develop cognitive flexibility in dealing with biological systems not through taking numerous biology classes, but through efforts to recast phenomena from disparate biological domains in terms of features of engineering systems, especially control and regulation. What biology classes they do take are usually theoretical or bioinformatics classes, without labs, so they have little understanding of how biological data are produced, which creates a major impediment to collaboration. We were told that our initial proposal—a full semester rotation in a biology research lab, which we still think a good strategy—would take too much time away from modeling work. We proposed, instead, an intensive "experimental summer camp" experience for beginning modelers, in which they spend a month in an experimental lab engaged in a real piece of research to learn hands-on what it takes to design and execute experiments, as well as something of the way experimentalists think about biological phenomena. The lab G director chose two modelers to spend the month at a laboratory working on yeasts with which he had a long collaboration. The modelers were not absolute novices to biology since they had been conducting the literature searches and building the pathways for about a year, as I discussed in chapter 5. However, they had no idea of the complex environment of a biosciences lab or sense of the nature and costs (time and money) of the experimental practices through which data are collected and analyzed. One was a telecommunications engineer (G16), the other a mechanical engineer (G5). G16 had no relation to the lab, but G5 had been collaborating with them for over a year. Both students came back excited about their experiences, each with a collaborative experimental paper under way from the research they had undertaken. It is instructive to quote highlights from our follow-up interviews from each about what they felt they had learned. Both were surprisingly reflective and articulate about this.

G16 contrasted her before and after understanding of the experimental procedures she had previously viewed as "*like recipes*": "*You are looking from far away. You just see this person is just going into the lab and pipetting, and that's not interesting and why would you do that? But then when you get it, you see there are a lot of reasonings going on and they are involved in their own sort of culture.*" She also expressed that her hands-on experiences, for instance, "*the stuff I saw—I actually pipetted a little bit,*" made her "*feel more self-confident in talking to biologists.*" In addition, she learned important things about her own practices. For instance, she had not understood why the lab director kept telling her to model trends in the data, not data points. She now understood why: "*Right now I would say there's a lot of human error in there. . . . It's both about the reliability of the data and the types of errors.*" Further, she was able to leverage her brief experiences to develop a more complex metalevel awareness of what she would need to know to collaborate better. As she told us in an extended reflection in which she appeared to talk to a potential experimental collaborator, "*We need to be able to communicate—we need to have an idea of what kind of experiments are done. . . . Their area of research is very limited. They just know some sorts of experiments they have in their labs with the equipment they have. And then you sometimes need to include someone else in the project, to do some other part for us to build a dynamic network of this pathway, this specific organism, we need this kind of data. If we don't have it, we can't. And then, 'I'm a modeler, you're a biologist, you don't do that type of experiment. Who do you think could do that?' And then 'how much do you think it will cost?' I can ask a question from you, but I need to have an idea that such a thing exists to be able to think about or suggest it at all.*" Perhaps most importantly, she had come to realize, "*So sometimes, like in a month, you just change inside. It's not about the exact things you learn—it's just knowing how to learn stuff.*"

G5 had already been collaborating with that experimental lab but found himself quite surprised by what he encountered when actually working in the lab. A major revelation for him was to understand that it is "*pretty hard to get time series data. Why they so focused on one pathway, one gene, one mutant—it's hard to imagine how hard it is to pull out."* He recounted: *"And it's like—oh yeah and I think that the techniques we have now are still very limited. So, I spent two years to realize that the gene has strong thermal tolerance phenotype, but it's like I know the gene is in the very bottom of the cells and when we knock it down the cell can't grow well under heat stress. I spent almost three weeks [of the four at the lab]—and what happened between the phenotype and the thermal tolerance behavior and the gene, I have totally no idea. And thousands or more than thousands of pathways or relationships between them—I spent three weeks to realize these things. That there are many, many things that are still unknown. And each step takes time."* He, too, discovered why the lab director had told him to model trends: *"They generate data so I can use it right. After this month, I need to reconsider the data I have because there are a lot of steps that might cause the inaccuracy of the data. I need to focus on the trend on the data rather than the exact value. The exact value is not that reliable, I think."* He also learned he needed to approach the experimental literature differently, using his new skills: *"[Before]I just find a paper and read it and usually believe the results—used to skip over the methods section—now I look at experimental design as part of evaluation."* Further, he felt he had a significant change in perspective on his own project—and possibly on what systems biology is about: *"So, before I, before this trip I totally focused on the mathematical problem—so how to make the model, how to process the data, all mathematical things. But now I can start to think about the links between my model and real world—it's not a quantitative behavior. It's like a property of the cell."* When we asked whether he thought the experience would make him a better modeler, he said there had not been enough time since returning to say, but, importantly, he now felt confident he could alter the way the collaboration had been going. Before, there was *"zero interaction—I only got the dataset they published in 2004 and that's it."* Now, he envisioned, *"Yes, it really changes things because, now if I have any problem about the data, I can just ask. Before I feel like I work alone."*

In sum, the modelers described the following gains: increased self-confidence, comfort with experiments, and some understanding of experimental procedure; enhanced ability to anticipate the needs and questions of experimentalists, to understand experimentalist reasoning processes, and to evaluate experimental literature; and new appreciation for the difficulties/constraints of experimentation and the possibilities of errors in the data. The latter helped them to relax their engineering values, which favor precision, and to understand why their adviser kept telling them to model trends, not every data point. The outcomes suggest it would be even more valuable to build the full semester experimental rotation into the curriculum.

On the experimentalist side, learning about model-building cannot be achieved by visiting a modeling lab. Hands-on experience requires a more structured approach. Fortunately, the department was interested in developing a new introductory graduate course in biosystems modeling. As envisioned, students from the biological sciences would develop conceptual understanding of modeling and basic modeling skills while working on systems biology problems with engineering or applied math students, who would be learning to adapt their engineering knowledge and skills to model biological systems. Wendy Newstetter worked with both lab directors through several iterations of the course. It took a while for a significant number of biology students to take the course, and during our study the only one was from lab C. It was C11, the biologist lab manager and research technologist, who was just transitioning to a PhD student. At this point she anticipated conducting only experimental research for her project. Given that all the other members of her lab were only modelers or bimodal, she said she took the course in order to get a better sense of what modeling was about. She cast this move as *"going over to the dark side,"* since her role as the sole biologist within the lab had positioned her as the staunch defender of biological practices when they were treated dismissively by the modelers. In an early interview as she was taking the course, she volunteered that she was beginning to rethink her PhD project in terms of adding a modeling component: *"I'm trying to stop myself from going to [lab director] and suggesting it actually (laughing). Every time I come out of the class and I'm like 'oh, this is fun, I learned something.' I want to go to her and go 'I want to do modeling,' but then I think I might regret it later, so I'm giving it some thought (chuckles)."* In the end (after our study concluded), she did develop a modeling dimension, becoming a bimodal researcher from the opposite direction from C9.

We interviewed her weekly as she was taking the course. It was interesting to see how, throughout the course, she experienced it through the lens of her earlier, unsuccessful research collaboration with a modeler in lab C. Even early in the course, she stated, *"You know I wish I had taken this class two years ago. I wish he and I had taken it together. Because we would have looked at each other and gone 'oh, I get it, I know what you're doing now.' And it would have been helpful for me to understand what kind of data he needed,* ***to understand what kind***

of questions he should have been asking of me. *. . . [I] didn't have insight into know exactly what kind of data would be useful to him. . . . And I think it is hard for him to explain it to me because he didn't know what I had to like go through to get the data. . . . It's funny because he's starting to do experiments now too, so I think he figured out the same thing from his end. It's easier if you have more knowledge on the other side."* The text I put in boldface is interesting from the perspective of interactional expertise because she now felt she understood what a modeler needs to elicit from an experimentalist with respect to the data requirements for a model. She continued to think about the requirements for collaboration throughout the course. Toward that end, she stressed the importance of having done coding herself now: *"I wasn't sure how he like converted what I gave him into something that could be put into code. . . . Now I'm going 'Oh, that's what he wanted. That's what he needed. Oh, OK, I wish I had known that. . . . I would have had better data for him'."*

She had coded only a little in MATLAB before, and had also taken math as an undergraduate, including calculus, and so had some experience with differential equations, but *"didn't realize what they were used for"* until she built ODE models. In her biological training, with a MS in ecology, she had thought of math not *"in terms of computers and math"* but in terms of what she called *"counting"* for instance, *"like how many birds do I have, how many bunnies, how many wolves. . . . We think of it as boring stuff you have to get through to get to the interesting, exciting bunnies and wolves."* Now, she described a new awareness of the affordances of models and a new understanding of math for biological analysis—as a flexible *"tool"* for *"actual real-world application."* She also reflected on the different epistemic norms and values of biology and modeling, and how it was a struggle to negotiate between them in her thinking: *"Biologists tend to think in a lot of details and it just seems there's no way you can build a model with all these details in it . . . it's hard for me. I have to like try and cut out a lot of things in my head when I think about how I would go about modeling something . . . because what I've been trying to do before is get all the details and not make any assumptions. It's like my whole training was like 'don't make any assumptions about what this will be.' And if you're modeling you have to start by making an assumption, 'cause otherwise you don't know where you're going to start. It's like 'I'm gonna assume this system is going to behave similar to this'."* She also noted that coming from biology to modeling might be the reason she was thinking of her models in terms of their biological subjects, for instance, the model she was building of cystic fibrosis for the class: *"I think of the model as a patient. . . . I don't think of it as trying to get the model to work as much as, 'but this would kill the patient, so I can't do that.' I think of [former collaborator's] things in the same way. I think of his model as a little bit of a cell."* So, now, with her intimate experience of model-building (cognitive partnership), she was anthropomorphizing not only the cells, but also the models. In an interview after the course, she reported having a split-brain experience with respect to her attitudes: *"I'm a changed woman. Now most of my brain is going 'what? Why are you getting rid of a data point?' and the other part's going 'look a smooth curve!'"* Finally, she felt it had been *"beneficial"* to have *"people from different points of view take the same class, because we get the other side and hopefully get some intuition about both. I had no intuition about what a mathematical function could be used for. And then there are some people who have no idea about what might possibly be going on in a biological system. You need both if you're going to model. You need to know both."* In sum, it was clear that by the end of the course she had developed sufficient familiarity with concepts, methods, and techniques for building systems biology models, even though the models were simple in comparison to what we saw in actual practice, to have a much more successful collaboration than prior to taking the course—and felt confident she could do so.

Admittedly, our samples are small, but at the very least they provide a "proof of concept" for the "small interventions, big payoffs" approach through which to help each side penetrate, however slightly, into the domain of the other. Although only a start, our findings suggest that specific, time-limited learning experiences are productive for cultivating characteristics needed for more effective collaborations, better reasoning, greater awareness of the affordances of the other's methods, and enhanced ability to reflect on both one's own perspective and that of the other. These researchers now had the capabilities to be effective boundary agents without needing the deep hybridization of the bimodal approach or that BME aspired to. In sum, even small, targeted learning interventions can have big payoffs to benefit collaboration, and thus effective problem-solving potential, in the ISB research space. It's conceivable that this approach would work in other interdisciplinary spaces as well.

7.3 Summary: "I Get It Now—I Know What You're Doing"

The goals of our project were multifaceted, but a major one was to investigate emerging interdisciplinary epistemic practices around problem-solving in frontier research laboratories in the bioengineering sciences, with an eye toward facilitating such research with situation-appropriate learning experiences. Cognitive ethnography provided the means to fathom both the nature of interdisciplinary problem-solving practices and how these are enacted in situ, in our cases, with respect to innovative model-building environments. Importantly, it provided the means to fathom ways in which the cognitive, social, material, and cultural dimensions of epistemic practices are integrated in model-building. We had no hypothesis about the nature of the interdisciplinarity we would encounter when we entered the BME labs. What we learned from our initial interviews, observations, and discussions with faculty constructing the fledgling BME program was both that in these labs engineers were tasked with conducting basic biological research through building living in vitro simulation models, composed partly of cells and cellular materials and partly of engineered materials, and that the faculty wanted to build an educational program that would require students to "integrate" all three dimensions of BME from the outset. In the chapters on the BME labs, we saw how hybridization is achieved for researchers and devices through processes of interlocking engineering and biological concepts, methods, and materials in order to build in vitro model-systems and conduct experiments that simulate selected biological processes. We participated in developing learning environments—classroom and instructional lab—for a novel curriculum that aimed to foster this kind of hybridization in BME problem-solving.

The preliminary research for preparing the grant proposal for our investigation of the ISB labs led to our hypothesis that "integration" in these labs would need a different characterization from the sort we had encountered in BME. Building systems-level computational models requires modelers to have both a sophisticated understanding of systems engineering concepts and the ability to adapt high-level computational and mathematical methods that have been developed for other purposes, as well biological collaborators who can conduct sophisticated, highly skilled wet-lab biological research to produce the requisite data. We found that the nature of the systems-level problems formulated in the emerging ISB field not only requires collaboration, but also that, at least to some extent, participants engage with the practices of the other fields, with little training to do either in the present state. The problem-solving situation in ISB creates an essential epistemic interdependence among the collaborating fields. The chapters on the ISB labs showed how it falls on the modeler to manage the complexity of the problem-solving process, which includes that the modeler determine how to adapt engineering concepts and methods to biological problems and how to find the biological data necessary to build, simulate, and test their models. Successful collaboration requires, at the very least, that experimentalists understand the data needs for building and testing systems models and that modelers understand the conditions and constraints under which data are collected. We developed minimalist strategies that would cultivate at least the chief interdisciplinary virtues we had determined to be required for effective collaboration in ISB.

Overall, the cognitive ethnographic research discussed in this book establishes that a method that was pioneered to examine cognitive practices in areas where problem-solving tasks and goals are largely well-defined can be extended fruitfully to investigate the open-ended problem-solving environments of emerging interdisciplinary sciences and engineering. Indeed, cognitive ethnography turns out to provide the primary means to develop nuanced, fine-structured analyses of the epistemic practices of varieties of interdisciplinarity as they are created and enacted in real-world, real-time situations. It provides a unique granularity for fathoming the nature and challenges of these exploratory, incremental, and nonlinear problem-solving practices, their development, and the epistemic principles guiding them. It can yield insight into how methods, norms, and standards come to be justified, and, thus, into why and to what extent it is reasonable to consider the fruits of such research to be trustworthy. And, as we have demonstrated, insights gleaned from intensive case studies can be used to develop strategies for facilitating specific varieties of interdisciplinary learning, integration, and collaboration. The research that has led to conclusions such as these started in a conversation with three visionary engineers who approached me with a wish for support to develop an educational program. Years after that conversation, this research has, in many respects, made me a "changed woman" as well.

Notes

Chapter 1

1. I use both "I" and "we" throughout the book. "I" indicates what I see as my individual contribution to the project, which consists primarily in its framing and in drawing on, and extending, analyses I have developed in previous research. "We" signals the inherently collaborative nature of the project, including data collection and analysis. As I said in the acknowledgments, it is impossible to disentangle specific contributions in our analyses. However, when another group member contributed significantly to a specific analysis and interpretation, I will note that they should be credited as co-analyst.

2. I discuss our rationale for investigating bioengineering sciences practices later, but here I note that I am including them under the rubric of "scientific practice" because I am concerned, mainly, with their epistemic practices. These hybrid fields aim at creating both biological knowledge and engineering applications, and so differ from conventional engineering fields in having epistemic goals (see also Boon 2011). One of our objectives has been to lay out the epistemic structure of biological engineering.

3. Woody (2014) locates the "practice turn" in philosophy in a concern with experimentation in the mid-1980s (e.g., Hacking 1983; Franklin 1989); however, there was also a simultaneous emerging concern with theoretical and conceptual practices (e.g., Gavroglu and Goudaroulis 1989; Gooding 1990; Nersessian 1984). The Science and Philosophy book series (Nijhoff/Kluwer) I created and began editing in 1984 was dedicated to the study of theoretical, conceptual, and experimental dimensions of practice.

4. Some sociocultural studies in the late 1990s also moved toward accounts that can be read as taking note of cognition, such as Peter Galison's (1997) concern with the "image" and "logic" traditions in the material culture of particle physicists, Karin Knorr Cetina's (Cetina 1999) analysis of scientific practices as part of "epistemic cultures," and Hans-Jörg Rheinberger's (1997) analysis of experimentation in molecular biology as producing "epistemic things."

5. Classic references for these perspectives are cognition as embodied (Barsalou 1999; Johnson 1987; Lakoff 1987); artifact-using (Clark 1998; Hutchins 1995a; Norman 1988); and situated (Greeno 1989a; Lave 1988; Suchman 1987).

6. Classic references for these perspectives are distributed cognition (Hutchins 1995a; Hollan et al. 2000; Kirsh 1995, 2001; Kirsh and Maglio 1994; Norman 1991); distributed intelligence (Pea 1993); activity theory (Cole and Engestrom 1993); situated action (Lave 1988; Greeno 1989a,b); and extended mind theory (Clark and Chalmers 1998; Clark 1998).

7. It is important to underscore that Hutchins's position is that distributed cognition is an analytical framework and not an ontological claim as advanced by Andy Clark with his "extended mind thesis" (Hutchins 2011).

8. For some time learning scientists have been examining science practices from a D-cog perspective; see especially Hall et al. (2002) and Hall et al. (2010). More recently, a handful of studies in the science studies fields and cognitive science have examined problem-solving in scientific research from a D-cog perspective (see, e.g., Alac and Hutchins 2004; Becvar et al. 2008; Charbonneau 2013; Giere 2002; Goodwin 1995), but they have not explicitly addressed the ways the framework itself needs extension.

9. Hutchins has stated in several talks I have heard that "embodied brains" are participants, without specifying what the brains contribute. Although I will not address this issue beyond what I say about the capacity for mental modeling/simulation here, such specification is an important open problem for the distributed cognition framework, since the necessity for there to be a human in the system to make it a *cognitive* system provides an important contrast with actor-network theory with which it is sometimes conflated (see also Giere 2002). Learning sciences research by Rogers Hall (Hall et al. 2002; Hall et al. 2010) and Charles Goodwin (1995), in particular, also seeks to elucidate the mental resources at work in problem-solving within the D-cog framework.

10. The literature on science listed in note 8 does attend to this dimension.

11. Kirsh and Malgio (1994), Kirsh (1995, 2001) Hall et al. (2010), and Chandrasekharan and Stewart (2007) provide important exceptions.

12. Classic references for this research are discourse and situation modeling (Johnson-Laird 1983; Perrig and Kintsch 1985; Zwaan 1999); mental animation (Hegarty 1992; Schwartz 1995; Schwartz and Black 1996); mental spatial simulation (Finke 1989; Kosslyn 1994; Shepard and Cooper 1982); and perceptual simulation in embodied mental representation (Barsalou 1999; Brass et al. 2002; Bryant and Tversky 1999; Glenberg 1997).

13. Cetina suggests that the collective knowledge in the Large Hadron Collider project could be considered as "a sort of distributed cognition" (1999, 173–174) with

no further explication. As Ronald Giere (2002) has pointed out, she seems to mean "collective cognition" in the sense of Durkheim, and not in the sense of cognitive science that we have been discussing. Her notion comprises people, but not the artifacts, which perhaps explains why she does not cast the microbiology lab as a D-cog system, which on my account it is.

14. Sociological and anthropological ethnographies of research labs took off in the late 1980s (see, e.g., Latour and Woolgar 1979; Knorr-Cetina 1983; Lynch 1985; Traweek 1988) and continue to the present day (see, e.g., Roosth 2017). Cognitive science investigations began in the 1990s. Kevin Dunbar (1995) pioneered what he called the in vivo/in vitro method of investigating scientific cognition (understood in the traditional sense of individual cognition) in research labs as a source of hypotheses to then be brought into the experimental psychology lab for rigorous investigation. Cognitively oriented anthropologists and learning scientists investigated lab practices more in line with our investigations relating cognition and culture (see, e.g., Goodwin 1995; Hall et al. 2002; Ochs and Jacoby 1997). Only quite recently have philosophers begun to investigate epistemic practices in research labs through observations, as embedded participants in the research, or with interviews (see, e.g., Andersen and Wagenknecht 2013; Bechtel and Abrahamsen 2013; Bursten 2015; Carusi et al. 2012; Green 2013, 2017; Hangel and Schickore 2017; Leonelli 2016; Loettgers 2007; Sheredos et al. 2013; Wagenknecht et al. 2015). We began our ethnographic investigations in 2000.

15. Kevin Dunbar, as noted earlier, has showed the fruitfulness of ethnographic/observational methods for even the traditional cognitive science perspective. He collected data on cognitive practices of scientists in research labs and then tested his hypotheses about cognitive processes (e.g., the role of analogy in problem-solving) in controlled experiments in the psychology laboratory on nonscientist subjects (see, e.g., Dunbar 1997; Dunbar and Blanchette 2001). Dunbar's research, primarily on research labs in industry, showed a considerable amount of concern with issues of priority, as well as conflict among researchers and among researchers and labs they viewed as competitors, as one also often finds in the STS analyses of research labs. Our research has been criticized as painting the atmosphere in the labs as "too nice" or "too harmonious." However, over fifteen years of nearly daily interaction with them, our ethnographers noted little conflict and drama among researchers in these labs, or in their attitudes toward other labs conducting similar research. There were, of course, the normal frustrations and annoyances of human interaction, but these did not appear to spill over into the research arena. In general, they showed respect, good will, and a cooperative spirit when engaging in research and discussion. One possible "explanation" for the difference that comes to mind is that these labs are populated by engineering scientists, and engineers tend to be pragmatic—a conclusion I have arrived at not only from this study but also from having spent most of my career in engineering environments (MIT, Case Western Reserve, and Georgia Tech).

16. The quote is from Hutchins's response to the highly favorable review of his 1995 book by Latour (in Keller et al. 1996). In it, Hutchins counters the claim made by Latour that with D-cog, cognition has been eliminated or reduced to sociocultural factors. I concur that Latour fundamentally misunderstands or misrepresents the central point of D-cog: cognitive processes comprise human cognitive capacities, material resources, and sociocultural practices, and so, in no sense has cognition been eliminated. Empirical research in environmental perspectives across the board establishes the inherently cultural nature of human cognition.

17. Although some ethnographers might object to such abstraction, including those in STS, Hutchins's position is in line with that of Geertz, who argued that ethnographic analysis has a "double task": "to uncover the conceptual structures that inform our subjects' acts . . . and to construct a system of analysis in whose terms what is generic to those structures, what belongs to them because they are what they are, will stand out" (Geertz 1983, 57).

18. The continual development of all dimensions of these research labs over many years is the reason why I have sometimes called the method "cognitive-historical ethnography."

19. Additional kinds of analyses, for example, with respect to social positioning and identity, gender, emotion and affect, epistemic identity, and learning can be found in the publications of the researchers in our group across a range of fields—for instance in the book, *Science as Psychology: Sense Making and Identity in Scientific Practice,* cowritten by Lisa Osbeck, Kareen Malone, Wendy Newstetter, and myself (Osbeck et al. 2011). The book addresses psychologists and philosophers and was awarded the William James Book Award by the American Psychological Association.

20. We had additional questions related to the learning practices and challenges in the different kinds of interdisciplinary fields as part of our NSF-funded research.

21. This research was conducted under an IRB protocol in which the participants are to remain anonymous. I have designated the labs by a letter and the researchers in each lab have been given a code, for example, lab A, researcher A10.

22. I discuss the BME labs in the past tense because both are now closed.

23. We are using the term "device" in the way the researchers in the BME labs referred to their in vitro simulation technologies. This notion differs from the notion of "inscription devices" that Latour and Woolgar (1979, 51) introduced. Their notion refers to technologies for creating figures or diagrams of phenomena. The BME devices are sites of in vitro simulation and experimentation. Further processing with instruments is needed to transform the information they provide into visual representations or quantitative measures.

24. Recently, the extent and variety of the current fields adapting ethnography and qualitative methods to their interests, goals, and epistemic norms and values led the

Journal of Qualitative Research to set up a task force to develop guidelines sufficient to evaluate methodological integrity in data collection and analysis across fields when reviewing journal submissions and grant proposals. Levitt et al. (2017) provides a useful conceptual analysis of the task force recommendations as they can be used to guide both the design and review of qualitative research.

25. Each of the researchers on this project already had an interdisciplinary background when they joined. As "instruments" of data collection and analysis, our group brought a wide range of perspectives to the project: philosophy of science and history of science (physics, biology, psychology), cognitive science (AI, cognitive psychology, philosophy), linguistic anthropology, physics, mathematics, learning sciences, human-centered computing, theoretical psychology, gender studies, psychoanalysis, architecture, and industrial design. We have analyzed data through many of these lenses and have published several contributions in fields represented by the interests of group members beyond philosophy of science and cognitive science.

Chapter 2

1. In the BME labs, the researchers used the word "device" to refer to their in vitro simulation models. This led us to believe that devices were a specific kind of model. Only later, in the context of a different usage of the word in an ISB lab, did we find out that, in bioengineering, "device," in general, means any engineered artifact that interfaces with biological entities, as discussed in chapter 6.

2. See Vermeulen et al. (2009) for their refinement of Goodman's definition.

3. To be clear, I am not saying researchers in other contexts have to know the historical processes through which a device or model-system has been developed and has attained its credibility in order to use it. By the time of our research, for instance, the flow-loop device that lab A had developed was in use in many other labs and, recall, the dish model-system was developed in a lab other than lab D, and a few other labs were also using it. These researchers do have to know what features it exemplifies (or can be made to) and why this selection of features is warranted (e.g., simulates first-order forces) and to evaluate the relevance of these for their own research goals.

4. I use italics throughout to indicate I am quoting a researcher from transcript material. I use the convention of ellipses (. . .) to indicate text that has been omitted, without changing the meaning, and em dashes (—) to indicate a pause in speaking.

5. That the researchers all use "*over*" instead of "*through*" the lumen (which is tubular in in vivo) is an interesting slip they made all the time. I suspect they made the mistake because they were thinking in terms of the in vitro simulation, in which, as we will see, the tubular constructs are cut open and laid flat in the flow chamber.

6. Although cumbersome, it is possible to cast this process using Hutchins's characterization of information flow in a D-cog system: the forces *generated* by the flow loop *represent* shear stresses (to a first-order approximation) as it *manipulates* the endothelial cells, which, *generates* conditioned cells that researchers *manipulate* with instruments that *generate* quantitative and qualitative information in various representational formats that *propagates* through the D-cog system as it performs a problem-solving task. We did not find Hutchins's characterization of a D-cog system generating, propagating, and manipulating representations a useful way to analyze the dimensions of the D-cog systems of the research laboratory we were interested to understand, but it was useful for constructing our diagrams of various model-systems, such as figure 2.6, which I have pared down for use here.

7. No doubt, by now the reader has noticed that researchers frequently speak of their models with anthropomorphic language. We found this to be the case across all of the labs, and not only with living entities and systems but also with computers and other technologies that performed as cognitive-cultural artifacts. And, contrary to what might be expected, novice researchers rarely used such language, but advanced researchers frequently did. We took this fact to indicate that such anthropomorphizing is not careless use of language or a sign of naivete, but rather signals a growing understanding of the artifact as a partner in problem-solving. Such utterances led us to develop the theme of "cognitive partnership," which can be formed with other researchers and with artifacts of particular salience to the research, that is, the cognitive-cultural artifacts. In the case of such artifacts, researchers often *attributed* agency to them. Such attribution by a researcher is different from the artifacts actually *having* agency (though of course they interact), as proposed by actor-network theory. We have developed the theme of cognitive partnering extensively in previous research, and I will not focus on it in this book (see Osbeck and Nersessian 2006; Osbeck and Nersessian 2013; Osbeck et al. 2011). There we also demonstrate the researchers' affective engagement with the entities and objects of their research we witnessed, namely cells and models, does not taint science, but rather helps to make it possible.

8. Anecdotally, I was seated next to a neuroscientist during a presentation by the lab D director to a cognitive science audience. Afterward I asked him what he thought of the research program. He responded along the lines that although the dish was a tremendously simplified model, if they could induce learning in the network without all the other parts of a brain neuroscientists believe are necessary for learning, they would have demonstrated something very important for the field.

9. MEAs currently in use in the field have around 26K electrodes and allow researchers to do more finely grained recording of the activity of individual neurons.

10. When MEArt was exhibited as mechanical art installation, the researchers called it "The Semi-living Artist," and described it as follows in the exhibits: "A geographically detached bio-cybernetic research and development project exploring aspects of creativity and artistry in the age of new biological technologies from both artistic

and scientific perspectives. The installation is distributed between two locations in the world. Its brain consists of cultured nerve cells that grow in a neuro-engineering lab in the US. Its body is a robotic drawing arm in Australia (or wherever exhibited) that is capable of producing drawings. They communicate via satellite. The brain and the body communicate in real time with each other for the duration of an exhibition" (https://www.symbiotica.uwa.edu.au/).

11. Hesse called these features "properties," but I use "features" to capture that properties, relations, and relational structures, as well as behaviors can be mapped.

12. Douglas Hofstadter provides an exception. Although his creative representation-building AI programs are quite simple, he does argue that these processes are a significant dimension of analogy in both mundane and scientific analogies (see, e.g., Hofstadter 1995). In my 2008 book, chapters 5 and 6 provide a discussion of the issue of representation-building in analogy with respect to the cognitive science literature.

13. As will be seen in chapters 3, 5, and 6, much of this account can be extended to building computational models of complex biological systems.

14. In my 2008 book I advocated that although the word "abstraction" is commonly used for separate processes alongside "idealization" and other abstractive notions, this is confusing. It is better to reserve "abstraction" for a comprehensive notion comprising various processes, including idealization, approximation, simplification, limiting case, and generic modeling. All of these processes can play a role in model-building as a means to manage the complexity of modeling biological systems.

15. This concentration was equally important to our educational research goals, since building devices for the purpose of model-based simulations is their primary epistemic practice.

Chapter 3

1. I introduced the view of conceptual change in science as a problem-solving process early in my research on the field concept (Nersessian 1984). Only later, in the course of my research into cognitive science, did I discover that the Russian psychologist Lev Vygotsky (1962) held a similar view about mundane concepts. In characterizing concept formation during learning (acquisition of culturally extant concepts), Vygotsky argued that a concept emerges and takes shape in the course of a complex operation aimed at the solution of some problem. He also advanced the notion that concept formation is an ongoing dynamic and sociocultural process in each use or acquisition of a concept. We both hold that concepts are neither completely fixed units of representation nor solely mental representations, but arise, develop, and live in the interactions among people as they create and use them.

2. Christopher Patton was trained as an ethnographer on this lab and had primary responsibility for data collection over the course of the years discussed here. He was the one who alerted us that something important seemed to be taking place with

the building of the computational model, which enabled us to capture additional relevant data as the case and our analysis of it unfolded. He and I did the first analysis of the case together. I later did a reanalysis, based on additional interview data, with Sanjay Chandrasekharan, and what I present here is an elaboration, which incorporates more data, on the work of the three of us.

3. See, especially, Chandrasekharan (2009), Chandrasekharan and Nersessian (2015), Chandrasekharan et al. (2012), Nersessian (1991b, 1992a,b, 2002, 2008).

Chapter 4

1. Gerson (2013) rightly cautions against unreflective appropriation of biological metaphors to analyze culture. However, Wimsatt (2013a) argues for the appropriateness of using the analogy of generative entrenchment in a biological ecosystem for cultural evolution because in those processes, too, there are "multiple evolving and interdependent lineages acting on different time and size scale" (564), which fits with my characterization of the research lab as an evolving, distributed cognitive-cultural system with epistemic goals.

2. We have published case studies of the building processes of the mechanical tester (Nersessian et al. 2005) and the compression bioreactor (Harmon and Nersessian 2008).

3. This discovery led us to apply to NSF for a small supplemental grant and also to the Spencer Foundation to investigate issues pertaining to gender and to race in BME epistemic cultures. Our research scientist, Kareen Malone, took the lead on that research, holding focus groups with BME students and faculty, as well as conducting targeted interviews with lab members on the topics. Some of our findings are discussed in Osbeck et al. (2011, chapter 6) and Malone et al. (2005).

4. Chapter 7, "The Learning Person," in *Science as Psychology* (Osbeck et al. 2011) provides an extended case study of her development as a BME researcher within lab A.

5. As a field, biomedical engineering had been in existence since at least the early 1960s, but there were a few established departments circa 2000, most notably at the Johns Hopkins University (est. 1962). Since this is not a historical account of the field, I present the situation as the founders of this new department expressed it to us. The notion that they aspired to become an interdisciplinary discipline, an "interdiscipline," came from us, but they embraced it immediately and started using it in their proposals and publicity. Indeed, in both studies, part of what we did was to provide them ways to conceptualize their educational and research aims and practices. One of the senior faculty called these ways "Nancy-speak" and "Wendy-speak."

6. Although I do not articulate it here, there are significant insights to be gained regarding the explicit formation of interdisciplinary research fields—characteristic of much late twentieth and early twenty-first-century science and engineering—by

examining them through the lens of research on "social movements" (Leonelli 2019). In the case at hand, the call of these researchers for a new breed of biomedical engineer suited for the twenty-first century came well in advance of an articulated means to carry out the objective. It was a collective normative vision, the broad outlines of which were announced to the administrations of the schools involved, the wider intellectual community, funding agencies, and prospective donors. Importantly, it made a bid to reshape the epistemic practices of a field, which the leaders felt needed to move beyond collaboration to hybridization in order to meet specified goals for twenty-first-century BME.

7. Interestingly, they did not get that NSF ERC, but decided to proceed with what they dubbed "a cognitively informed" educational program anyway. The gamble paid off in that in approximately five years they went from nonexistent to the number-two-ranked BME department in the *US News and World Report* rankings, eventually taking over the number one spot, over such rivals as the long-established department at Johns Hopkins, as well as departments at MIT and Stanford. Twelve years after the program started, the program received a State Regents' Award for the best university educational program in the state. The program also received the 2019 Bernard M. Gordon Prize for Innovation in Engineering and Technology Education from the US National Academies of Engineering. These prestigious awards provide validation for the "translational approach" pioneered in our research, as well as for the educational program itself, which has conducted longitudinal evaluations of its outcomes.

8. The BME-dedicated building was under construction as we began to plan the implementation of the PDL approach. Because of the envisioned introductory PDL course, they decided to construct five specially designed classrooms with a seating structure and floor-to-ceiling whiteboards surrounding the room to facilitate interaction among the participants. Since we were doing research on the courses, two rooms were equipped with a separate observational window and recording compartment for us. They recognized that the plan for students to work in groups of eight with a facilitator was costly from the outset, but the educational experiment is seen as so successful by the administration that it has continued to support the model despite the significant growth in the student population. In recent years more than 160 undergraduate students are enrolled in a semester, with facilitators needed for twenty teams, plus for the graduate courses.

9. The department has continued to hire its own cognitive and learning scientists to provide support for ongoing curriculum development.

Chapter 5

1. The analysis developed in this chapter draws significantly on research conducted together (individually) with Miles MacLeod and with Sanjay Chandrasekharan, who were postdoctoral researchers on the project.

2. Some of these strategies are in widespread use in all kinds of modeling. However, the modelers we studied transferred and adapted them from their use in the engineering fields in which they were trained or had developed familiarity with.

3. See Stuart and Nersessian (2019) for a discussion of a novel attempt by modelers, in a different systems biology lab that Michael Stuart and I investigated, to mitigate the collaboration problem by developing computational visualizations of what is going on inside the black box that would be comprehensible to experimentalists, as well as other modelers not involved in the building process.

4. There are significant parallels between what these modelers are doing in mathematizing causal relational structures in biological networks and what Maxwell did in constructing a mathematical representation of the causal relational structure between electricity and magnetism that produces the dynamical behavior of the electromagnetic field, without specifying the underlying causal mechanisms. This strategy proved especially productive in the Maxwell case, as it was later understood that the field is a nonmechanical dynamical structure. For a detailed analysis, see Nersessian (2008, chapter 2).

5. BST was developed, initially, by Michael Savageau (1969a, 1969b, 1970), who is a pioneer in the application and adaptation of systems engineering concepts and methods to biological systems.

6. Most members of lab G are not native English speakers, and so many of the quotes I use are ungrammatical.

7. Quite recently, Lenhard (2020) has discussed some of these roles in physics-based simulation. The account he develops relies on publications and anecdotal evidence, and some post hoc interviews, and not in situ examination of modeler practices.

8. Here I elaborate on a case study analysis originally developed and written with Sanjay Chandrasekhran, who should be considered its coauthor.

9. See Wendy S. Parker (2013) for a discussion of ensemble models in climate science, which comprise different models rather than the same model fitted with different parameters.

10. There are other fields in which model-building often does not have a theoretical starting point. For example, Parker (2013) has pointed this out with respect to modeling in behavioral and social sciences, and Peck (2008) has also shown this for modeling in ecology. Our characterization of "modeling from the ground up" should be distinguished from Keller's (2003) notion of "modeling from above." As Keller describes it, the latter is a strategy that aims to simulate the phenomenon itself, not by trying to map its underlying causal structure or dynamics, but rather by generating the phenomenon from a simple yet artificial system of interactions and relations. What we call modeling from the ground up is what she would call "modeling from below" in that it relies on information about the causal structure

and dynamics of a system's compositional elements. Both however can begin from nontheoretical starting points.

11. Concurrent with our research, some philosophers of biology have begun to examine how modeling works differently in systems biology than in physics. This research largely focuses on issues of mechanistic explanation and template development (see, e.g., Bechtel 2011; Brigandt 2013; Levy and Bechtel 2013; Serban and Green 2019). It is clear that systems biology provides a rich domain with which to expand philosophical understanding of computational modeling and simulation. The kind of mesoscopic model-building practices we have studied, especially in lab G, for instance, tend to provide understanding that, while making use of mechanistic information, does not provide mechanistic explanation (MacLeod and Nersessian, 2015).

12. I, by contrast, have been arguing (since Nersessian 1984) that such considerations need to inform philosophical accounts of scientific practices, generally.

13. Scientific thought experimenting is one form of possible-worlds thinking. We have posited that, with the advent of computational simulation, in many scientific fields what are customarily called thought experiments can largely be supplanted, or reduced to a minimal role, by in silico simulation models (Chandrasekharen et al. 2012).

14. See MacLeod and Nersessian (2019) for a detailed analysis of the mesoscopic strategy.

15. Similar observations, however, have been made with respect to agent-based modeling in ecology. This is not surprising, given the comparable complexity of the problems and lack of domain theory that characterize both fields. As with Peck's point that "there are no formal methodological procedures for building these types of models suggesting that constructing an ecological simulation can legitimately be described as an art" (2008, 393), our modelers, too, describe their modeling practices as, in part, "art." Likewise the ISB modeling we have observed is an individual project in which each modeler chooses the methods and strategies he or she thinks best resolve the problem without any formal procedure governing the process, though often, for novices, in discussion with someone with greater expertise. A major benefit of an ethnographic approach is that it exposes the often hidden, creative choices that are "rarely disclosed in formal descriptions of model-building" (Peck 2008, 395). These parallels with ecology suggest that there is a deeper commonality among the methodologies employed across these kinds of simulation-building contexts. Empirical investigations such as ours help to broaden understanding of the range of scientific practices involved in the methodologies of computational modeling and simulation.

16. Although I am not attending to it here, our investigations show that the affective experiences of researchers in relation to one another and to the entities and artifacts that are part of their research are important dimensions of the cognitive-cultural

system that create and sustain epistemic practices. Thinking about this issue led us to note the affective dimension of our own language of "coupling" and cognitive "partnering" (Osbeck and Nersessian 2006). We have used "coupling" guided consciously by previous work in cognitive science, but perhaps unconsciously by cultural conceptions of partnering and coupling as the "joining of two persons into one." The "two as one" notion does come close to expressing the kind of relationship of cognitive and cultural domains that enables these to be understood as a single system, each intimately implicated in the other.

Chapter 6

1. It appears to have been a risk worth taking. I have followed the work of the lab and career of the director at a distance for over five years since we ended our investigation, out of curiosity to see how things were turning out. The lab has produced some significant discoveries, has numerous high-level publications, is well-funded, has many more students and postdocs, and now the lab director, who had just received tenure when we left, has been promoted to full professor with an endowed chair.

2. Although the phenomenon of cell signaling was discovered in 1855 by Claude Bernard when he found that certain "secretions" released into the bloodstream had effects on distant cells, the process was not conceptualized as "cell signaling" until the 1970s. According to Nair et al. (2019) "the word 'signal transduction' appeared in biological literature in the 1970s, further elucidation of which was provided by Martin Rodbell in 1980 who postulated that 'individual cells were cybernetic systems made up of three distinct molecular components: discriminators, transducers and amplifiers'" (2). Although biologists use a variety of terms today (reception, transduction, response), this early terminology shows the role of cybernetic and control engineering in the formation of the biological concept.

3. Modelers in both labs complained about the fact that molecular biologists often see computational model representations as too abstract, while they interpret Michaelis-Menten kinetics as providing direct representations of mechanistic reality. In fact, as modelers point out, Michaelis-Menten is itself an abstraction based on various simplifications and assumptions applied to a mass-action representation. It is a mathematical model of the rate at which enzymes catalyze a particular reaction. As C7 expressed the complaint, *"For example, there's a very famous equation that's just called the Michaelis-Menten equation. It's supposed to represent [enzyme] kinetics. But that is an approximation. Most biologists, you know, do not realize that. And they have it in their subconscious that, well, that this is a precise representation of the exact kinetics, and if I were to use a more basic representation of which Michaelis-Menten is an approximation, they would not trust that."*

4. Our research goals required that we understand the research conducted by the labs we investigated in sufficient detail so as to be able to discuss the science,

modeling, and technology in our interviews. Lab C's intensive and complex biological research, complete with what I deemed its "alphabet soup" terminology, provided our group the most significant challenge of all the labs. I had had sufficient background in physics, computer science, and neuroscience to be able to guide our group in learning what was necessary to probe the participants in the other labs on their research. My lack of knowledge of biology enabled me to understand firsthand some of the challenge lab C and lab G researchers faced as they encountered the alphabet soup. We were fortunate to have in our group, at that time, a graduate student, Vrishali Subramanian, with a background in biosciences, who wanted to be trained to do ethnography to use in her research on environmental policy. She helped us understand the biological content of the specific research, as well as make sense of what we were reading in the recommended immunology textbook, during the first two years of our investigations.

5. After hearing a talk by us that mentioned how they used such language to talk about the cells and some technologies with which the cells interfaced (e.g., Bio-Plex), they discussed how they were unaware that they did so. One researcher called it "*creepy*," and said this was treating the cells "*like [she did] the Muppets*," but she also stated that she found herself "*unable to stop doing it.*"

6. In an interview following the meeting, C11 explained that she thought she had been able to come up with the solution to the problem that was stumping C10 because biologists think differently from engineers. As she phrased it, "*She's like 'I calculated, I did the calculations correctly!' Whereas I thought 'yes, but the cells going through this pathway, in this path [zigzag], that would make the cells unhappy' because I think like a biology* [sic]*.*"

7. Here I present a case study analysis originally developed and written with Miles MacLeod, who should be considered its coauthor.

8. C9's paper-writing strategy was that she would begin to develop a paper as she was in the process of conducting experiments or building a model.

9. There has been some recognition of the system-like nature of modeling, simulation, and experimentation in computational biology. In particular, Carusi et al. (2012) argue that these should be viewed as forming an "MSE" system, since each interacts with the other in the discovery process. Although they are correct about the system-like nature, theirs is a generic methodological account that could be used to describe either a unimodal or bimodal strategy. They do not provide detailed examination of such interactions in case studies and do not discuss the novel bimodal strategy we examine here.

10. Although there is no cognitive science literature on embodied engagement by scientists in conducting wet-lab experimentation, there is a substantial literature on how embodiment in mundane experience informs conceptual understanding that I think relevant (see, e.g., Barsalou 1999; Glenberg 2010; Johnson 1987; Lakoff 1987;

Prinz 2002). See also the research of David Kirsh (2010) on dance movements as a form of thinking and the study by the anthropologist Natasha Meyers (2015) on the bodily enactments by protein crystallographers of the molecules for which they are constructing three-dimensional models, from a cultural theory perspective.

11. Note our use of "localization" here differs from the use of the term by Bechtel and Richardson (2010), Wimsatt (2007), and others. Our emphasis is on localization of errors or inaccuracies in a network rather than localization of function, although the former often serves to help reveal the latter, as it did for C9.

Chapter 7

1. A fictional character created by Patton who interjects stories and insights to lighten the task of wading through more than six hundred pages on qualitative methods. The name is pronounced "How come."

2. There are some notable exceptions in philosophy and cognitive science, most of them recent. These include Andersen (2010), Andersen and Wagenknecht (2013), Brigandt (2013), Christensen and Schunn (2007), Dunbar (1995, 1999), Goodwin (1995), Grüne-Yanoff (2011), Hall et al. (2002), Hall et al. (2010), O'Malley et al. (2007), O'Malley and Soyer (2012). Additionally, philosophers have begun to attend specifically to what interdisciplinary "integration" means in cases of contemporary and historical science; see, for instance, Andersen (2016), Green and Andersen (2019), Griesemer (2013), Leonelli (2013), Love and Lugar (2013), O'Malley and Soyer (2012), O'Rourke et al. (2016), Plutynski (2013).

3. A broad characterization of interdisciplinary research, as proposed by the US National Research Council, is considered standard: "A mode of research by teams or individuals that integrates information, data, techniques, tools, perspectives, concepts, and/or theories from two or more disciplines or bodies of specialized knowledge to advance fundamental understanding or to solve problems whose solutions are beyond the scope of a single discipline or field of research practice" (NAS, NAE, and IM 2005, 26).

4. An invitation by the Council of Graduate Schools to give a presentation at their annual meeting (2012) about what characteristics foster creativity in interdisciplinary research and how these might be cultivated in graduate education provided the initial context for my thinking about what I now call "interdisciplinary epistemic virtues."

5. Kevin Dunbar's (1995) findings about the relations between creative analogy use and innovative outcomes in biology research labs provides a nice example of cognitive flexibility as a feature of a problem-solving system. He found that in labs where researchers had homogeneous backgrounds, they made fewer analogies in collaborative problem-solving sessions than those where researchers had more heterogeneous

backgrounds. As he tracked the research, the heterogeneous labs produced more innovative outcomes.

6. How to distinguish the notions of interactional and contributory expertise has been the subject of an extensive debate in the literature that I need not consider for my purposes here (see, e.g., Andersen 2016; Collins and Evans 2015; Collins et al. 2007; Collins et al. 2016; Goddiksen 2014).

7. When we began our research, the faculty talked about interdisciplinary integration but later adopted our language of "interdiscipline" for their aspirations for the field. As they phrased it in a proposal submitted for an award, "Many educational programs in BME might be described as 'engineering with a little biology thrown in.' We maintain that practitioners for the twenty-first century need to be trained in a truly integrative fashion. BME is best understood as an "interdiscipline"—that is, a field that is inherently interdisciplinary. BME is situated at the intersection of three disciplines: biology, engineering, and medicine. All three are essential to the practice of a biomedical engineer."

8. Although I do not elaborate here, in our in-depth analyses of affordances and challenges in interdisciplinary practice in both BME and ISB, we found it useful to draw from *positioning theory* as developed in social psychology to analyze the ways in which participants talk about themselves and one another. "Positioning theory is a contribution to the cognitive psychology of social action. It is concerned with revealing the explicit and implicit patterns of reasoning that are realized in the ways that people act toward others" (Harré et al. 2009, 5). Our analyses have extended the theory to consider positioning in relation to epistemic identity and the epistemic effects of positioning strategies in scientific communities (see, e.g., Osbeck and Nersessian 2010, 2012, 2017; Osbeck et al. 2011).

References

Alac, M., and Hutchins, E. (2004). I see what you are saying: Action as cognition in fMRI brain mapping practice. *Journal of Cognition and Culture* 4 (3–4): 629–662.

Andersen, H. (2010). Joint acceptance and scientific change: A case study. *Episteme* 7:248–265.

Andersen, H. (2016). Collaboration, interdisciplinarity, and the epistemology of contemporary science. *Studies in History and Philosophy of Science Part A* 56:1–10.

Andersen, H., and Wagenknecht, S. (2013). Epistemic dependence in interdisciplinary groups. *Synthese* 190 (11): 1881–1898. doi:10.1007/s11229-012-0172-1

Arabatzis, T. (2006). *Representing Electrons: A Biographical Approach to Theoretical Entities*. Chicago: University of Chicago Press.

Aurigemma, J., Chandrasekharan, S., Nersessian, N. J., and Newstetter, W. (2013). Turning experiments into objects: The cognitive processes involved in the design of a lab-on-a-chip device. *Journal of Engineering Education* 102 (1): 117–140.

Bailer-Jones, D. M. (2009). *Scientific Models in Philosophy of Science*. Pittsburgh: University of Pittsburgh Press.

Barsalou, L. W. (1999). Perceptual symbol systems. *Behavioral and Brain Sciences* 22:577–609.

Batterman, R. W., and Green, S. (2020). Steel and bone: Mesoscale modeling and middle-out strategies in physics and biology. *Synthese* 199 (4): 1–26.

Bazerman, C. (1988). *Shaping Written Knowledge: The Genre and Activity of the Experimental Article in Science*. Rhetoric of the Human Services, vol. 356. Madison: University of Wisconsin Press.

Bechtel, W. (2011). Mechanism and biological explanation. *Philosophy of Science* 78 (4): 533–557.

Bechtel, W., and Abrahamsen, A. A. (2013). Thinking dynamically about biological mechanisms: Networks of coupled oscillators. *Foundations of Science* 18 (4): 707–723. doi:10.1007/s10699-012-9301-z

Bechtel, W., and Richardson, R. C. (2010). *Discovering Complexity: Decomposition and Localization as Strategies in Scientific Research*. Cambridge, MA: MIT Press.

Becvar, A., Hollan, J., and Hutchins, E. (2008). Representing gestures as cognitive artifacts. In *Resources, Co-evolution, and Artifacts: Theory in CSCW*, edited by M. S. Ackerman, C. Halverson, T. Erickson, and W. A. Kellog, 117–143. New York: Springer.

Beisbart, C. (2018). Are computer simulations experiments? And if not, how are they related to each other? *European Journal for Philosophy of Science* 8 (2): 171–204.

Bertolaso, M. (2011). Hierarchies and causal relationships in interpretative models of the neoplastic process. *History and Philosophy of the Life Sciences* 33 (4): 515–535.

Bertolaso, M., Giuliani, A., and Filippi, S. (2014). The mesoscopic level and its epistemological relevance in systems biology. In *Recent Advances in Systems Biological Research*, edited by A. Valente, A. Sarkar, and Y. Gao, 19–36. Hauppauge, NY: Nova Science Publishers.

Black, M. (1962). *Models and Metaphors*. Ithaca: Cornell University Press.

Bloor, D. (1991). *Knowledge and Social Imagery*. 2nd ed. Chicago: University of Chicago Press.

Boon, M. (2011). In defense of engineering sciences: On the epistemological relations between science and technology. *Techné: Research in Philosophy and Technology* 15:49–71.

Boon, M. (2017). An engineering paradigm in the biomedical sciences: Knowledge as epistemic tool. *Progress in Biophysics and Molecular Biology* 129 (Supplement C): 25–39. doi:https://doi.org/10.1016/j.pbiomolbio.2017.04.001

Boon, M., and Knuuttila, T. (2009). Models as epistemic tools in engineering sciences: A pragmatic approach. *Handbook of the Philosophy of Technological Sciences* 9:687–719.

Brass, M., Bekkering, H., and Prinz, W. (2002). Movement observation affects movement execution in a simple response task. *Acta Psychologica* 106 (1–2): 3–22.

Braun, V., and Clarke, V. (2006). Using thematic analysis in psychology. *Qualitative Research in Psychology* 3: 77–101. Brigandt, I. (2013). Systems biology and the integration of mechanistic explanation and mathematical explanation. *Studies in the History and Philosophy of Biological and Biomedical Sciences* 44:477–492.

Brown, A. L. (1992). Design experiments: Theoretical and methodological challenges in creating complex interventions in classroom settings. *Journal of the Learning Sciences* 2 (2): 141–178.

Bruggeman, F. J., and Westerhoff, H. V. (2007). The nature of systems biology. *TRENDS in Microbiology* 15 (1): 45–50.

Bryant, D. J., and Tversky, B. (1999). Mental representations of perspective and spatial relations from diagrams and models. *Journal of Experimental Psychology: Learning, Memory, and Cognition* 25:137–156.

Bursten, J. (2015). Surfaces, Scales, and Synthesis: Scientific Reasoning at the Nanoscale. Diss., University of Pittsburgh.

Calvert, J. (2010). Systems biology, interdisciplinarity and disciplinary identity. *Collaboration in the New Life Sciences*. Aldershot, UK: Ashgate.

Calvert, J., and Fujimura, J. H. (2011). Calculating life? Duelling discourses in interdisciplinary systems biology. *Studies in History and Philosophy of Science Part C: Studies in History and Philosophy of Biological and Biomedical Sciences* 42:155–163.

Caporael, L. R. (2014). Evolution, groups, and scaffolded minds. In *Developing Scaffolds in Evolution, Culture, and Cognition*, edited by L. R. Caporael, J. R. Griesemer, and W. C. Wimsatt, 57–76. Cambridge, MA: MIT Press.

Carey, S. (2009). *The Origin of Concepts*. Oxford: Oxford University Press.Carusi, A., Burrage, K., and Rodríguez, B. (2012). Bridging experiments, models and simulations: An integrative approach to validation in computational cardiac electrophysiology. *American Journal of Physiology—Heart and Circulatory Physiology* 303 (2): H144–H155.

Cetina, K. K. (1999). *Epistemic Cultures: How the Sciences Make Knowledge*. Cambridge, MA: Harvard University Press.

Chandrasekharan, S. (2009). Building to discover: A common coding approach. *Cognitive Science* 33 (6): 1059–1086.

Chandrasekharan, S., and Stewart, T. C. (2007). The origin of epistemic structures and proto-representations. *Adaptive Behaviour* 15:329–353.

Chandrasekharan, S., and Nersessian, N. J. (2015). Building cognition: The construction of external representations for discovery. *Cognitive Science* 39:1727–1763.

Chandrasekharan, S., Nersessian, N. J., and Subramanian, V. (2012). Computational modeling: Is this the end of thought experiments in science? *Thought Experiments in Philosophy, Science, and the Arts* 11:239.

Chandrasekharan, S., and Nersessian, N. J. (2017). Rethinking correspondence: how the process of constructing models leads to discoveries and transfer in the bioengineering sciences. *Synthese* 198 (Suppl. 21):1–30 doi:http://dx.doi.org/10.1007/s11229-017-1463-3

Chang, H. (2004). *Inventing Temperature: Measurement and Scientific Progress*. Oxford: Oxford University Press.

Chang, H. (2012). *Is Water H2O?: Evidence, Realism and Pluralism*. Dordrecht: Springer.

Charbonneau, M. (2013). The cognitive life of mechanical molecular models. *Studies in History and Philosophy of Science Part C: Studies in History and Philosophy of Biological and Biomedical Sciences* 44 (4, pt. A): 585–594.

Christensen, B. T., and Schunn, C. D. (2007). The relationship between analogical distance to analogical function and pre-inventive structure: The case of engineering design. *Memory and Cognition* 35:29–38.

Christensen, B. T., and Schunn, C. D. (2008). The role and impact of mental simulation in design. *Applied Cognitive Psychology* 22:1–18.

Clark, A. (1998). *Being There: Putting Brain, Body, and World Together Again*. Cambridge, MA: MIT Press.

Clark, A., and Chalmers D. (1998). The extended mind. *Analysis* 58 (1):7–19.

Cole, M., and Engeström, Y. (1993). A cultural-historical approach to distributed cognition. In *Distributed Cognitions: Psychological and Educational Considerations*, edited by G. Salomon, 1–46. Cambridge, UK: Cambridge University Press.

Collins, A. (1992). Toward a design science of education. In *New Directions in Educational Technology*, edited by Eileen Scanlon and Tim O'Shea, 15–22. Berlin: Heidelberg. doi: 10.1007/978-3-642-77750-9_2

Collins, H., and Evans, R. (2015). Expertise revisited, part I: Interactional expertise. *Studies in History and Philosophy of Science Part A* 54:113–123. doi:http://dx.doi.org/10.1016/j.shpsa.2015.07.004

Collins, H. M., and Evans, R. (2002). The third wave of science studies: Studies of expertise and experience. *Social Studies of Science* 32 (2): 235–296.

Collins, H., Evans, R., and Gorman, M. (2007). Trading zones and interactional expertise. *Studies in History and Philosophy of Science Part A* 38 (4): 657–666. doi:http://dx.doi.org/10.1016/j.shpsa.2007.09.003

Collins, H., Evans, R., and Weinel, M. (2016). Expertise revisited, part II: Contributory expertise. *Studies in History and Philosophy of Science Part A* 56:103–110. doi:http://dx.doi.org/10.1016/j.shpsa.2015.07.003

Corbin, J., and Strauss, A. (2008). *Basics of Qualitative Research: Techniques and Procedures for Developing Grounded Theory*. 3rd ed. Los Angeles: Sage.

Craik, K. (1943). *The Nature of Explanation*. Cambridge, UK: Cambridge University Press.

DeKleer, J., and Brown, J. S. (1983). Assumptions and ambiguities in mechanistic mental models. In *Mental Models*, edited by D. Gentner and A. Stevens, 155–190. Hillsdale, NJ: Lawrence Erlbaum.

Dennett, D. C. (2000). Making tools for thinking. In *Metarepresentations: A Multidisciplinary Perspective*, edited by D. Sperber, 17–29. New York: Oxford University Press.

Donald, M. (1991). *Origins of the Modern Mind: Three Stages in the Evolution of Culture and Cognition*. Cambridge, MA: Harvard University Press.

Dunbar, K. (1995). How scientists really reason: Scientific reasoning in real-world laboratories. In *The Nature of Insight*, edited by R. J. Sternberg and J. E. Davidson, 365–395. Cambridge, MA: MIT Press.

Dunbar, K. (1997). How scientists think: On-line creativity and conceptual change. In *Conceptual Structures and Processes*, edited by T. Ward, S. M. Smith, and J. Vaid, 461–493. Washington, DC: APA Press.

Dunbar, K. (1999). How scientists build models: In vivo science as a window on the scientific mind. In *Model-Based Reasoning in Scientific Discovery*, edited by L. Magnani, N. J. Nersessian, and P. Thagard, 85–99. New York: Kluwer Academic/Plenum Publishers.

Dunbar, K., and Blanchette, I. (2001). The *in vivo/in vitro* approach to cognition: The case of analogy. *TRENDS in Cognitive Science* 5:334–339.

Eisner, E. W. (2003). On the art and science of qualitative research in psychology. In *Qualitative Research in Psychology: Expanding Perspectives on Methodology and Design*, edited by P. Carnic, J. Rhodes, and L. Yardley, 17–29. Washington, DC: American Psychological Association.

Elgin, C. Z. (2009). Exemplification, idealization, and understanding. In *Fictions in Science: Essays on Idealization and Modeling*, edited by M. Suárez, 77–90. London: Routledge.

Elgin, C. Z. (2018). *True Enough*. Cambridge, MA: MIT Press.

Feest, U., and Steinle, F. (2012). *Scientific Concepts and Investigative Practice*. Berlin: De Gruyter.

Finke, R. A. (1989). *Principles of Mental Imagery*. Cambridge, MA: MIT Press.

Franklin, A. (1989). *The Neglect of Experiment*. Cambridge, UK: Cambridge University Press.

Frigg, R., and Reiss, J. (2009). The philosophy of simulation: Hot new issues or same old stew? *Synthese* 169 (3): 593–613.

Galison, P. (1997). *Image and Logic: A Material Culture of Microphysics*. Chicago: University of Chicago Press.

Gavroglu, K., and Goudaroulis, Y. (1989). *Methodological Aspects of the Development of Low Temperature Physics 1881–1957: Concepts out of Context(s)*. Dordrecht: Nijhoff.

Geertz, C. (1973). *The Interpretation of Cultures*. New York: Basic Books.

Geertz, C. (1983). Thick description: Toward an interpretive theory of culture. In *Contemporary Field Research*, edited by R. Emerson, 37–59. Prospect Heights, IL: Waveland Press.

Gentner, D. (1983). Structure-mapping: A theoretical framework for analogy. *Cognitive Science* 7:155–170.

Gentner, D., Rattermann, M. J., and Forbus, K. D. (1993). The roles of similarity in transfer: Separating retrievability from inferential soundness. *Cognitive Psychology* 25:524–575.

Gerson, E. M. (2013). Integration of specialties: An institutional and organizational view. *Studies in History and Philosophy of Science Part C: Studies in History and Philosophy of Biological and Biomedical Sciences* 44 (4): 515–524.

Giere, R. N. (2002). Scientific cognition as distributed cognition. In *The Cognitive Basis of Science*, edited by P. Carruthers, S. Stich, and M. Siegal, 285–299. Cambridge, UK: Cambridge University Press.

Glaser, B., and Strauss, A. (1967). *The Discovery of Grounded Theory: Strategies for Qualitative Research*. Piscataway, NJ: Aldine Transaction.

Glenberg, A. M. (1997). Mental models, space, and embodied cognition. In *Creative Thought: An Investigation of Conceptual Structures and Processes*, edited by T. Ward, S. M. Smith, and J. Vaid, 495–522. Washington, DC: American Psychological Association.

Glenberg, A. M. (2010). Embodiment as a unifying perspective for psychology. *WIREs Cognitive Science* 1:586–596. doi:https://doi.org/10.1002/wcs.55

Goddiksen, M. (2014). Clarifying interactional and contributory expertise. *Studies in History and Philosophy of Science Part A* 47:111–117. doi:http://dx.doi.org/10.1016/j.shpsa.2014.06.001

Gooding, D. (1990). *Experiment and the Making of Meaning: Human Agency in Scientific Observation and Experiment*. Dordrecht: Kluwer.

Goodman, N. (1968). *Languages of Art*. Indianapolis, IN: Hackett.

Goodwin, C. (1995). Seeing in depth. *Social Studies of Science* 25:237–274.

Gorman, M. (1997). Mind in the world: Cognition and practice in the invention of the telephone. *Social Studies of Science* 27:583–624.

Green, S. (2013). When one model is not enough: Combining epistemic tools in systems biology. *Studies in History and Philosophy of Science Part C: Studies in History and Philosophy of Biological and Biomedical Sciences* 44 (2): 170–180. doi:http://dx.doi.org/10.1016/j.shpsc.2013.03.012

Green, S. (2017). *Philosophy of Systems Biology*. New York: Springer.

Green, S., and Andersen, H. (2019). Systems science and the art of interdisciplinary integration. *Systems Research and Behavioral Science* 36 (5): 727–743. doi:https://doi.org/10.1002/sres.2633

Greeno, J. G. (1989a). A perspective on thinking. *American Psychologist* 44:134–141.

Greeno, J. G. (1989b). Situations, mental models, and generative knowledge. In *Complex Information Processing*, edited by D. Klahr and K. Kotovsky, 285–318. Hillsdale, NJ: Lawrence Erlbaum.

Griesemer, J. (2013). Integration of approaches in David Wake's model-taxon research platform for evolutionary morphology. *Studies in History and Philosophy of Science Part C: Studies in History and Philosophy of Biological and Biomedical Sciences* 44 (4): 525–536.

Grüne-Yanoff, T. (2011). Models as products of interdisciplinary exchange: Evidence from evolutionary game theory. *Studies in History and Philosophy of Science Part A* 42 (2): 386–397.

Guba, E. G. (1978). *Toward a Methodology of Naturalistic Inquiry in Educational Evaluation*. Berkeley: University of California Press.

Guba, E. G. (1981). Criteria for assessing the trustworthiness of naturalistic inquiries. *Educational Research Technology and Development,* 29(2): 75–91.Hacking, I. (1983). *Representing and Intervening*. New York: Cambridge University Press.

Hall, R., Stevens, R., and Torralba, T. (2002). Disrupting representational infrastructure in conversation across disciplines. *Mind, Culture, and Activity* 9:179–210.

Hall, R., Wieckert, K., and Wright, K. (2010). How does cognition get distributed? Case studies of making concepts general in technical and scientific work. In *Generalization of Knowledge: Multidisciplinary Perspectives,* edited by M. Banich and D. Caccmise, 225–246. New York: Psychology Press.

Hangel, N., and Schickore, J. (2017). Scientists' conceptions of good research practice. *Perspectives on Science* 25 (6): 766–791. doi:10.1162/POSC_a_00265

Harmon, E., and Nersessian, N. J. (2008). *Cognitive Partnerships on the Bench Top: Designing to Support Scientific Researchers*. Paper presented at the Proceedings of the 7th ACM Conference on Designing Interactive Systems. doi:10.1145/1394445.1394458

Harré, R. (1970). *The Principles of Scientific Thinking*. London: Macmillan.

Harré, R., Moghaddam, F. M., Cairnie, T. P., Rothbart, D., and Sabat, S. R. (2009). Recent advances in positioning theory. *Theory & Psychology* 19 (1): 5–31.

Hegarty, M. (1992). Mental animation: Inferring motion from static diagrams of mechanical systems. *Journal of Experimental Psychology: Learning, Memory, and Cognition* 18 (5): 1084–1102.

Hegarty, M. (2004). Mechanical reasoning by mental simulation. *Trends in Cognitive Science* 8:280–285.

Hesse, M. (1963). *Models and Analogies in Science*. London: Sheed and Ward.

Hofstadter, D. (1995). *Fluid Concepts and Creative Analogies: Computer Models of the Fundamental Mechanisms of Thought*. New York: Basic Books.

Hollan, J., Hutchins, E., and Kirsh, D. (2000). Distributed cognition: Toward a new foundation for human-computer interaction research. *ACM Transactions on Computer-Human Interaction (TOCHI)* 7 (2): 174–196.

Holyoak, K. J., and Koh, K. (1987). Surface and structural similarity in analogical transfer. *Memory & Cognition* 15 (4):332–340.

Holyoak, K., and Thagard, P. (1989). Analogical mapping by constraint satisfaction: A computational theory. *Cognitive Science* 13:295–356.

Hood, L., Heath, J. R., Phelps, M. E., and Lin, B. (2004). Systems biology and new technologies enable predictive and preventative medicine. *Science* 306 (5696): 640–643.

Humphreys, P. (2004). *Extending Ourselves: Computational Science, Empiricism, and Scientific Method*. New York: Oxford University Press.

Humphreys, P. (2009). The philosophical novelty of computer simulation methods. *Synthese* 169:615–626. doi:10.1007/s11229-008-9435-2

Hutchins, E. (1995a). *Cognition in the Wild*. Cambridge, MA: MIT Press.

Hutchins, E. (1995b). How a cockpit remembers its speed. *Cognitive Science* 19:265–288.

Hutchins, E. (1996). Response to reviewers. *Mind, Culture, and Activity* 3 (1): 64–68. doi:10.1207/s15327884mca0301_6

Hutchins, E. (2011). Enculturating the supersized mind. *Philosophical Studies* 152 (3): 437–446.

Johnson, M. (1987). *The Body in the Mind: The Bodily Basis of Meaning, Imagination, and Reason*. Chicago: University of Chicago Press.

Johnson-Laird, P. N. (1983). *Mental Models*. Cambridge, MA: MIT Press.

Keller, E. F. (2002). *Making Sense of Life: Explaining Biological Development with Models, Metaphors, and Machines*. Cambridge, MA: Harvard University Press.

Keller, E. F. (2003). Models, simulation, and computer experiments. In *The Philosophy of Scientific Experimentation*, edited by H. Radder, 198–215. Pittsburgh: University of Pittsburgh Press.

Keller, J. D., Bazerman, C., and Latour, B. (1996). Cognition in the Wild (book). *Mind, Culture, and Activity* 3 (1): 46–63.

Kirsh, D. (1995). *Complementary Strategies: Why We Use Our Hands when We Think*. Paper presented at the Proceedings of the Seventeenth Annual Conference of the Cognitive Science Society. https://www.researchgate.net/publication/236864140_Complementary_Strategies_-_Why_we_use_our_hands_when_we_think

Kirsh, D. (2001). The context of work. *Human-Computer Interaction* 16 (2–4): 305–322.

Kirsh, D. (2010). Thinking with external representations. *AI and Society* 24:441–454.

Kirsh, D., and Maglio, P. (1994). On distinguishing epistemic from pragmatic action. *Cognitive Science* 18:513–549.

Kitano, H. (2002). Looking beyond the details: A rise in system-oriented approaches in genetics and molecular biology. *Current Genetics* 41 (1): 1–10.

Klein, J. T. (2010). A taxonomy of interdisciplinarity. In *The Oxford Handbook of Interdisciplinarity*, edited by R. Frodeman, J. T. Klein, and C. Mitcham, 15–30. New York: Oxford University Press.

Knorr-Cetina, K. D. (1983). *The Ethnographic Study of Scientific Work: Towards a Constructivist Interpretation of Science.* http://nbn-resolving.de/urn:nbn:de:bsz:352-opus-80624http://nbn-resolving.de/urn:nbn:de:bsz:352-opus-80624https://kops.uni-konstanz.de/bitstream/handle/123456789/11543/knorrethnographic.pdf?sequence=1

Knuuttila, T. (2005). Models, representation, and mediation. *Philosophy of Science* 72 (5):1260–1271.

Knuuttila, T. (2011). Modelling and representing: An artefactual approach to model-based representation. *Studies in History and Philosophy of Science Part A* 42 (2): 262–271.

Knuuttila, T., and Loettgers, A. (2011). Synthetic modeling and the functional role of noise. https://www.researchgate.net/publication/50854247_Synthetic_Modeling_and_the_Functional_Role_of_Noise

Knuuttila, T., and Loettgers, A. (2014). Varieties of noise: Analogical reasoning in synthetic biology. *Studies in History and Philosophy of Science Part A* 48:76–88.

Kosslyn, S. M. (1994). *Image and Brain.* Cambridge, MA: MIT Press.

Krohs, U., and Callebaut, W. (2007). Data without models merging with models without data. In *Systems Biology: Philosophical Foundations*, edited by F. C. Boogerd, F. J. Bruggeman, J.-H. S. Hofmeyer, and H. V. Westerhoff, 181–213. New York: Elsevier.

Kuhn, T. (1962). *The Structure of Scientific Revolutions: International Encyclopedia of Unified Science.* Vol. 2. Chicago: University of Chicago Press.

Kurz-Milcke, E., Nersessian, N. J., and Newstetter, W. (2004). What has history to do with cognition? Interactive methods for studying research laboratories. *Journal of Cognition and Culture* 4:663–700.

Lakoff, G. (1987). *Women, Fire, and Dangerous Things: What Categories Reveal about the Mind.* Chicago: University of Chicago Press.

Latour, B. (1987). *Science in Action.* Cambridge, MA: Harvard University Press.

Latour, B., and Woolgar, S. (1979). *Laboratory Life: The Construction of Scientific Facts.* Princeton, NJ: Princeton University Press.

Lave, J. (1988). *Cognition in Practice: Mind, Mathematics, and Culture in Everyday Life.* New York: Cambridge University Press.

Lave, J., and Wenger, E. (1991). *Situated Learning: Legitimate Peripheral Participation.* Cambridge, UK: Cambridge University Press.

Lenhard, J. (2006). Surprised by a nanowire: Simulation, control, and understanding. *Philosophy of Science* 73 (5): 605–616.

Lenhard, J. (2007). Computer simulation: The cooperation between experimenting and modeling *Philosophy of Science* 74 (2): 176–194.

Lenhard, J. (2020). *Calculated Surprises: A Philosophy of Computational Simulation.* Oxford: Oxford University Press.

Leonelli, S. (2013). Integrating data to acquire new knowledge: Three modes of integration in plant science. *Studies in History and Philosophy of Science Part C: Studies in History and Philosophy of Biological and Biomedical Sciences* 44 (4): 503–514.

Leonelli, S. (2016). *Data-Centric Biology: A Philosophical Study.* Chicago: University of Chicago Press.

Leonelli, S. (2019). Scientific agency and social scaffolding in contemporary data-intensive biology. In *Beyond the Meme: Development and Structure in Cultural Evolution.* Minneapolis: University of Minnesota Press. https://manifold.umn.edu/read/untitled-efb05e81-47d9-4015-a30f-1b133318b43a/section/88c7360c-5cb0-4d7f-9661-2e71fca23ae5

Levitt, H. M., Motulsky, S. L., Wertz, F. J., Morrow, S. L., and Ponterotto, J. G. (2017). Recommendations for designing and reviewing qualitative research in psychology: Promoting methodological integrity. *Qualitative Psychology* 4:2–22. doi:10.1037/qup0000082

Levy, A., and Bechtel, W. (2013). Abstraction and the organization of mechanisms. *Philosophy of Science* 80 (2): 241–261.

Lincoln, Y., and Guba, E. (1985). *Naturalistic Inquiry.* Newberry Park, CA: Sage.

Loettgers, A. (2007). *The Hopfield Model and Its Role in the Development of Synthetic Biology.* Paper presented at the International Joint Conference on Neural Networks. doi:10.1109/IJCNN.2007.4371175

Longino, H. (1990). *Science as Social Knowledge.* Princeton, NJ: Princeton University Press.

Longino, H. (2001). *The Fate of Knowledge.* Princeton, NJ: Princeton University Press.

Love, A. C., and Lugar, G. L. (2013). Dimensions of integration in interdisciplinary explanations of the origin of evolutionary novelty. *Studies in History and Philosophy of Science Part C: Studies in History and Philosophy of Biological and Biomedical Sciences* 44 (4): 537–550.

Love, A. C., and Wimsatt, W. (2019). *Beyond the Meme: Development and Structure in Cultural Evolution*. Minnesota Studies in the Philosophy of Science, vol. 22. Minneapolis: University of Minnesota Press.

Lynch, M. (1985). *Art and Artifact in Laboratory Science: A Study of Shop Work and Shop Talk in a Research Laboratory*. London: Routledge and Kegan Paul.

MacLeod, M., and Nersessian, N. J. (2013). Coupling simulation and experiment: The bimodal strategy in integrative systems biology. *Studies in the History and Philosophy of Biological and Biomedical Sciences* 44 (4, pt. A): 572–584. doi:10.1016/j.shpsc.2013.07.001

MacLeod, M., and Nersessian, N. J. (2015). Modeling systems-level dynamics: Understanding without mechanistic explanation in integrative systems biology. *Studies in History and Philosophy of Science Part C: Studies in History and Philosophy of Biological and Biomedical Sciences* 49:1–11.

MacLeod, M., and Nersessian, N. J. (2016). Interdisciplinary problem-solving: Emerging modes in integrative systems biology. *European Journal for Philosophy of Science* 6 (3): 401–418.

MacLeod, M., and Nersessian, N. J. (2018). Modeling complexity: Cognitive constraints and computational model-building in integrative systems biology. *History and Philosophy of the Life Sciences* 40 (1):1–28.

MacLeod, M., and Nersessian, N. J. (2019). Mesoscopic modeling as a cognitive strategy for handling complex biological systems. *Studies in the History and Philosophy of Biological and Biomedical Sciences* 78:101201. doi:10.1016/j.shpsc.2019.101201

Macleod, M., and Nersessian, N. J. (2020). Bounded rationality, distributed cognition, and the computational modeling of complex systems. In *The Routledge Handbook of Bounded Rationality*, edited by R. Viale, 120–130. London: Routledge.

Malone, K. R., Newstetter, W. C., and Nersessian, N. J. (2005). Gender writ small: Gender enactments and gender narratives about lab organization and knowledge transmission in a biomedical engineering research setting. *Journal of Women and Minorities in Science and Engineering* 11 (1): 61–82.

Mansnerus, E., and Wagenknecht, S. (2015). Feeling with the organism: A blueprint for empirical philosophy of science. In *Empirical Philosophy of Science: Introducing Qualitative Methods into Philosophy of Science*, edited by S. Wagenknecht, N. J. Nersessian, and H. Andersen, 37–64. Cham: Springer.

Morgan, M. S., and Morrison, M. (1999). *Models as Mediators: Perspectives on Natural and Social Science*. Ideas in Context, vol. 52. Cambridge, UK: Cambridge University Press.

Myers, N. (2015). *Rendering Life Molecular: Models, Modelers, and Excitable Matter*. Durham, NC: Duke University Press.

Nair, A., Chauhan, P., Saha, B., and Kubatzky, K. F. (2019). Conceptual evolution of cell signaling. *International Journal of Molecular Science* 20 (13). doi:10.3390/ijms20133292

NAS, NAE, and IM. (2005). *Facilitating Interdisciplinary Research*. Washington, DC: National Academies Press.

Nersessian, N. J. (1984). *Faraday to Einstein: Constructing Meaning in Scientific Theories*. Dordrecht: Martinus Nijhoff/Kluwer.

Nersessian, N. J. (1987). A cognitive-historical approach to meaning in scientific theories. In *The Process of Science*, edited by N. J. Nersessian, 161–179. Dordrecht: Kluwer.

Nersessian, N. J. (1991a). Discussion: The method to "meaning": A reply to Leplin. *Philosophy of Science* 58:678–687.

Nersessian, N. J. (1991b). Why do thought experiments work? In *Proceedings of the Cognitive Science Society* 13:430–438. Hillsdale, NJ: Lawrence Erlbaum.

Nersessian, N. J. (1992a). How do scientists think? Capturing the dynamics of conceptual change in science. In *Minnesota Studies in the Philosophy of Science*, edited by R. Giere, 3–45. Minneapolis: University of Minnesota Press.

Nersessian, N. J. (1992b). In the theoretician's laboratory: Thought experimenting as mental modeling. In *PSA 1992* 2, edited by D. Hull, M. Forbes, and K. Okruhlik, 291–301. East Lansing, MI: PSA.

Nersessian, N. J. (1995). Should physicists preach what they practice? Constructive modeling in doing and learning physics. *Science & Education* 4:203–226.

Nersessian, N. J. (2002). The cognitive basis of model-based reasoning in science. In *The Cognitive Basis of Science*, edited by P. Carruthers, S. Stich, and M. Siegal, 133–153. Cambridge, UK: Cambridge University Press.

Nersessian, N. J. (2005). Interpreting scientific and engineering practices: Integrating the cognitive, social, and cultural dimensions. In *Scientific and Technological Thinking*, edited by M. Gorman, R. D. Tweney, D. Gooding, and A. Kincannon, 17–56. Hillsdale, NJ: Lawrence Erlbaum.

Nersessian, N. J. (2006). *Boundary Objects, Trading Zones, and Adaptive Spaces: How to Create Interdisciplinary Emergence?* Paper presented at the Invited Address, National Science Foundation Sciences of Learning Centers PI meeting, Washington, DC.

Nersessian, N. J. (2008). *Creating Scientific Concepts*. Cambridge, MA: MIT Press.

Nersessian, N. J. (2009). How do engineering scientists think? Model-based simulation in biomedical engineering research laboratories. *Topics in Cognitive Science* 1:730–757.

Nersessian, N. J. (2012a). Engineering concepts: The interplay between concept formation and modeling practices in bioengineering sciences. *Mind, Culture, and Activity* 19 (3): 222–239.

Nersessian, N. J. (2012b). Modeling practices in conceptual innovation: An ethnographic study of a neural engineering research laboratory. In *Scientific Concepts and Investigative Practice,* edited by U. Feest and F. Steinle, 245–269. Berlin: DeGruyter.

Nersessian, N. J. (2017). Systems biology modeling practices: Reflections of a philosopher-ethnographer. In *Philosophy of Systems Biology,* edited by S. Green, 215–225. New York: Springer.

Nersessian, N. J., and Chandrasekharan, S. (2009). Hybrid analogies in conceptual innovation in science. *Cognitive Systems Research* 10:178–188.

Nersessian, N. J., Kurz-Milcke, E., and Davies, J. (2005). Ubiquitous computing in science and engineering research laboratories: A case study from biomedical engineering. In *In-Use Knowledge,* edited by G. Kouzelis, M. Pournari, M. Stšppler, and V. Tselfes, 167–195. Berlin: Peter Lang

Nersessian, N. J., Kurz-Milcke, E., Newstetter, W., and Davies, J. (2003). Research laboratories as evolving distributed cognitive systems. In *Proceedings of the Cognitive Science Society 25,* edited by D. Alterman and D. Kirsch, 857–862. Hillsdale, NJ: Lawrence Erlbaum Associates.

Nersessian, N. J., and Macleod, M. (2022, in press). Rethinking ethnography for philosophy of science. *Philosophy of Science.*

Nersessian, N. J., and Newstetter, W. C. (2013). Interdisciplinarity in engineering research and learning. In *Cambridge Handbook of Engineering Education Research,* edited by J. Aditya and B. Olds, 713–730. Cambridge, UK: Cambridge University Press.

Newell, A., and Simon, H. A. (1972). *Human Problem Solving.* Englewood Cliffs, NJ: Prentice Hall.

Newstetter, W., Kurz-Milcke, E., and Nersessian, N. J. (2004). Cognitive partnerships on the bench tops. *Proceedings of the Sixth International Conference on Learning Sciences.* https://www.researchgate.net/publication/234799965_Cognitive_partnerships_on_the_bench_tops

Newstetter, W. C. (2005). Designing cognitive apprenticeships for biomedical engineering. *Journal of Engineering Education* 94 (2): 207–213. doi:10.1002/j.2168-9830.2005.tb00841.x

Newstetter, W. C. (2006). Fostering integrative problem solving in biomedical engineering: The PBL approach. *Annals of Biomedical Engineering* 34 (2): 217–225.

Newstetter, W. C., Behravesh, E., Nersessian, N. J., and Fasse, B. B. (2010). Design principles for problem-driven learning laboratories in biomedical engineering education. *Annals of Biomedical Engineering* 38 (10): 3257–3267.

Newstetter, W. C., Kurz-Milcke, E., and Nersessian, N. J. (2004). Agentive learning in engineering research labs. In *Frontiers in Education, 2004.* Frontiers in Education Conference 2004, T2F/7–T2F12. IEEE Xplore. doi:10.1109/FIE.2004.1408503

Noble, D. (2006). *The Music of Life: Biology beyond the Genome.* Oxford: Oxford University Press.

Norman, D. A. (1988). *The Psychology of Everyday Things.* New York: Basic Books.

Norman, D. A. (1991). Cognitive artifacts. In *Designing Interaction,* edited by J. M. Carroll. Cambridge, UK: Cambridge University Press.

NRC. (2015). *Enhancing the Effectiveness of Team Science.* Washington, DC: National Academies Press.

Ochs, E., and Jacoby, S. (1997). Down to the wire: The cultural clock of physicists and the discourse of consensus. *Language in Society* 26:479–505.

O'Malley, M. A., Calvert, J., and Dupré, J. (2007). The study of socioethical issues in systems biology. *American Journal of Bioethics* 7 (4): 67–78.

O'Malley, M. A., and Dupré, J. (2005). Fundamental issues in systems biology. *BioEssays* 27 (12): 1270–1276.

O'Malley, M. A., and Soyer, O. S. (2012). The roles of integration in molecular systems biology. *Studies in History and Philosophy of Science Part C: Studies in History and Philosophy of Biological and Biomedical Sciences* 43 (1): 58–68.

O'Rourke, M., Crowley, S., and Gonnerman, C. (2016). On the nature of cross-disciplinary integration: A philosophical framework. *Studies in History and Philosophy of Science Part C: Studies in History and Philosophy of Biological and Biomedical Sciences* 56:62–70.

Osbeck, L., and Nersessian, N. J. (2006). The distribution of representation. *Journal for the Theory of Social Behaviour* 36:141–160.

Osbeck, L. M., and Nersessian, N. J. (2010). Forms of positioning in interdisciplinary science practice and their epistemic effects. *Journal for the Theory of Social Behaviour* 40 (2): 136–161.

Osbeck, L. M., and Nersessian, N. J. (2012). The acting person in science practice. In *Psychology of Science: Implicit and Explicit Processes,* edited by R. W. Proctor and E. J. Capaldi, 86–111. Oxford: Oxford University Press.

Osbeck, L. M., and Nersessian, N. J. (2013). Beyond motivation and metaphor: "Scientific passions" and anthropomorphism. In *EPSA11 Perspectives and Foundational Problems in Philosophy of Science,* 455–466. New York: Springer.

Osbeck, L. M., and Nersessian, N. J. (2015). Prolegomena to an empirical philosophy of science. In *Empirical Philosophy of Science,* edited by S. Wagnekencht, N. J. Nersessian, and H. Andersen, 13–35. Cham: Springer.

Osbeck, L. M., and Nersessian, N. J. (2017). Epistemic identities in interdisciplinary science. *Perspectives on Science* 25:226–260.

Osbeck, L. M., and Nersessian, N. J. (2019). "Groping for trouts in a peculiar river": Challenges in exploration and application for ethnographic study of interdisciplinary science. In *Psychological Studies of Science and Technology*, edited by K. C. O'Doherty, L. M. Osbeck, E. Schraube, and J. Yen, 103–126. Cham: Springer.

Osbeck, L. M., Nersessian, N. J., Malone, K. R., and Newstetter, W. C. (2011). *Science as Psychology: Sense-Making and Identity in Science Practice*. Cambridge, UK: Cambridge University Press.

Parker, W. S. (2009). Does matter really matter? Computer simulations, experiments, and materiality. *Synthese* 169 (3): 483–496.

Parker, W. S. (2010a). Comparative process tracing and climate change fingerprints. *Philosophy of Science* 77 (5): 1083–1095.

Parker, W. S. (2010b). Predicting weather and climate: Uncertainty, ensembles and probability. *Studies in History and Philosophy of Science Part B: Studies in History and Philosophy of Modern Physics* 41 (3): 263–272.

Parker, W. S. (2013). Ensemble modeling, uncertainty and robust predictions. *Wiley Interdisciplinary Reviews: Climate Change* 4:213–223.

Patton, M. Q. (2002). *Qualitative Research & Evaluation Methods*. 3rd ed. London: Sage.

Pea, R. D. (1993). Practices of distributed intelligence and designs for education. *Distributed Cognitions: Psychological and Educational Considerations* 11:47–87.

Peck, S. L. (2008). The hermeneutics of ecological simulation. *Biology & Philosophy* 23 (3): 383–402.

Perrig, W., and Kintsch, W. (1985). Propositional and situational representations of text. *Journal of Memory and Language* 24:503–518.

Plutynski, A. (2013). Cancer and the goals of integration. *Studies in History and Philosophy of Science Part C: Studies in History and Philosophy of Biological and Biomedical Sciences* 44 (4, pt A): 466–476.

Prinz, J. J. (2002). *Furnishing the Mind: Concepts and Their Perceptual Basis*. Cambridge, MA: MIT Press.

Quine, W. V. O. (1969). Naturalized epistemology. In *Ontological Relativity and Other Essays*, 69–91. Cambridge, MA: Harvard University Press.

Rahaman, J., Agrawal, H., Srivastava, N., and Chandrasekharan, S. (2018). Recombinant enaction: Manipulatives generate new procedures in the imagination, by extending and recombining action spaces. *Cognitive Science* 42 (2): 370–415.

Rheinberger, H.-J. (1997). *Toward a History of Epistemic Things: Synthesizing Proteins in the Test Tube*. Stanford, CA: Stanford University Press.

Roessner, D., Porter, A. L., Nersessian, N. J., and Carley, S. (2013). Validating indicators of interdisciplinarity: Linking bibliometric measures to studies of engineering research labs. *Scientometrics* 94 (2): 439–468.

Roosth, S. (2017). *Synthetic: How Life Got Made*. Chicago: University of Chicago Press.

Roschelle, J., and Greeno, J. G. (1987). *Mental Models in Expert Physics Reasoning*. GK-2. Retrieved from ERIC Document ED, Office of Naval Research, pp. 285–736. Washington, DC.

Ross, L. N. (2018). Causal concepts in biology: How pathways differ from mechanisms and why it matters. *British Journal for the Philosophy of Science* 72 (1): 1–30. doi:10.1093/bjps/axy078

Rouse, J. (1996). *Engaging Science: How to Understand Its Practices Philosophically*. Ithaca, NY: Cornell University.

Savageau, M. A. (1969a). Biochemical systems analysis: I. Some mathematical properties of the rate law for the component enzymatic reactions. *Journal of Theoretical Biology* 25 (3): 365–369.

Savageau, M. A. (1969b). Biochemical systems analysis: II. The steady-state solutions for an N-pool system using a power-law approximation. *Journal of Theoretical Biology* 25 (3): 370–379.

Savageau, M. A. (1970). Biochemical systems analysis: III. Dynamic solutions using a power-law approximation. *Journal of Theoretical Biology* 26 (2): 215–226.

Schank, J. C., and Wimsatt, W. C. (1986). Generative entrenchment and evolution. *PSA: Proceedings of the Biennial Meeting of the Philosophy of Science Association* 1986 (2): 33–60. doi:10.1086/psaprocbienmeetp.1986.2.192789

Schwartz, D. L. (1995). Reasoning about the referent of a picture versus reasoning about a picture as the referent. *Memory and Cognition* 23:709–722.

Schwartz, D. L., and Black, J. B. (1996). Analog imagery in mental model reasoning: Depictive models. *Cognitive Psychology* 30:154–219.

Schwartz, D. L., and Martin, T. (2006). Distributed learning and mutual adaptation. *Pragmatics and Cognition* 14:313–332.

Serban, M., and Green, S. (2019). Biological robustness: Design, organization, and mechanisms. http://philsci-archive.pitt.edu/id/eprint/16251

Shepard, R. N., and Cooper, L. A. (1982). *Mental Images and Their Transformations*. Cambridge, MA: MIT Press.

Sheredos, B., Burnston, D., Abrahamsen, A., and Bechtel, W. (2013). Why do biologists use so many diagrams? *Philosophy of Science* 80 (5): 931–944.

Shore, B. (1997). *Culture in Mind: Cognition, Culture and the Problem of Meaning*. New York: Oxford University Press.

Simon, H. A. (1957). *Models of Man: Social and Rational*. Hoboken, NJ: John Wiley and Sons.

Spiro, R. J., Coulson, R. L., Feltovich, P. J., and Anderson, D. K. (1994). *Cognitive Flexibility Theory: Advanced Knowledge Acquisition in Ill-Structured Domains*. Paper presented at the 10th Annual Conference of the Cognitive Science Society. doi:10.1017/CBO9780511529863.023

Spiro, R. J., Feltovich, P. L., Jacobson, M. J., and Coulson, R. L. (1992). Cognitive flexibility, constructivism, and hypertext: Random access for advanced knowledge acquisition in ill-structured domains. In *Constructivism and the Technology of Instruction: A Conversation*, edited by T. M. Duffy and D. Jonassen. Hillsdale, NJ: Lawrence Erlbaum.

Star, S. L., and Griesemer, J. G. (1989). Institutional ecology, "translations" and boundary objects: Amateurs and professionals in Berkeley's Museum of Vertebrate Zoology, 1907–39. *Social Studies of Science* 19:387–420.

Strauss, A., and Corbin, I. (1998). *Basics of Qualitative Research Techniques and Procedures for Developing Grounded Theory*. 2nd ed. London: Sage.

Stuart, M. T., and Nersessian, N. J. (2019). Peeking inside the black box: A new kind of scientific visualization. *Minds and Machines* 29 (1): 87–107.

Suchman, L. A. (1987). *Plans and Situated Actions: The Problem of Human-Machine Communication*. Cambridge, UK: Cambridge University Press.

Tomasello, M. (1999). *The Cultural Origins of Human Cognition*. Cambridge, MA: Harvard University Press.

Trafton, J. G., Trickett, S. B., and Mintz, F. E. (2005). Connecting internal and external representations: Spatial transformations of scientific visualizations. *Foundations of Science* 10:89–106.

Traweek, S. (1988). *Beamtimes and Lifetimes: The World of High Energy Physics*. Cambridge, MA: Harvard University Press.

Trewavas, A. (2006). A brief history of systems biology. *Plant Cell* 18 (10): 2420–2430.

Trickett, S. B., and Trafton, J. G. (2007). "What if. . . .": The use of conceptual simulations in scientific reasoning. *Cognitive Science* 31:843–876.

Vermeulen, I., Brun, G., and Baumberger, C. (2009). Five ways of (not) defining exemplification. In *From Logic to Art: Themes from Nelson Goodman*, edited by G. Ernst, J. Steinbrenner, and O. Scholz, 219–250. London: Ontos.

Voit, E. O., Qi, Z., and Kikuchi, S. (2012). Mesoscopic models of neurotransmission as intermediates between disease simulators and tools for discovering design principles. *Pharmacopsychiatry* 45 (1): 22.

Vygotsky, L. S. (1962). *Thought and Language*. Cambridge, MA: MIT Press.

Wagenknecht, S., Nersessian, N. J., and Andersen, H. (2015). Empirical philosophy of science: Introducing qualitative methods into philosophy of science. In *Empirical Philosophy of Science*, edited by S. Wagnekencht, N. J. Nersessian, and H. Andersen, 1–10. Cham: Springer.

Westerhoff, H. V., and Kell, D. B. (2007). The methodologies of systems biology. *Systems Biology: Philosophical Foundations*, edited by F. C. Boogerd, F. J. Bruggeman, J.-H. S. Hofmeyer, and H. V. Westerhoff, 23–70. New York: Elsevier.

Westerhoff, H. V., Kolodkin, A., Conradie, R., Wilkinson, S. J., Bruggeman, F. J., Krab, K., . . . Moné, M. J. (2009a). Systems biology towards life in silico: Mathematics of the control of living cells. *Journal of Mathematical Biology* 58 (1): 7–34.

Westerhoff, H. V., Winder, C., Messiha, H., Simeonidis, E., Adamczyk, M., Verma, M., . . . Dunn, W. (2009b). Systems biology: The elements and principles of life. *FEBS Letters* 583 (24): 3882–3890.

Wimsatt, W. C. (1974). Complexity and organization. *PSA 1972: Proceedings of the Biennial Conference of the Philosophy of Science Association*, 67–80. Dordrecht: D. Reidel.

Wimsatt, W. C. (2007). *Re-engineering Philosophy for Limited Beings: Piecewise Approximations to Reality*. Cambridge, MA: Harvard University Press.

Wimsatt, W. C. (2013a). Articulating Babel: An approach to cultural evolution. *Studies in History and Philosophy of Science Part C: Studies in History and Philosophy of Biological and Biomedical Sciences* 44 (4): 563–571.

Wimsatt, W. C. (2013b). Entrenchment and scaffolding: An architecture for a theory of cultural change. In *Developing Scaffolding in Evolution, Cognition and Culture*, edited by L. R. Caporael, J. R. Griesemer, and W. C. Wimsatt, 77–105. Cambridge, MA: MIT Press.

Winsberg, E. (2001). Simulations, models, and theories: Complex physical systems and their representations. *Philosophy of Science* 68 (3): 442–454.

Winsberg, E. (2009). A tale of two methods. *Synthese* 169 (3): 575–592.

Winsberg, E. (2010). *Science in the Age of Computer Simulation*. Chicago: University of Chicago Press.

Woods, D. D. (1997). Towards a theoretical base for representation design in the computer medium: Ecological perception and aiding human cognition. In *The*

Ecology of Human-Machine Systems, edited by J. Flach, P. Hancock, J. Caird, and K. Vincente, 157–188. Hillsdale, NJ: Lawrence Erlbaum.

Woody, A. I. (2014). Chemistry's periodic law: Rethinking representation and explanation after the turn to practice. *Science after the Practice Turn in Philosophy, History, and the Social Studies of Science*, edited by L. Soler, S. Zwart, M. Lynch, and V. Israël-Jost. London: Routledge.

Zagzebski, L. T. (1996). *Virtues of the Mind: An Inquiry into the Nature of Virtue and the Ethical Foundations of Knowledge*. Cambridge, UK: Cambridge University Press.

Zhang, J., and Norman, D. A. (1995). A representational analysis of numeration systems. *Cognition* 57:271–295.

Zwaan, R. A. (1999). Situation models: The mental leap into imagined worlds. *Current Directions in Psychological Science* 8:15–18.

Index

Note: Figures and tables are indicated by "f" and "t" respectively, following page numbers